ELECTROMAGNETIC COMPATIBILITY DESIGN GUIDE

for

AVIONICS AND RELATED GROUND SUPPORT EQUIPMENT

TABLE OF CONTENTS

TABLE OF CONTENTS (cont'd)

TABLE OF CONTENTS (cont'd

TABLE OF CONTENTS (cont'd

TABLE OF CONTENTS (cont'd)

LIST OF ILLUSTRATIONS

LIST OF ILLUSTRATIONS (cont'd)

LIST OF ILLUSTRATIONS (cont'd)

LIST OF ILLUSTRATIONS (cont'd)

LIST OF ILLUSTRATIONS (cont'd)

LIST OF ILLUSTRATIONS (cont'd)

LIST OF ILLUSTRATIONS (cont'd)

LIST OF TABLES

PREFACE

This handbook was originally published as NAVAIR AD1115 and was developed over a period of nine years through the support of the Naval Surface Weapons Center in Dahlgren, Virginia. Over these many years, there have been many comments and suggestions received from engineers and technicians in the field. These have been evaluated and appropriate changes incorporated where we felt it was appropriate. The handbook has been a long term project of the staff of Sachs/Freeman Associates, Inc.

The support and cooperation of many organizations and individuals in connection with the preparation of the document is appreciated. In particular the assistance and guidance of H. Shechtel and T. Butrim of the Avionics Support Service Division, NASC, and of C. Gallaher, E. Tolive and B. Hilbish was critical to the program. The second edition of this guide was prepared under the direction of C. Caldwell, W. Lenzi, R. Wasilko, and B. Zander. A special appreciation is extended to the many individuals at the Naval Avionics Facility, Indianapolis for many inputs to the guide.

During the course of preparing this guide, many manufacturers of EMC-associated components and hardware were contacted. Their cooperation in providing information about their products is considered to enhance the utility of the guide.

It should be pointed out that any reference to particular contractor's and manufacturer's products in this publication is not to be construed as an endorsement. The references are made for illustrative purposes only.

1.0 INTRODUCTION

1.1 PURPOSE

The purpose of this publication is to provide electromagnetic compatibility guidance in connection with the design and development of Navy Ground Support Equipment. In terms of this purpose, it is appropriate to define both Ground Support Equipment and Electromagnetic Compatibility. Ground Support Equipment (GSE) is defined in OPNAVINST 4790.2 as follows:

GSE includes all equipment on the ground required to make a weapons system, support system, subsystem, or end-item of equipment operational in its intended environment. This includes all equipment required to install, launch, arrest (except Navy shipboard and shorebased launching and arresting equipment, less missile target launching equipment), guide, control, direct, inspect, test, adjust, calibrate, appraise, gauge, measure, assemble, disassemble, handle, transport, safeguard, store, actuate, service, repair, overhaul, maintain, or operate the system, subsystem, end item, or component....

Specifically excluded from this GSE definition are handtools, housekeeping items, office equipment, common production tools, and ancillary equipment such as headsets and microphones.

Electromagnetic Compatibility (EMC) is defined in MIL-STD-463[1] as follows:

The capability of electronic equipments or systems to be operated with a defined margin of safety in the intended operational environment at designed levels or efficiency without degradation due to interference.

The definition is established in an electromagnetic (EM) context, so it can be interpreted in terms of an electromagnetic (rather than a physical or tactical) environment. Physical or tactical environments identify where systems or equipments are located, but do not generally include descriptions of EM emission or susceptibility effects.

1.2 OBJECTIVE

It is probably not necessary to justify a statement that representative Navy shipboard and shore EM environments are becoming more complex and of higher intensity. Many examples can be cited of the increased use of electromagnetic emitters within or in the vicinity of shore maintenance shops, or near base flight-lines, or on ships of the fleet. These equipments have output levels ranging from fractions of milliwatts to tens of megawatts, with waveforms having noiselike properties as well as those of a very complex coherent nature. They often are developed to intentionally radiate these power levels within selected portions of the radio spectrum, but equally often may be emitting EM energies into parts of the spectrum as by-products of some non-radiation or limited radiation application.

In a like manner, more equipments are appearing in these same operational locations that have relatively sensitive

1 MIL-STD-463, Military Standard Definitions, and System of Units, Electromagnetic Interference Technology, 9 June 1966.

EM degradation thresholds. When these thresholds are exceeded (due to either unintended or intended emissions), they can result in reduced equipment performance capabilities. This, in turn, can cause a reduction in overall effectiveness of the mission in which that equipment is used.

Ground-support equipment operate in these environments. They are subject to the diverse energy levels and waveforms created by the fleet and other proximate activities. They also can influence the performance of other environmental equipments, should they couple power to susceptible devices. A very critical consideration throughout the life-cycle of GSE must therefore be its ability to operate compatibly in this environment. That is the objective of this design manual. More specifically, this publication is intended to provide GSE engineering and management personnel with an awareness that a potential of interference exists between ground-support equipment and other electromagnetic equipment. It hopes to convey the philosophy that, if this potential is not recognized and taken into account during GSE design, degradation to testing or operational equipment and systems (or both) can result.

The EMI problems encountered by GSE, and the interference reduction techniques that must be employed to achieve EM compatibility, are often unique to this class of equipment. This is brought about by features of the GSE itself (typically small, portable, and with limited RF circuitry), and by the specialized EM environments in which the equipment may operate (e.g., high narrow-band electric fields). A major objective of the design guide is to provide the ground-support equipment designer with EMC guidance that concentrates on those areas of control dictated by these types of equipments and environments.

The design guide provides extensive technical data and information to assist avionics-oriented GSE design and development engineers in creating a product that will be compatible with its operational environment. While it is not possible to anticipate all of the interface situations that can arise during the course of equipment development, the guidelines. approaches and techniques covered in this manual are expected to furnish cognizent engineers that have system EMC responsibility with the necessary tools for identifying, attacking and resolving EMI problem areas.

It is hoped that use of this guide will benefit GSE hardware at all stages in its design/development cycle. Recognition of the characteristics of a proposed GSE, and the features of its expected operational environment, can influence choices of conceptual approaches and point out design tradeoffs dictated by EM compatibility considerations. Early identification of EMI system reduction requirements can prevent extensive hardware redesign and its attendant additional costs. Application of the correct EMI reduction measures can result in a GSE that is compatible with its environment, and meets its other performance and physical (size, weight, volume) objectives.

1.3 ORIENTATION

An individual who has worked previously in the area of interference suppression of electrical/electronic equipment and systems is likely to comment that there are already several EMI design guides available on this subject. He would be correct; there are not only several good compilations of EMC design techniques available, but there have been innumerable books and technical papers prepared on various facets of equipment EM interaction and interference reduction, and in such related areas as electromagnetic pulse (EMP) effects, EM hazards to ordnance, and biological effects of EM radiation. A few of the more extensive unclassified EMC design guides, handbooks, and manuals published by the government in recent years are listed below:

Many of the interference reduction techniques covered by the listed publications are also applicable to the types of EMI situations that a GSE design/development engineer might encounter, and these techniques have therefore been incorporated into this publication as well. However, additional material has been provided in this guide to take into account two unique aspects of EMC control in GSE.

First, the class of avionics-oriented GSE performs selected operational

Electromagnetic Compatibility, DH1-4
Air Force Systems
Series 1-0 General
Second Edition 10 January 1972

Interference Reduction Guide for Design Engineers
U.S. Army Electronics Laboratories
Ft. Monmouth, N.J. August 1964

Electromagnetic Compatibility Principles and Practices
Office of Manned Space Flight
Apollo Program Design Reliability Series
National Aeronautics and Space Administration May 1965

Engineering Design Handbook
Hardening Weapon Systems Against RF Energy
AMC Pamphlet AMCP 706-235
Headquarters, U.S. Army Material Command February 1972

Military Standardization Handbook
Electro-Magnetic (Radiated) Environment Considerations for Design and Procurement of Electrical and Electronic Equipment
MIL-HDBK-235-1, Unclassified; -2, Confidential; -3, Secret
23 June 1972

Handbook of Instructions for Aerospace Systems Design (HIASD)
AFSC 80-9, Vol. IV, EMC
Headquarters, Air Force Systems Command 20 April 1967

Electromagnetic Compatibility Manual
NAVAIR 5335
Naval Air Systems Command
Washington, D.C. 1972

Electronics Installation and Maintenance Book
Electromagnetic Interference Reduction
NAVSHIPS 0967-000-0150
Naval Ship Engineering Center June 1970

functions within the Navy, and constitutes a specialized group of devices as far as interference suppression requirements are concerned. This design guide recognizes the somewhat homogenous nature of this group of equipments, and concentrates on potential EMI problem areas and interference reduction techniques that are most directly applicable to GSE.

Second, the EM environments (particularly the shipboard environments) in which avionics-oriented GSE must operate are often very specialized. The design guide pays particular attention to these environments and the unique EMI problems they can create.

Since this guide is directed toward experienced system designers, fundamental or basic engineering discussions have been avoided. The reader is referred to the above-listed documents as well as those contained in the chapter appendices) for expansions of many of the topics covered in this publications.

1.4 SCOPE

The eleven chapters of this publication are intended to provide technical and management personnel concerned with GSE design and development with the necessary inputs so that they can evaluate, analyze and resolve potential EMI problems, and maintain system-to-environment compatibility throughout the life of the equipment. The scope of these chapters are as follows:

Chapter 2, Sources and Coupling of EM Energy - An unclassified discussion of the environments in which GSE must operate, with particular attention to radiated and conducted energies on flight and hangar decks, along base flightlines, and in shops and rework facilities.

Chapter 3, GSE Design Considerations - A summary of general EMC design guidelines; factors influencing GSE design; cost, maintenance, and safety considerations; examples of GSE and avionics design.

Chapter 4, Control and Test Planning - Guidance in organizing and implementing effective interference regulations, through EMC Control and Test Plans.

Chapter 5, Shielding - A discussion of applicable shielding techniques, and emphasis on case and cable shielding, and the cable/connector interface.

Chapter 6, Bonding - A summary of bonding techniques of interest to GSE designers.

Chapter 7, Grounding - A summary of single-point and multiple-point grounding, including the grounding of shielded cables.

Chapter 8, Filtering - A detailed discussion of the various types of filters that can be employed to reduce conductive EMI, and their applications in GSE design.

Chapter 9, Testing Requirements and Techniques - Documentation of EMI test procedures and guidelines not contained in previous chapter of the guide.

Chapter 10, Specifications - A summary of government EMC-related specifications and standards applicable at component, subsystem, equipment, and system levels.

Chapter 11, Tables and Nomographs - A compilation of additional tables, charts, nomographs and other design aids useful to GSE design engineers in investigating EMI effects.

The major EMI problems previously encountered by the GSE community have centered around the effects of the operational environment on ground support equipment performance. As a result, the emphasis of the design guide is in this direction, rather than on possible GSE interference to the environment, or internal interference effects within particular GSE. This is not to imply that these other areas have been excluded in the guide, but merely points out that the material in the guide concentrates on coupling problems known to have been encountered in the past.

2.0 SOURCES AND COUPLING OF EM ENERGY

2.1 THE ENVIRONMENT

The operational electromagnetic environments in which GSE must operate are essentially created by the multitude of man-made electrical and electronic equipments that will be operating in the vicinity of the testing devices. Except in very rare cases, non-man-made sources that contribute to the environment (such as solar noise, galactic noise, atmospheric noise, etc.) can be disregarded.

From an EM point of view, these environments can be quite complex. As indicated in Chapter 1, they may contain discrete or very narrow-band sources (such as the outputs of communication transmitter), or sources with relatively complex modulation techniques that tend to produce a broad-band output (such as radars), or impulsive or noiselike sources whose emissions cover a wide frequency range (such as gasoline engines or flourescent lights). They may include emissions that operate almost continuously, as well as others that are on only intermittently. Some of the sources may be under the responsibility of activities other than the Navy, and in some circumstances, the sources may include civilian as well as government equipments.

The portion of the electromagnetic spectrum representing environments of concern to GSE extends from frequencies of a few Hertz, to frequencies in the tens of Gigahertz range. The portion of the band between 10 kHz and 275 MHz is allocated under international radio regulations and national rules and procedures for communications and navigation purposes. Frequency allocations and assignments for mobile radio, radar, radio astronomy, space telemetry, radiosonde, and other diverse types of equipments for both commercial and U.S. government use are made in conformance with these rules. A brief summary of the manner in which this part of the spectrum is allocated to various services is shown in Table 2-1, with special emphasis on U.S. government allocations.

Many other types of emissions exist that are not under the strict control of the above regulations, but could influence the performance of GSE. They occur at the lower end of the spectrum, and include emissions at power-line frequencies, ignition interference, brush noise from rotating machinery, arc discharge radiation from welders, and many others.

The designer of ground-support equipment must recognize that generally he has very limited control over the environment. Whether the equipment he is developing is a portable unit to be operated on a carrier flight deck, or a large testing device that will be connected to 3-phase, 60 Hz power mains at a base installation, he must expect that it will have to operate satisfactorily in the EM fields created by other operational equipments, without the application of many constraints to those other equipments. In addition, he must control the emissions from his GSE so that equipments in the operational environment will not be degraded.

In order to properly design avionics-oriented GSE to meet these constraints, a good understanding of what these environments can look like is necessary. The remainder of this chapter will discuss the general characteristics of environments, and will also identify specific features of

Table 2-1

Allocation of the Radio Spectrum[1]

Frequency, MHz	Service
0.010 - 0.014	Radionavigation
0.014 - 0.200	Fixed, Maritime Mobile
0.200 - 0.285	Aeronautical Radionavigation
0.285 - 0.325	Maritime Radionavigation
0.325 - 0.405	Aeronautical Radionavigation
0.405 - 0.415	Maritime Radionavigation
0.415 - 0.490	Maritime Mobile
0.490 - 0.535	Mobile
0.535 - 1.605	Broadcasting
1.605 - 1.800	Aeronautical Radionavigation, Fixed, Mobile, Radiolocation
1.800 - 2.000	Radionavigation
2.000 - 2.850	Maritime Mobile
2.850 - 3.155	Aeronautical Mobile
3.155 - 3.400	Fixed, Land Mobile, Maritime Mobile
3.400 - 3.500	Aeronautical Mobile
3.500 - 4.000	Amateur
4.000 - 4.063	Fixed
4.063 - 4.438	Maritime Mobile
4.438 - 4.750	Fixed, Aeronautical Mobile
4.750 - 5.450	Fixed
5.450 - 5.730	Aeronautical Mobile
5.730 - 5.950	Fixed
5.950 - 6.200	Broadcasting
6.200 - 6.525	Maritime Mobile
6.525 - 6.765	Aeronautical Mobile
6.765 - 7.000	Fixed
7.000 - 7.300	Amateur
7.300 - 8.195	Fixed
8.195 - 8.815	Maritime Mobile
8.815 - 9.040	Aeronautical Mobile
9.040 - 9.500	Fixed
9.500 - 9.775	Broadcasting
9.775 - 11.700	Fixed, Aeronautical Mobile
11.700 - 11.975	Broadcasting
11.975 - 12.330	Fixed
12.330 - 13.200	Maritime Mobile
13.200 - 13.360	Aeronautical Mobile
13.360 - 14.000	Fixed
14.000 - 14.350	Amateur
14.350 - 15.100	Fixed, Aeronautical Mobile
15.100 - 15.450	Broadcasting
15.450 - 16.460	Fixed
16.460 - 17.360	Maritime Mobile
17.360 - 17.700	Fixed
17.700 - 17.900	Broadcasting
17.900 - 18.030	Aeronautical Mobile
18.030 - 21.000	Fixed
21.000 - 21.450	Amateur
21.450 - 21.750	Broadcasting
21.750 - 22.000	Fixed, Aeronautical Mobile
22.000 - 22.720	Maritime Mobile
22.720 - 25.000	Fixed, Aeronautical Mobile
25.000 - 25.600	Fixed, Land Mobile, Aeronautical Mobile
25.600 - 26.100	Broadcasting
26.100 - 28.000	Fixed, Land Mobile, Citizens
28.000 - 29.700	Amateur
29.700 - 29.800	Land Mobile
29.800 - 30.560	Fixed, Mobile
30.560 - 50.000	Land Mobile
50.000 - 54.000	Amateur
54.000 - 72.000	Broadcasting
72.000 - 76.000	Fixed, Radio Astronomy, Aeronautical Radionavigation
76.000 - 108.000	Broadcasting
108.000 - 118.000	Aeronautical Radionavigation

Frequency, MHz	Service
118.000 - 136.000	Aeronautical Mobile
136.000 - 144.000	Space, Meteorological, Fixed Mobile
144.000 - 148.000	Amateur
148.000 - 150.800	Fixed Mobile
150.800 - 156.250	Land Mobile
156.250 - 157.450	Maritime Mobile
157.450 - 161.575	Land Mobile
161.575 - 161.625	Maritime Mobile
161.625 - 161.775	Land Mobile
161.775 - 162.0125	Maritime Mobile
162.0125 - 174.000	Fixed, Land Mobile, Public Safety
174.000 - 216.000	Broadcasting
216.000 - 225.000	Radiolocation, Fixed, Mobile
225.000 - 399.900	Fixed Mobile, Radionavigation
399.900 - 420.000	Satellite, Meteorological, Fixed Mobile
420.000 - 450.000	Radiolocation
450.000 - 470.000	Land Mobile
470.000 - 806.000	Broadcasting
806.000 - 947.000	Land Mobile
947.000 - 960.000	Fixed
960.000 - 1215.000	Aeronautical, Radionavigation
1215.000 - 1300.000	Amateur
1300.000 - 1400.000	Aeronautical, Radiolocation
1400.000 - 1542.500	Fixed, Mobile, Space
1542.500 - 1660.000	Aeronautical, Maritime Mobile
1660.000 - 2300.000	Meteorological, Fixed Mobile Space
2300.000 - 2450.000	Amateur
2450.000 - 2500.000	Fixed Mobile
2500.000 - 2690.000	Broadcasting, Satellite
2690.000 - 3300.000	Radio Astronomy, Maritime, Radiolocation
3300.000 - 3700.000	Radiolocation, Amateur
3700.000 - 4200.000	Satellite, Fixed
4200.000 - 5925.000	Aeronautical Radionavigation, Fixed Mobile, Amateur
5925.000 - 6425.000	Satellite, Fixed
6425.000 - 7250.000	Fixed Mobile
7250.000 - 7750.000	Satellite
7750.000 - 7900.000	Fixed Mobile
7900.000 - 8400.000	Satellite
8400.000 - 11700.000	Radiolocation, Space, Aeronautical, Amateur, Fixed Mobile
11700.000 - 12200.000	Satellite, Broadcasting
12200.000 - 14000.000	Fixed Mobile, Aeronautical
14000.000 - 14500.000	Satellite, Fixed, Radionavigation
14500.000 - 22000.000	Fixed Mobile, Aeronautical, Satellite
22000.000 - 41000.000	Space Research, Fixed Mobile Radiolocation, Amateur
41000.000 - 43000.000	Satellite, Broadcasting
43000.000 - 48000.000	Aeronautical, Maritime
48000.000 - 66000.000	Space, Fixed Radiolocation
66000.000 - 71000.000	Aeronautical, Maritime
71000.000 - 86000.000	Radiolocation, Fixed, Mobile
86000.000 - 95000.000	Space
95000.000 -101000.000	Aeronautical, Maritime
101000.000 -142000.000	Space
142000.000 -150000.000	Aeronautical, Mobile
150000.000 -190000.000	Space
190000.000 -200000.000	Aeronautical, Mobile
200000.000 -250000.000	Space
250000.000 -265000.000	Aeronautical, Mobile
265000.000 -275000.000	Fixed, Mobile, Satellite
275000.000 -300000.000	Fixed, Mobile
Above 300	Not allocated

environments in which GSE equipments may be expected to operate. These discussions will be limited in detail so that the design guide can be maintained as an unclassified document. For more complete descriptions of the environments, the reader is referred to MIL-HDBK-235( ) entitled "Electromagnetic (Radiated) Environment Considerations for Design and Procurement of Electrical and Electronic Equipment."[2]

2.2 ENVIRONMENT SOURCES

2.2.1 General Signal Characteristics

From the standpoint of spectrum occupancy, electromagnetic signals are often classified as being narrowband or broadband types. Although many signals cannot be categorized this conveniently (since "narrow" and "wide" are relative terms), the narrowband/broadband division is a useful one for understanding and analyzing signal effects.

A narrowband signal is obviously one that occupies a very small portion of the radio spectrum. The ultimate in such a signal is a continuous sinewave, since the energy in a sinusoidal wave is concentrated at a single frequency. However, most communications transmitters in which a carrier is modulated by voice sidebands, or in which suppressed-carrier techniques are employed, are usually classed as narrowband emitters, since most of their emission is confined to only tens or hundreds of kilohertz of bandwidth. Single-channel AM, FM and SSB transmitters fall in this category, as well as some multiplex analog and digital systems.

In addition to the fundamental outputs of communications transmitters, other types of signals can be classed as narrowband emissions. The harmonic outputs of narrowband communication transmitters are also narrowband. So are 60 Hertz and 400 Hertz prime power signals and their harmonics, communication receiver local oscillator emissions, fire-control system CW illuminators, radio altimeters, and many other functional man-made sources.

In contrast to narrowband emissions, a broadband signal may disperse its energy across tens or hundreds of Megahertz or more. This type of distribution results from signals that are composed of narrow pulses having relatively short rise and fall times. The randomness with which the pulses occur, and the individual characteristics of the pulses involved, dictate the spectral distribution that will result.

Broadband signals can be further divided into random and impulsive sources. Random sources consist of closely spaced electromagnetic impulses that are not clearly distinguishable from one another. The impulses are frequent and overlap, with sharp peaks exceeding the average level. This type of output may be intentionally generated by a noise jammer, may be unintentionally produced by a communication or radar transmitter at frequencies far removed from its operating frequency, or may be contributed by nature in such forms as galactic and solar noise. Impulse sources are characterized by sharp pulses that are relatively infrequent and clearly separated. This type of output may be generated by internal combustion engine ignition systems, power line discharges, motor brush sparking, gas tube discharges, and other electrical or electromechanical devices.

In general, the frequency spectrum associated with either random or impulse energies is extremely broad. Because the energy is spread over such a wide bandwidth, the magnitude is generally specified in terms of micro-volts per kilohertz of bandwidth. For random interference, both peak and average levels are proportional to the square root of the receiver bandwidth. For impulse interference, peak voltages are proportional to the receiver bandwidth, whereas average voltages are independent of bandwidth.

There are a variety of natural and man-made sources of broadband interference. Natural interference is referred to as static and it produces the familiar noises that can sometimes be heard in a radio receiver. This natural static originates in thunder-storms, galactic sources, and the sun. Man-made broadband noise arises from numerous electrical devices and therefore is especially significant in cities and industrial areas, although it also may be particularly severe aboard ships as well.

Table 2-2 provides a list of examples of narrowband and broadband man-made interference sources[3]. The table has been subdivided to give a general indication of whether the source involved is in operation for long periods of time (continuous), for short time periods (intermittent), or for very

Table 2-2

Typical Man-Made Interference Sources

Broadband			Narrowband	
Transient	Intermittent	Continuous	Intermittent	Continuous
Mechanical Function switches Motor starters Thermostats Timer units Thyratron trigger circuits	Electronic computers Motor speed controls Poor or loose ground connections Arc Welding equipment Electric Drills	Commutation noise Electric typewriters Ignition systems Arc and vapor lamps Pulse generators Pulse Radar transmitters Sliding contacts Teletype-writer equipment Voltage regulators	cw-Doppler radar Radio transmitters and their harmonics Signal generators, oscillators and other types of test equip-ment Transponders Diathermy Equipment	Power-line hum Receiver local oscillators

short and infrequent time periods, (transient). The levels of those sources on this list that can cause interference to GSE will be discussed later in this design guide.

2.2.2 Pulse Signal Characteristics

One particular type of broadband source that is often encountered in a GSE topside environment is pulsed radar equipment. Pulsed emissions contain spectrum components that can encompass a very wide frequency range.

Figure 2-1 illustrates this effect. The figure shows a typical time representation of a pulse train, and its corresponding power spectrum as might be observed on an ideal spectrum analyzer. In this case, the envelope of the pulse in the time domain is a rectangle, and the resultant frequency domain presentation follows a $\sin^2 x/x^2$ shape.

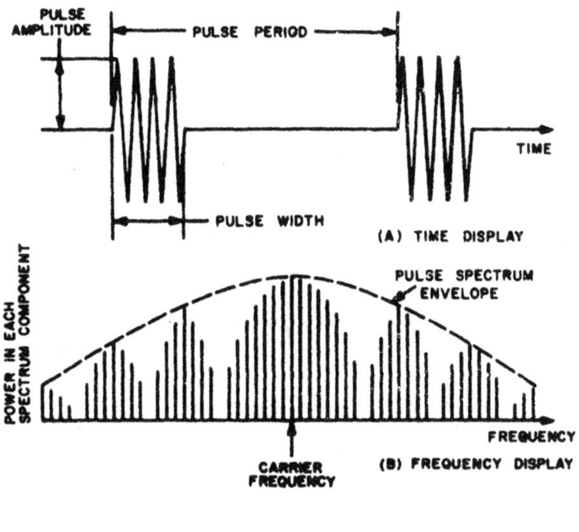

Figure 2-1

Pulse Spectrum Representations

Fourier analysis techniques can be employed to describe the spectrums of pulse signals having a variety of shapes. A usual way of describing the spectrum is in power in a given bandwidth (such as dBm per kilohertz). It is also often convenient to describe the envelope of the spectrum (see Figure 2-1(b)), and not the detailed lobe structure of the spectrum output.

Table 2-3 summarizes pulse spectrum envelope analytical representations applicable to the rectangular pulse case just discussed, as well as for trapezoidal, cosine-squared, and rectangular with linear pulse compression cases. Pulse compression is a radar technique to improve range resolution, and involves intentionally shifting the carrier frequency of the radar during its pulse interval.

Other pulse spectra data are provided in Figure 2-2[3]. In this figure, the spectrum falloff curves have been normalized on the basis of area under each pulse. The advantages to be gained in terms of spectral occupancy by increasing pulse rise and decay times is evident, with pulses having shapes approximating a Gaussian curve being relatively efficient in conserving spectrum space. For any pulse shape, emission levels at low frequencies (frequencies close to the carrier) depend primarily on the area under the pulse, while at high frequencies the levels depend on the number and steepness of the slopes.

It should be pointed out that actual spectrums generated by pulsed systems or transient signals often are considerably different from those created using Fourier spectrum synthesis techniques. For example, Figure 2-3 is a spectrum analyzer photograph of a magnetron output for a 4.5 microsecond, 300 pulses-per-second pulse modulation signal[5]; note its asymmetry and the

Table 2-3

Representative Pulse Spectrums

PULSE CHARACTERISTICS	SPECTRUM ENVELOPE (positive side only shown)	POWER SPECTRAL DENSITY VALUES	FREQUENCY INTERCEPT VALUES
RECTANGULAR 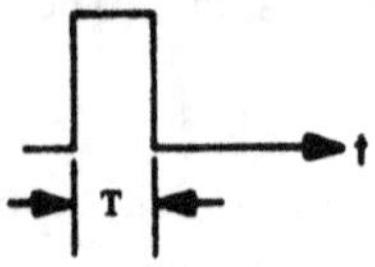	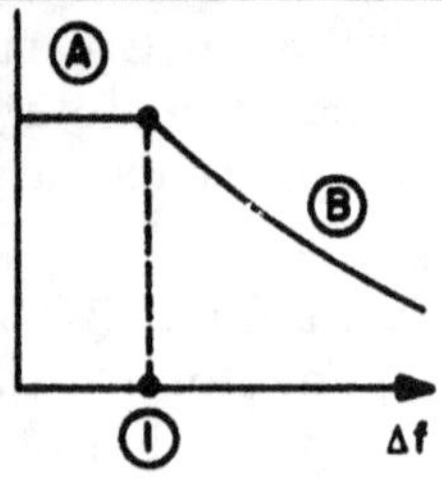	Ⓐ $P = \hat{P}\tau^2/T$ Ⓑ $P = \hat{P}/\pi^2 T\Delta f^2$	① $1/\pi\tau$ P = Power per unit bandwidth, in watts per kilohertz $\hat{P}$ = Peak pulse power, in watts τ = Pulse width, in sec. T = Pulse period, in sec. f = Frequency displacement from carrier frequency, in kHz
TRAPEZOIDAL 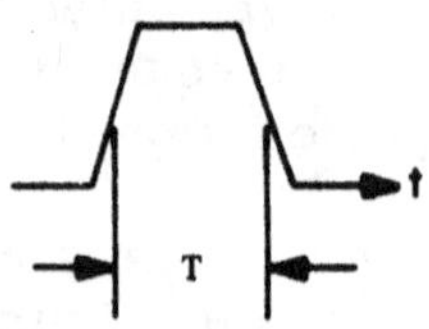	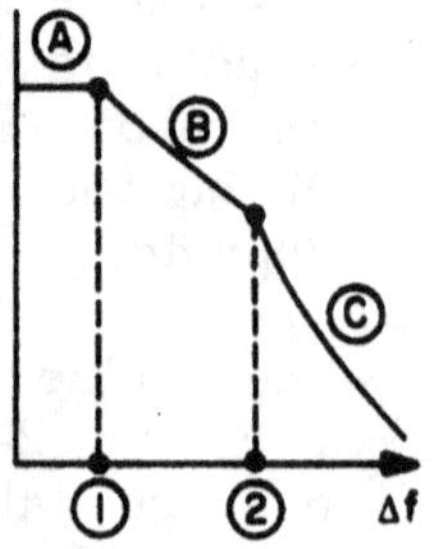	Ⓐ $P = \hat{P}\tau^2/T$ Ⓑ $P = \hat{P}/\pi^2 T\Delta f^2$ Ⓒ $P = \dfrac{\hat{P}\left(\frac{1}{\delta_1} + \frac{1}{\delta_2}\right)}{4\pi^4 \Delta f^4}$	① $1/\pi\tau$ ② $\frac{1}{2\pi}\left(\frac{1}{\delta_1} + \frac{1}{\delta_2}\right)$ δ_1 = Pulse risetime, usec. δ_2 = Pulse decay time, usec.
COSINE-SQUARED 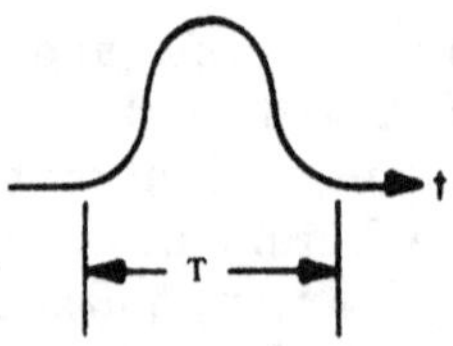	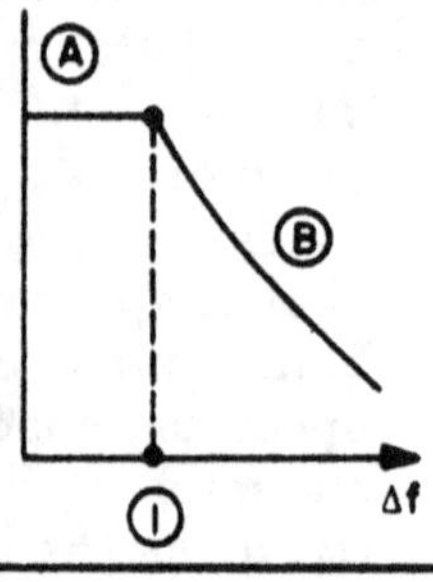	Ⓐ $P = \hat{P}\tau^2/T$ Ⓑ $P = \dfrac{\hat{P}}{4T\pi^2\Delta f^2(1-4\ {}^2\Delta f^2)^2}$	① $\dfrac{\sqrt{\frac{9}{4} - \frac{\pi^2}{6} + 48\pi^2\tau^3}}{12\pi\tau^3}$
LINEAR PULSE COMPRESSION 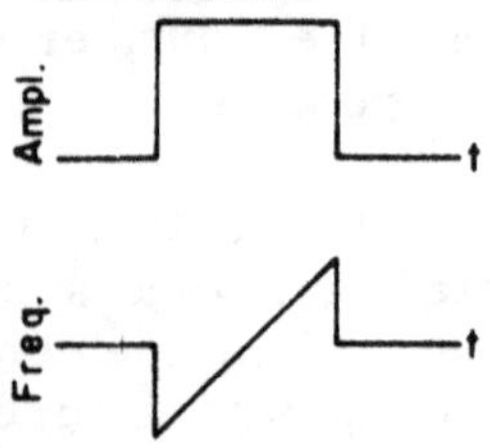	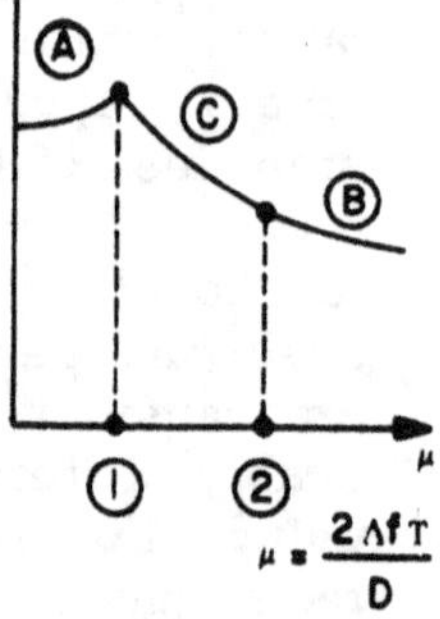	Ⓐ $P = \dfrac{\hat{P}\tau^2}{TD} \cdot \left(1 + \dfrac{1}{\pi\sqrt{D}}\ \dfrac{2}{1-\mu^2}\right)^2$ Ⓑ $P = \dfrac{4\hat{P}\tau^2}{\pi^2 D^2 T} \cdot \left(\dfrac{1}{\left[\mu - \frac{1}{\mu}\right]^2}\right)$ Ⓒ Log-linear interpolation	① $\mu = 1 - \sqrt{\frac{2}{D}}$ ② $\mu = 1 + \sqrt{\frac{2}{D}}$ D = Pulse compression ratio

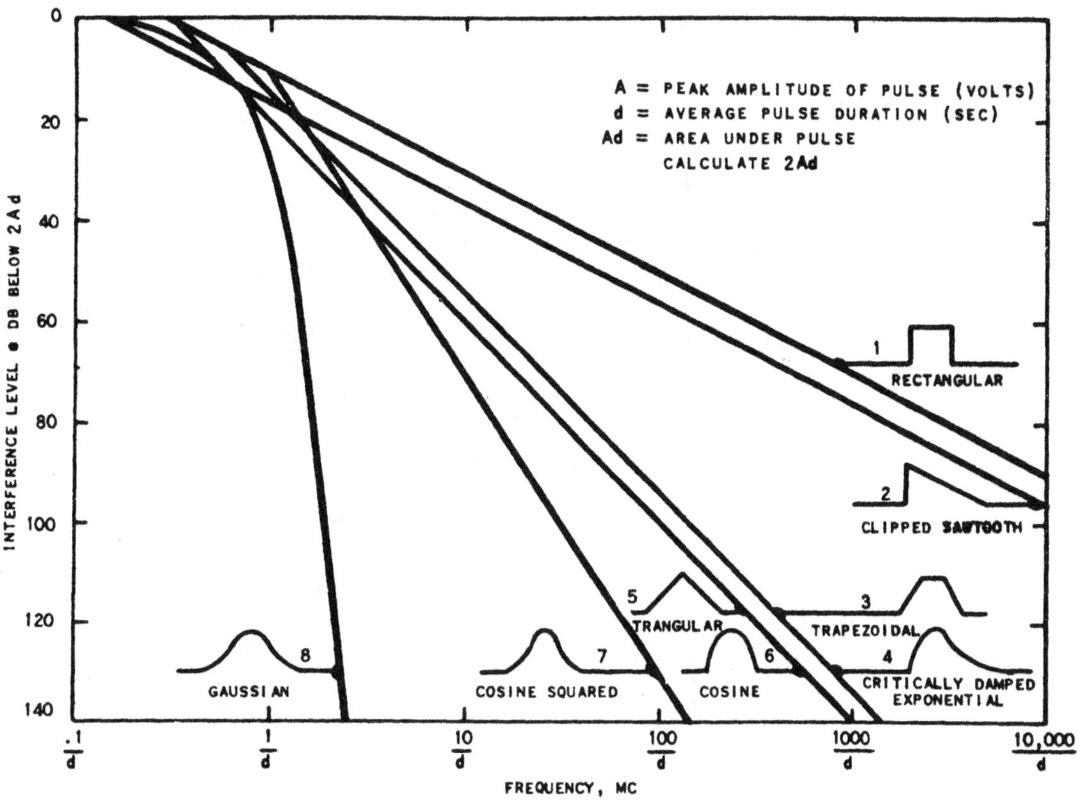

Figure 2-2

Interference Levels for Various Pulse Shapes

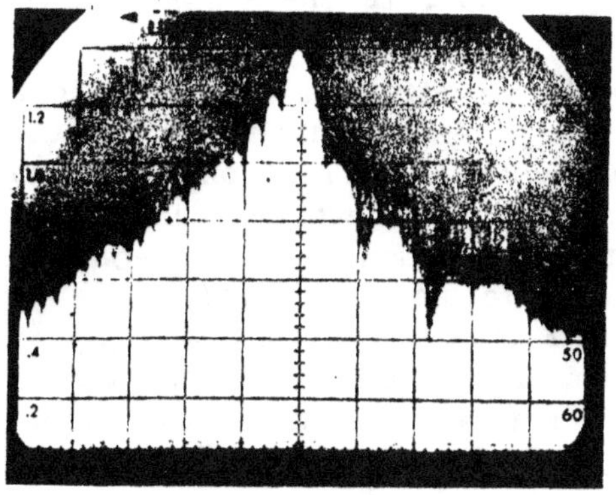

Figure 2-3

Spectrum Analyzer Photograph of Magnetron output when modulated by a 4.5 μs, 300 pps pulse. Ordinate is linear in frequency; abscissa is logarithmic in level.

breakup of lobes on the high side of the spectrum. Such differences are due to approximations that are made of the actual waveshapes in the analysis, combinations of pulse amplitude and carrier frequency changes that actually occur and are not taken into account analytically, and noise outputs that exist in the pulse generators. Measurements of operational environment characteristics, if such measurements are possible, provide a much more reliable indication of the environment than attempting to describe the environment analytically.

2.2.3 Signal Transmission

Interference signals may be transferred

from one point to another by one of two means, either by conduction or by radiation. In general, Conducted interference is transferred over common interconnections between a source and a receiver, and its propagation conforms to conventional circuit and transmission-line theory. Radiated interference is transferred via an electromagnetic field produced by a signal source, and its distribution conforms to field theory relationships. Figure 2-4 shows the conceptual differences between conducted and radiated paths[6].

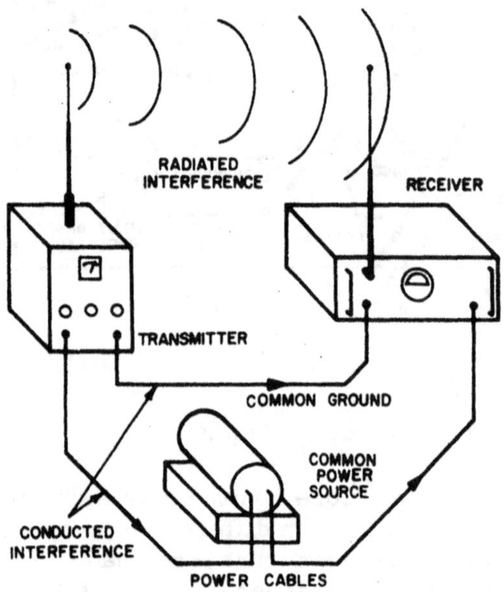

Figure 2-4

Propagation of Conducted and Radiated Interference

Some transfer of signals may occur over a combination of conducted and radiated paths. For example, a source might couple energy onto a power cable by radiated means, and the power system could then transfer this energy to various equipments tied to the power system by conduction. Similarly, coupling between two cables and attendent propagation of energy into and out of the cables may involve both radiation and conduction effects.

The next section of this guide will discuss radiated and conducted coupling in more detail. It will describe the factors and devices contributing to each, and will provide representative radiated and conducted environmental data peculiar to GSE.

2.3 RADIATED ENERGY

2.3.1 The Radiation Field

The term radiated interference is commonly used to mean any interfering signal transferred through a medium by an electromagnetic field. Technically speaking, however, the radiation field represents the energy which actually escapes from a source and spreads out in space according to the laws of wave propagation.

A varying electric field is always accompanied by a magnetic field, and, conversely, a varying magnetic field is always accompanied by an electric field. The ratio of electric-to-magnetic fields is the wave impedance.

In the vicinity of a source of electromagnetic energy, the ratio of electric-to-magnetic field is strongly dependent upon the point where it is observed. One common source (Fig. 2-5a) is the "loop" formed by a conductor carrying large current variations into a load from a low-impedance source, with the return current flowing through a ground plane. This produces a strong H field, and unless the source voltage is exceptionally high, a relatively weak E field. In the immediate vicinity of this loop, the field impedance is very low ... because E/H is low (small E, large H). Away from the loop, however, the H field weakens more rapidly than does the E field, so that their ratio changes, until a free-space value for E/H of 377 is reached, beyond which, unless disturbed by interfering objects or a change in medium, the field impedance remains constant, at 377 ohms.

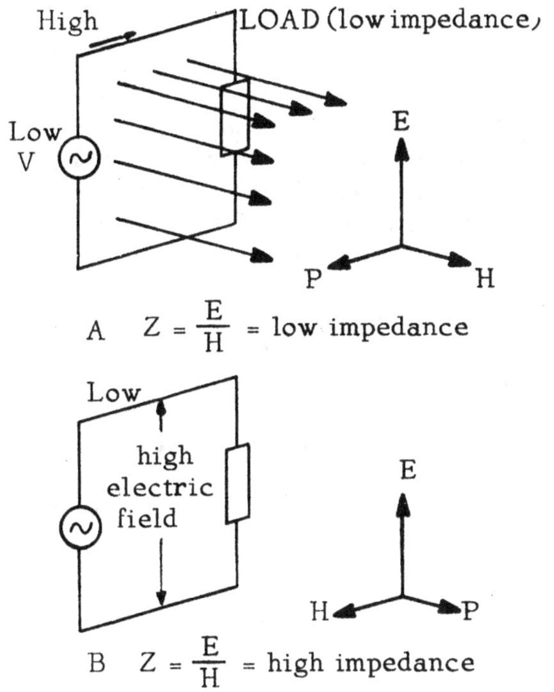

E = electric field
H = magnetic field
P = power flow

Figure 2-5
Sources and Their Fields

Fig. 2-5B shows a conductor connecting a high-voltage, low-current AC-generator to its load. Here again, the return current flows through the ground plane. However, in this example, the H field is weak, because the current is low, and the E field is strong, because the voltage per meter (the voltage gradient) between the conductor and the ground plane is high. In the immediate vicinity of this source, the field impedance is very high, because E/H is high (large E, small H). Away from this source, however, the E field weakens faster than does the H field, and E/H approaches 377 ohms.

The distance at which the wave impedance approaches 377 ohms is approximately one wavelength.

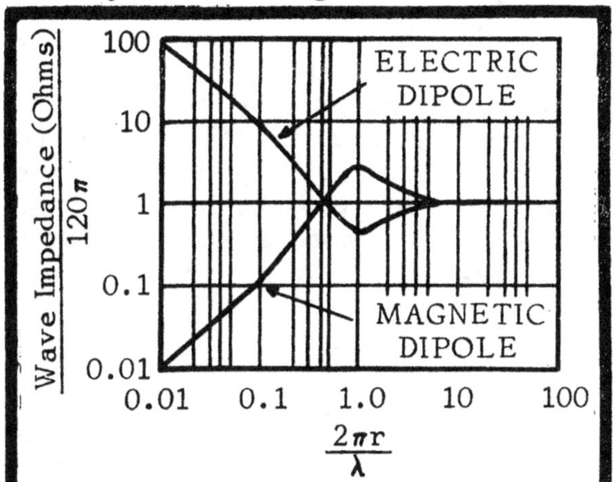

where: r = distance between source and monitoring point in same units as λ

Figure 2-6
Wave Impedance vs. Distance of Dipole From Source

Figure 2-6 shows how the wave impedance for the cases of Fig. 2-5 varies with distance r from the source. At distances beyond one wavelength, λ, from the source, the values are close to the free space value of 377 ohms.

For the radiation field, the ratio of electric-to-magnetic field is equal to the impedance of the medium in which the wave propagates. The impedance of free-space is approximately 377 ohms. In the radiation field, the electric- or magnetic-field intensity is an inverse function of the distance from the source. Emission that is in the radiation field of a source is essentially a plane wave.

The field intensity at any point in a radiation field is also a function of the size and orientation of the radiator. In order for any body to be an efficient radiator, its electrical dimensions must be at least of the order of one-half wave-length. Thus, although some sources have copious amounts of energy within the low frequency spectrum, little of it maybe radiated because of short, ineffective antenna lengths. Whether or not any radiated energy will degrade the performance of a device that must operate in the vicinity of the source is a function of the directivity of the energy as it leaves the source, the losses incurred in propagating to the device, and the susceptibility of the device to the particular characteristics of the energy.

The problem of determining the radiated signal level at a point in space some distance from a source is not a simple task for several reasons. When the source drives an antenna, it is often necessary to establish the effective gain of the antenna in the direction of the observer when the observer is not in the antenna mainbeam. In addition, the characteristics of the antenna change with frequency, so that the effective gain of the antenna at a frequency of interest outside of the design range of the antenna (such as at the third harmonic of the source) may have to be considered if a receiver in the environment may be susceptible to that frequency. The difficulty may be further compounded by the fact that many antennas either physically or electrically change their directional characteristics (such as a scanning antenna) so that the effective gain in a given direction varies as a function of time. And finally, if there are any conducting materials in the neighborhood of the antenna, such as ship superstructure, these materials can introduce reflection effects that may alter the radiation field.

When the radiator is not a well-defined structure like an antenna, but is an equipment case, or a cable located in a harness, then the problem of determining the radiated level at some point in space becomes even more complex. With very limited exceptions, it is necessary to perform measurements to obtain even approximate indications of expected levels.

2.3.2 Distinction Between Induction and Radiation Fields

Except in very simple cases, the mathematical description of the total fields around a conductor carrying a varying current is very complicated and cannot be solved. It can be shown, however, that the fields at a remote point due to the uniform current flow in a short, thin wire (or a very small, circular loop) have three types of terms. There are terms that vary as $1/r^3$, $1/r^2$, and $1/r$ (where r is the distance from the source to the point of observation). The relative importance of the various terms depends upon the ratio of r to the wave-length of the radiation, as shown in Figure 2-7.

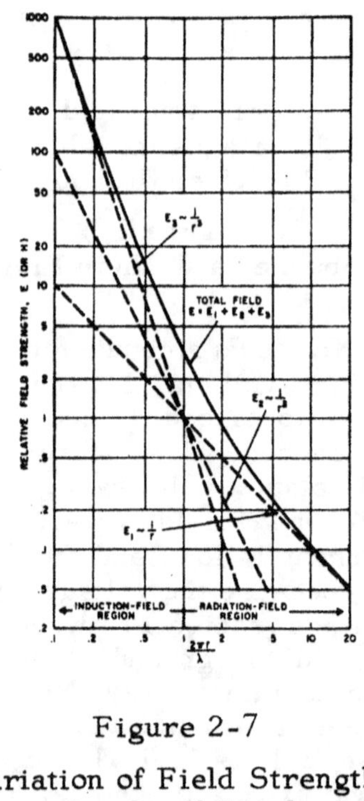

Figure 2-7

Variation of Field Strength with Distance for Small Dipole Antenna[5]

From Figure 2-7, it can be seen that, when r/λ is much greater than $1/2\pi$, the radiation ($1/r$) field is predominant. When r/λ is equal to $1/2\pi$, the induction components ($1/r^2$, $1/r^3$) and the radiation ($1/r$) field are equal. When r/λ is much smaller than $1/2\pi$, the radiation ($1/r$) term is negligible.

A general rule that is often used in the case of antenna radiators, is that one is in the radiation ($1/r$) field of the antenna main beam when r is greater than $2D^2/\lambda$, where D is the maximum aperture dimension of the antenna. D is thus the diameter of a circular aperture or the diagonal of a rectangular aperture. For interference analysis purposes, it appears that r can be considerably less than this value (perhaps as low as $r \leq D^2/\lambda$) before significant error is introduced by using the $1/r$ relationship[6]. As one moves out of the antenna main beam, D becomes much smaller (because it is actually a function of the effective antenna aperture in the direction of the point of interest), and the $1/r$ relationship becomes valid at much closer separations.

2.3.3 Nonlinear Environmental Effects

The characteristics of high rf power and close equipment proximity in a shipboard environment have resulted in selected elements of the ship acting as unique EM interference sources. These sources of interference are located topside, and consist of those portions of the ship that, under high signal conditions, act as nonlinear electrical devices. Some GSE may have to operate in environments containing contributions from these sources.

Environmental nonlinear sources aboard ship consist of nonlinear resistance junctions created at metallic mating surfaces. These junctions may be formed by salt or chemical deposition, by corrosion, or through a number of other means. The result of the nonlinear junction is that it acts like a mixer device and has the potential of creating intermodulation.

The process of intermodulation causes sum and difference products during transmissions involving two or more signals of different frequencies. For example, if n frequencies are transmitted, the frequencies of intermodulation products formed in a nonlinear element can be determined by the relationship

$$f_i = p f_1 \pm q f_2 \pm r f_3 \pm \cdots \pm z f_n$$

where

f_i is an intermodulation frequency

p, q, r...z are integers

and the sum (p + q + r +... + z) defines the intermodulation order. Due to the symmetrical nature of the nonlinearities, odd-order intermodulation products and harmonics are much more pronounced than even order. Therefore, the third-order intermodulation products and the third harmonic are normally the strongest signals generated by environmental nonlinearities.

The effects of nonlinear environmental sources are most pronounced in the HF (3-30 MHz) frequency range.

As the number of transmitters in the environment are increased, the number of possible intermodulation products generated rises rapidly. With three signals being radiated, 10 third-order products are generated. With ten signals being radiated, approximately 800 third-order products are generated.

In a fairly recent study of this effect[7],

metallic life lines were found to be one of the major non linear environmental sources. The junction itself was formed either at the turnbuckles in the line, or at the eyelets in the line, and the line acted as a fairly efficient radiating element.

Another series of sources were associated with the deck expansion joints on an AVT (World War II carrier). These included a vertical copper boot bridging the gap created by the joint, as well as cables and other structures crossing underneath the joint's covering plates. Other sources included mooring and anchor chains, slap-down plates, pipe and bracket joints, hatch hinges, life-raft hangers, ladders, armored cables, antenna guy wires, guard rails, booms, metallic frames and supports, gang planks, and similar structures.

2.3.4 Propagation Effects

Radiation involves the transmission of electrical signals through space. Since the space above the earth's surface is used as the transmission medium, the propagation characteristics are affected by the phenomena which occur in this region.

This section of the design guide provides a summary of these propagation characteristics. The various modes of propagation that are discussed are applicable for the most part to situations where an interference source is a considerable distance away from its potential victim. This will not be the problem with most GSE (whether they are used below decks, topside, or at shore installations), since EMI situations that occur are expected to involve close-in equipments and devices. However, the summary is intended as introductory material for the limited number of cases where the effects of distant equipments are of concern.

In the LF and VLF frequency bands (3-300 kHz) wave propagation takes place via ground-wave; that is, EM radiation propagated along the earth's surface. At these frequencies, the ionosphere absorbs rather than reflects any radiation, while ground wave paths may be used over long distances. The MF band (0.3-3 MHz) is transitional in nature, with ground-wave capabilities at the lower end and some skywave (waves reflected from the ionosphere) at the other. In the HF band (3-30 MHz) propagation may be either by ground or sky waves, depending upon the equipment physical configuration, frequency, and ionospheric conditions. In this band, ground-wave propagation proves the most reliable, but the distance coverage is limited to approximately 100 miles. On the other hand, skywave propagation has an unlimited range, depending upon the parameters outlined above.

As frequencies increase above 30 MHz, the energy reflected from the ionosphere decreases and eventually disappears. Thus, above the HF band, transmission depends mainly upon free-space-propagation. Propagation beyond the horizon is possible in the VHF and UHF bands (30 MHz-3 GHz) from atmospheric scattering. Propagation, particularly sky wave propagation, is a complex subject concerned with many variables. These include (1) the focusing effect of the ionospheric layers, (2) fading due to multipath propagation, (3) polarization fading, (4) loss of energy due to ground and ionospheric absorption, and (5) losses due to dust, water, gasses, and the dielectric of the media.

The general characteristics of the most commonly encountered propagation modes are listed in Table 2-4, along with information on effective frequencies and ranges, fading and other variational characteristics, and link reliability.

Table 2-4 [8] Summary of Characteristics of Selected Propagation Modes

PROPAGATION MODE	FREQUENCY (MHz)	DISTANCE (Mi)	MECHANISM	FADING	DIURNAL VARIATIONS	SEASONAL VARIATIONS	AVAILABILITY & RELIABILITY
Line-of-sight links	No definite limits. Most useful above 50	Up to 30-50 per hop; no fixed upper limit on system length	"Free-space" (1/r) propagation, modified by multipath	1. Fast, due to multipath 2. Slow, due to ray bending	Day: Std. Atmos. Night: can get temp. inversion, fading	Less fading in winter	Approaching 100%
Obstructed-path links	No definite limits. Most useful (est.) 50-5000	Up to 200 per hop (est.)	Knife-edge diffraction	Similar to that for line-of-sight paths, but perhaps slightly less severe	±3 dB for no reflection multipath; +10 dB for reflection multipath		Approaching 100%
Tropospheric Scatter	40-10,000	<700-800	Scattering by blobs or reflection from layers in the atmosphere	1. Fast, due to multipath in atmospher. 2. Slow, due to change in average atmospheric properties	Slightly stronger in morning	Hourly median varies with $\sigma \cong 3$ to 8 dB about yearly median. Signal strongest in summer, weaker in winter.	Approaching 100%
Ionospheric Reflection	0.03-80 Most useful: 0.08-0.2; 2-30	No limit	Reflection from layers of high ionization density in the ionosphere; coupled with reflection from the ground for ranges greater than about 2500 miles.	1. Rapid multipath fading 2. Polarization fading ≅ 3 dB 3. Skip fading (when f is close to Max. Unable Frequency 4. Absorption	Max. Usable Frequency varies greatly being lower at night, highest at noon. Absorption highest during day, noise highest during night.	Max. Usable Frequency, absorption, and noise levels shift over the earth's surface with the seasons.	Typically ≅ 90% or less, considering absorption, noise, multipath and skip fading, and shift of Max. and Lowest Usable frequency. Ranges from about 60% in the auroral zones to about 98% in the equatorial zone.
Ionospheric Scatter	30-60	< 1400 Most useful: 600-1200	Scattering by D layer of ionosphere	1. Fast due to multipath (.2-5 Hz) 2. Slow due to variation in height of reflecting layer	10-12 dB variation. Strongest at noon; weakest at 8 p.m.	Hourly median varies with $\sigma \cong 6$-8 dB about yearly median. Signal strongest in June, July, weakest in Sept.	≅ 95%
Meteor-burst	25-75	200-1000	Electron trails due to ionized meteor material	Extremely rapid (up to ≅ 500 dB/sec)	Maximum possible information rate may vary by 3 to 1; Max. at 6 a.m., Min. at 6 p.m.	Meteor rate may vary by 10 to 1. Max in Summer, Min. in Spring (Northern Hemisphere)	≅ 95% availability. When mode is available, duty cycle is approx. 5 to 10%.
Surface Wave	0.010-0.300 for long range (up to ≅ 500 mi or more). Up to 3 short ranges	< 1200 (greatest over sea water, least over poor soil	Wave guided by interface between earth and air	Multipath with skywaves can occur at ranges beyong about 400 miles	Only very slight due to slight variations of earth's electrical constants (e.g., due to freezing)	Only slight due to slight variation of earth's electrical constants	≅ 95% (suffers from atmospheric noise, especially at low latitudes)
Active Satellites	100-10,000	No limit with multi-satellite system	Receiver-Transmitter in satellite	1. Movement of antenna beam 2. Doppler shift 3. Faraday fading (≅ 0, f >1000 MHz)	None except due to motion in orbit	None except due to motion in orbit	Approaching 100%
Passive Satellites	1000-10,000	No limit with multi-satellite system	Passive reflector	1. Movement of antenna beam 2. Doppler shift	None except due to motion in orbit	None except due to motion in orbit	Approaching 100%
ELF and VLF	Most useful: 0.01-0.03	No limit	Ducting within the waveguide formed by the ionosphere and the surface of the earth	Very slow fading due to 1. Diurnal or Abnormal changes in ionospheric conditions 2. Multipath interference with surface waves	Slightly stronger at night	Slightly stronger in winter	90% to 99.9%

2.3.5 The GSE Radiated Environments

The major potential sources of radiated interference to GSE located topside on Navy vessels obviously comes from the array of communications and radar equipments aboard that ship. Interference from nearby ships must also be considered when the GSE must be operating during refueling operations, while in port, or in other situations where other vessels would be close by and might have their electronic gear operating.

A look at the complement of electronic radiators aboard a particular ship will provide an immediate indication of those frequency bands where high field strengths will probably be encountered. Because of the dimensional restrictions and general structure of the ship, it can be expected that cases may exist where the GSE must operate in the induction field region of an antenna, or at distances well below the $r > D^2/2\lambda$ condition discussed in 2.3.2. Moreover, the direction of maximum energy from a particular source may not correspond to the optical direction to the source, due to super-structure masking, and reflections from masts, stacks, and enclosed above-decks spaces.

More specifically, field strength levels on carrier flight decks and weather decks of other type ships could exceed 2 volts/meter in communication equipment operating bands, with levels as high as 100 volts/meter and more from equipments operating in the HF band, the specific level depending on location. Flight deck average power density levels in acquisition and tracking radar bands might reach 10 milliwatts per square centimeter average levels and higher, while peak power density levels of hundreds to thousands of times these values are possible. These signals will have the characteristics of either plane waves or high-impedance waves, so that electric field rejection techniques in GSE design will apply.

The hangar deck and avionics support equipment spaces EM environments can be expected to be somewhat lower than the flight deck environment, particularly above the VHF/UHF range. Nevertheless, these levels must also be taken into account when developing GSE that must operate in this location. Levels in the HF range can exceed 1 volt/meter, while microwave levels could reach one milliwatt per square centimeter.

The major documents currently used for electromagnetic interference control are standards MIL-STD-461A and MIL-STD-462. Conducted and radiated emission and susceptibility limits for a wide range of electrical/electronic equipment are defined in MIL-STD-461A, while MIL-STD-462 documents the test procedures to be used. These standards are usually prescribed in GSE design contracts or agreements.

The GSE designer should appreciate that MIL-STD-461A limits often are not compatible with the environment in which the GSE must operate. This is particularly true in the case of susceptibility requirements for carrier flight deck equipment, which should be established based upon the intended emission levels that can be found in MIL-HDBK-235-2 (NAVY).[2] The designer must also recognize there may be need for conducted susceptibility tests on signal lines, or for broadband radiated susceptibility testing, neither of which are included in MIL-STD-461A at present.

Figure 2-8 contains sample E-field data on a CVA hangar bay EM environment.[9] The information in the figure is not representative of worst case conditions, and is provided only to show the general trend of the environment at that location. Each dot represents the maximum level narrowband signal located over the frequency band in the vicinity of the dot. The level of the maximum data point in the HF range corresponds to a level of

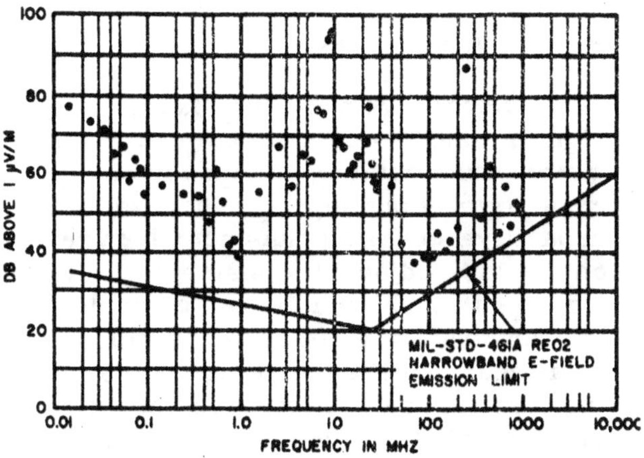

No measurements made over 1-10 GHz range.

Figure 2-8

Narrowband Radiated Emissions Measured in Forward Hangar Bay aboard CVA-62, Port Side, Amidships Section

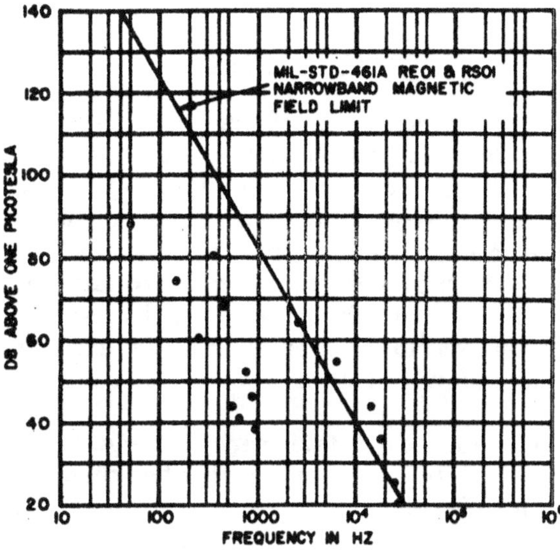

Figure 2-10

Narrowband Radiated Emissions Measured in Avionics Workshop #1 Aboard CVA-62

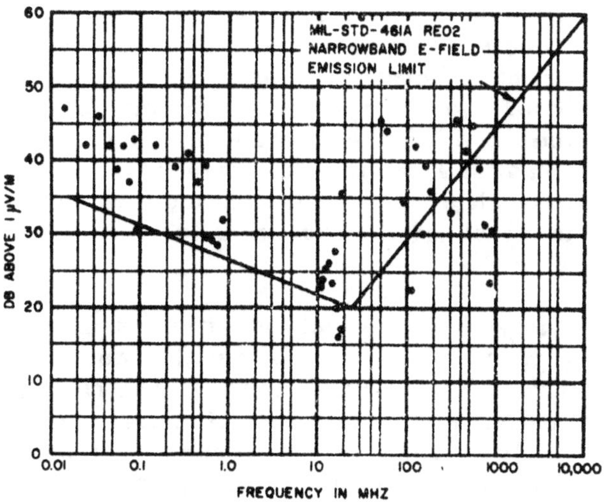

No measurements made over 1-10 MHz and 1-10 GHz ranges

Figure 2-9

Narrowband Radiated Emissions Measured in Avionics Workshop #1 aboard CVA-62

0.6 volts/meter. The MIL-STD-461A narrowband radiated emission limit (test RE02) is also shown in Figure 2-8 for comparative purposes.

Similar data is provided in Figure 2-9 for an avionics shop environment. Note the generally lower levels involved, as compared with Figure 2-8. The level of the maximum data point in this case is 200 microvolts/meter.

Radiated narrowband magnetic field data in the avionics workshop is plotted in Figure 2-10. For the case illustrated, it will be noted that GSE compliance to the radiated susceptibility threshold of MIL-STD-461A, test RS01, would not necessarily assure satisfactory performance in this environment, due to signal levels exceeding the threshold.

2.4 CONDUCTED ENERGY

2.4.1 Conducted Routes

Conducted interference is the interference that propagates through a metallic conductor such as wiring or any metallic structure. The conductor itself can directly transfer interference from one point in a circuit to another, or can be the pickup mechanism for converting radiated interference to conducted effects.

Conducted interference requires a complete circuit path between an interference source and a receiver, such as may be formed by wiring, chassis, structure, or mutual inductances or capacitances. A common path for conducted interference is a ground plane. Such interference is especially severe when several electronic devices are connected to a common power source that has insufficient RF isolation. Under these circumstances, interference may be transferred to the input terminals of each device because of mutual impedance in the ground path.

Whenever there is a direct connection between two circuits, and in addition a return path, a conduction current may flow between the circuits. The return path may be another metallic lead, a mutual capacitance or inductance, or a common ground return. In general, the magnitude of resulting current depends on the difference of potential to ground between the points of exit and entry in the exciting circuit, and on the total loop impedance between these two points. When the dimensions of the circuits involved become large compared with the wavelength of interest (a situation which may occur when the prime power distribution system is involved), then positions along the line where interference can enter and exit from the system become important.

There are three predominant routes for conduction currents. They are power supply leads, control and accessory cables, and ground returns. These paths may introduce interference to GSE, particularly those that must operate from ship or base prime power, or those that connect to aircraft or weapon system power.

In addition to a common impedance path, signals can be coupled onto equipment wires and cables by inductive or capacitive coupling. Inductive or capacitive coupling can occur between any two adjacent unshielded or inadequately shielded wires or cables, or between any internal wiring in equipment, where the return paths for the system may be through the same ground system. Whenever currents or voltages in one circuit induce corresponding voltages or currents in another circuit, these two circuits are said to be mutually coupled.

Inductive coupling occurs when a conductor is present in an electromagnetic field set up by interference energy in a nearby conductor. It is important to again note that mutual inductance exists only between two complete circuits, each a complete loop. However, the return portion of the loop of the circuit may be far removed from the inductive field, and in such a case, it is common practice to speak of the voltage induced into a wire.

Analysis shows that inductive coupling varies in a complicated fashion dependent upon the geometrical configuration of the circuits. At fairly close separations, mutual inductive coupling is comparable to ordinary transformer action. A varying current in one loop of wire sets up a varying field around that loop. In turn, the varying field can induce a varying voltage and resulting current in a second loop of wire.

There are two predominant methods of inductive coupling of interference, through transformers and through parallel wire runs. Because of core losses, a transformer normally will pass LF a-c signals and act as a low-performance filter against HF interference.

The parallel wire arrangement represents a special case of a transformer consisting of single turn primary and secondary windings. When two such lines are feeding different pieces of equipment from different sources, interference voltages present in either line can be induced into the other line. This situation enables interference to affect any susceptible part of a system, receiver, or circuit. For example, primary power leads located near a B+ lead may induce a-c hum into the receiver via the d-c circuit.

Capacitive coupling is the linking of one circuit with another by means of the capacitance existing between them. The degree of coupling depends on the voltage and the frequency of the interference as well as the value of the interconnecting capacitance. At low frequencies, an extremely large capacitor is required to couple the signal; at radio frequencies, only a small capacitance is required. Again, it should be noted that a complete path is necessary for this type of coupling, and it is normally the common ground connection for both leads which provides the return path.

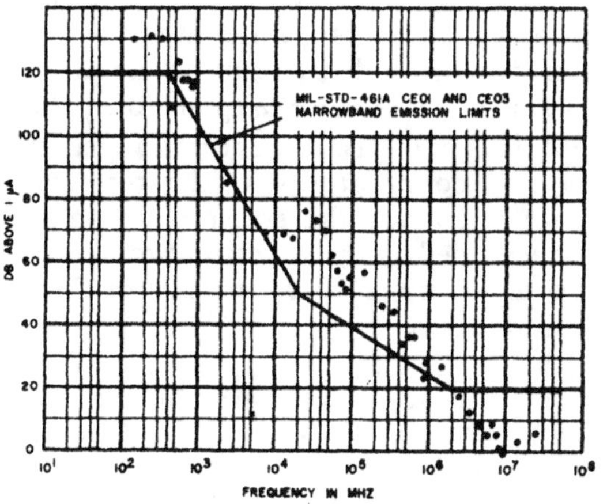

Figure 2-11

Narrowband Conducted Emissions Measured in Avionics Workshop #1 Aboard CVA-62

Multiconductor cables act as especially good capacitive couplers at radio frequencies. If interference is present on a single lead of a multiconductor cable, the interference may appear on all conductors of the cable.

2.4.2 The GSE Conducted Environment

The major mechanism of transmitting conducted EMI energy to GSE is via the prime power distribution system. The significance of this type of interference is therefore a function of the types and designs of other equipments that are also operating from these lines, and their locations on the lines.

Some indication of the conducted interference currents that might be encountered aboard ship is shown in Figure 2-11. The data represents the average of the three current maximums measured on a three-wire 60 Hz power system terminating in an avionics shop; significantly greater current levels were obtained on individual lines. Similar data taken on the same power buses at other locations aboard this same ship indicate that the results obtained are location-dependent, particularily over the lower ranges of measurement.

2.5 COMBINED EFFECTS

A common case of interference to GSE located on carrier flight decks results from a combination of radiated and conducted effects.

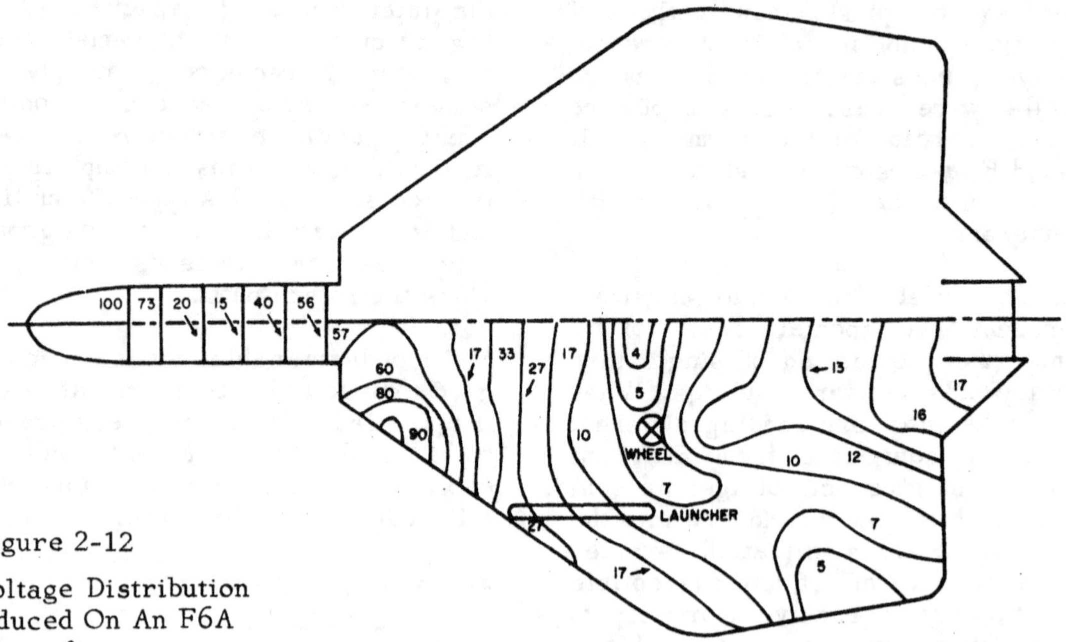

Figure 2-12

Voltage Distribution Induced On An F6A Aircraft

The wavelengths associated with the frequencies in the HF band (3-30) vary from 100 meters to 10 meters. Aircraft dimensions at these frequencies are a significant portion of a wavelength; the aircraft structure may act as a reasonably good antenna. This antenna-like behavior produces an RF voltage with respect to the carrier deck on the aircraft structure. The voltage to deck in a 100 v/m field may be as high as 250 volts RF[10]. Figure 2-12 shows the relative voltage distribution induced on an F6A[11] aircraft.

This voltage pattern varies with such factors as frequency, aircraft type, aircraft configuration and aircraft orientation with respect to the antenna. Attaching a ground strap from the aircraft body to the deck would not eradicate the problem. This would reference only one point of the aircraft to the deck. The electrical length of the rest of the aircraft would still be an appreciable fraction of a wavelength. A series of ground straps along the aircraft structure would cut down the effective electrical length presented by the aircraft to the HF. This is impractical due to the difficulties in achieving a low impedance RF bond between the aircraft and the deck.

The sources of the fields producing the pickup are the whip antennas used for communications and located on the carrier island and around the periphery of the flight deck. Aircraft undergoing organizational level maintenance may be parked in close proximity to these antennas. These whip antennas are non-directional and can be operated in a vertical or horizontal position. While interference from radar transmitters may be correlated with the rotation of their antennas, the source of interference caused by these whip antennas may not be obvious to operators of the affected GSE.

Figure 2-13 diagrams the coupling mechanisms involved. Many avionic systems use aircraft ground in their circuits. The aircraft structure serves as a ground return for the power supply in most aircraft; this "ground" may have an RF potential with respect to the deck "ground". A GSE item placed on the carrier deck and connected by cabling to the aircraft avionics will appear as a relatively low impedance to deck to the RF energy on the aircraft. RF currents will flow down the cable through the GSE into the deck. Even if the avionics circuit is isolated from aircraft ground or if there is no DC path to the deck, RF currents may flow due to capacitive coupling.

If the cabling is shielded and well terminated at both the aircraft and the GSE, the shield will provide a low impedance path to the deck and the RF currents will be confined to the outer portion of the shield due to skin effect. No coupling to the circuit lines will occur if the shield presents a low impedance to the RF energy. Discontinuities in the shielding such as anodized connectors or inadequate cable braid will cause voltage drops due to the RF currents flowing on the shield. Coupling to GSE and avionics systems may then occur because of these discontinuities in the cable shield.

Mounting GSE on the aircraft can provide one means of reducing the possibility of EMI in the HF range. By mounting the GSE on the aircraft the difference of potential between equipment and the aircraft is greatly reduced or eliminated. However, the mounting surfaces must be clean, and all interconnecting cabling must be kept as short as possible.

Most shielding effectiveness measurement methods use radiated fields to measure the combined reflection and penetration loss. In the case described, where large currents exist on the surface of the shields, only the penetration loss of a shield is effective. The most informative measure of shielding effectiveness for this type of coupling is the Surface Transfer Impedance (STI) test described in MIL-STD-1377.

If threshold signal levels are low, interference may result even if good shielding techniques are used. In this case and in cases where proper shielding techniques cannot be readily implemented, filtering may be necessary.

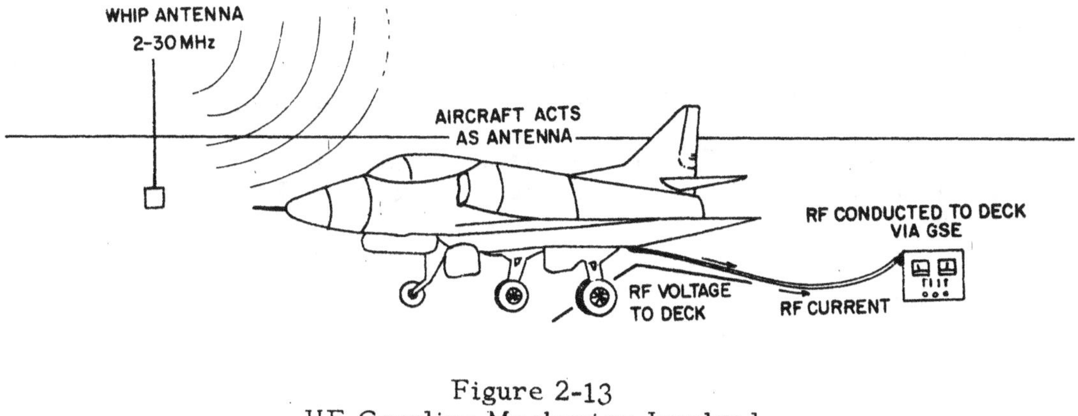

Figure 2-13
HF Coupling Mechanism Involved

REFERENCES

1. Manual of Regulations and Procedures for Radio Frequency Management, Office of Telecommunications Policy, January 1, 1973.

2. MIL-HDBK-235( ), "Military Standardization Handbook, Electro-Magnetic (Radiated) Environment Considerations for Design and Procurement of Electrical and Electronic Equipment," 23 June 1972.
   MIL-HDBK-235-1, Unclassified
   -2, Confidential
   -3, Secret

3. Includes material from "Interference Reduction Guide for Design Engineers," Volumes I and II, U.S. Army Electronics Laboratories, 1 August 1964.

4. Application Note 63, "Spectrum Analysis," Hewlett-Packard Co., August 1968.

5. Department of the Air Force, "Mutual Electromagnetic Interference," AFM-100-35, USAF Communications-Electronics Doctrine, 1 Sept 1960, p. 35-24.

6. R.C. Hausen, and L.L. Bailin, "A New Method of Near Field Analysis," IRE Transactions on Antennas and Propagation, Vol AP-7, Special Supplement, Dec. 1959, pp. S458-S467.

7. R.F. Elsner, et. al., "Environmental Interference Study Aboard A Naval Vessel," 1968 IEEE Electromagnetic Compatibility Symposium Record, July 23-25, 1968, pp. 330-338.

8. Armour Research Foundation, "Methods of Comparison and Evaluation of Communications Systems," Vol. I, Contract AF 30(602)-2074, 31 March 1960.

9. Data was reduced from W.E. Tate et al., "Shipboard Below Decks Electromagnetic Environment Study," Final Report Contract N00024-70-C-1357, with Honeywell Test Instruments Division.

10. H. K. Wolfe, "Measured Antenna Characteristics of A-4 and F-6 Aircraft with respect to HERO", Naval Weapons Laboratory TR-2569, Aug 1970

11. H. A. Schwab, and R. W. Carson, "A Method of Measuring the Relative Intensity of Electromagnetic Environment with respect to HERO", Naval Weapons Laboratory TR-2177, June 1968

3.0 DESIGN CONSIDERATIONS

3.1 GENERAL GUIDELINES

Chapter 2 outlined the factors that must be considered in the design of ground support equipment (GSE) to achieve electromagnetic compatibility (EMC).

Following the definition of requirements that a given piece of GSE must fulfill, its design normally begins with a delineation of the electrical circuits needed to interconnect its components. Such components may consist of black-box units, integrated circuits, or individual components or piece parts. After the electrical aspects of the design have been done, the mechanical configuration of the equipment is established and detailed.

3.1.1 Electrical Design

An early step in any electrical design is the selection of signal levels. In selecting signal levels, consideration must be given to the radiated EM environments in which the equipment must operate. If high level radiated EM environments are expected to be encountered, such as those found on a carrier flight deck, large signal levels and low impedances should be used in the design. If high level radiated EM environments are not to be encountered, signal levels should be the lowest that are consistent with the achievement of the needed signal-to-system noise ratio throughout the equipment. Generally low voltage, low impedance design is desirable from the viewpoint of minimizing capacitive coupling problems, but may aggravate inductive coupling problems if current flows are too high.

Because interference often has a different spectral content from that of the desired signals, much can be done by the proper use of filters, both in constraining signals to their desired paths within the equipment, such as through common power supply leads. Chapter 8 discusses filtering in detail.

After the basic circuit design has been accomplished and the necessary filtering has been specified, attention must be given to conductive coupling that may occur in leads carrying currents from differing sources. Since even a short piece of wire has an impedance, voltages are developed along wires and can cause effects not intended by the designer. This is especially true when attempts are made to return signal and power circuits through a common lead. Chapter 7, Grounding, discusses this aspect of design.

3.1.2 Physical Layout of Components

After the basic circuit design has been completed and grounding requirements have been established, attention must be given to the physical layout of the components. Capacitive and radiative coupling are of most concern here, but inductive coupling also must be considered. Often the desired amount of physical separation between sources of energy and sensitive elements cannot be achieved within the physical size constraints of the equipment, or the entire piece of equipment may have to operate in the presence of a strong external source. These problems of physical proximity make shielding necessary, especially the shielding of cables and connectors. Special design considerations with respect to cables and connectors are the following:

Cables

(A) Cabling penetrating the case should be shielded and the shield terminated in a peripheral bond to the case at the point of entry.

(B) Cable shield grounds should be maintained separate from any signal grounds or circuitry grounds.

(C) Cable shields should be bonded peripherally to adapter and connector shells; cable shields should not be "pig-tailed off" and run through on connector pins.

Connectors

(A) Connectors should be of the type which make peripheral shield (shell) ground before the pins mate during the process of connection. The pins should disconnect before the shield (shell) separates.

(B) Pins of connectors leading to electronic circuitry should be, wherever possible, female. Otherwise, they should be recessed male pins so as to exclude contact with any portion of the shell of the mating connector or with fingers of the operator.

Shielding techniques, including techniques applicable to cables and connectors, are discussed in Chapter 5.

3.1.3 Mechanical Factors

The construction of equipment cases, as well as component installation, both internal and external to the case, are important in the successful EMC design of GSE. Significant factors with respect to case construction are:

(A) The use of material with good shielding properties.

(B) The use of welded or overlapped seams.

(C) Secure bonding of panels, cover plates and aperture covers to the case, using gaskets with adequately spaced and properly tightened screws or bolts.

(D) The cleaning of mating surfaces immediately prior to assembly to ensure good electrical contact and minimize corrosion. The subject of bonding is discussed in Chapter 6.

Significant factors with respect to the installation of components internal to the case are:

(A) Isolation (electrical and physical) of all incoming circuitry (power lines, signal lines, and signal processing circuitry, etc.) from all internal circuitry should be accomplished by shielding entry areas and circuitry and by utilizing pi-type or tee-type feed-through filtering for lines into the internal circuitry.

(B) Electrical and physical isolation of internal interference generating circuitry from other circuitry should be accomplished.

With respect to the installation of components external to the case, the following items warrant consideration:

(A) EMI boots should be installed on toggle switches.

(B) EMI rotary shaft seals should be incorporated.

(C) Screening and shielding of meter faces and indicators should be accomplished.

(D) Shielded lights should be used.

(E) Shielded fuseholders should be used.

(F) Application of waveguide-below-

-cutoff techniques should be employed.

Other factors of importance to the GSE Designer are the following:

Recognize that high level EM environments can significantly aggravate the personnel and equipment safety situation, through the charging of isolated structures, through the performance of equipment and associated cables that were not designed to operate in that environment, and through mechanical configurations that can result in the generation of high current levels. Section 3.2 discusses Safety.

Appreciate the potential safety hazards that this type of environment can represent, and take these factors into account during hardware design. Particular attention should be given to grounding the test equipment/aircraft or weapon/flight or hangar deck configuration, and to establishing the signal path interfaces.

Apply appropriate consideration to the protection of components that may be susceptible to or degraded by high-level fields.

Be sure that high reliability interference suppression components are employed.

Develop an EMC design plan that is comprehensive enough to identify the potential operational problems that must be resolved, and is flexible enough to allow consideration of alternative remedial approaches. Recognize that the limited nature of most GSE production may make it most cost-effective to overdesign the EMC aspects of the equipment.

In, general, avionics GSE protection against a high-power, electromagnetic environment should be included in the design of the equipment from its inception and directed toward the elimination and exclusion of energy which will affect the operation of the equipment from the standpoint of both radiated and conducted interference. To achieve this objective, the techniques just outlined should be used. Chapters 5, 6, 7, and 8 provide detailed information about the application of these techniques.

3.2 SAFETY

3.2.1 The Effects of High RF Fields

High ambient electromagnetic levels present on a flight or hangar deck not only introduce unique design requirements on the equipments involved, but they also can create operator and equipment hazards as well.

Two classes of safety problems can be identified. One type is due to arc discharges occurring between GSE, the flight deck, and/or an aircraft. The other type is caused by large currents being induced in a GSE.

The arc discharges result because aircraft and GSE are often isolated from the flight or hangar deck, or from each other, and build up different levels of electric potential. These potential differences may arise under normal circumstances because of the incompatibilities of equipment ground systems, or because of different charge accumulations on different assemblies, but the problem can be severely aggravated by a high level EM environment. Under these circumstances, when an attempt is made to ground the test equipment to the deck, or to connect a test cable to an aircraft, a breakdown of the gap between the two charged assemblies occurs, and an arc results. Similarily, when test cables between these same units are removed, an arc can be drawn in the process. Considerable damage can be caused to GSE or an aircraft due to the resultant surge current effects, and injury to the equipment operator could take place. The transient might also be capable of triggering ordnance devices such as fuses or detonators.*

Environmental electric fields can couple to cables, discontinuities in shields, and other elements of GSE and GSE-interfaces. While a cable is not considered as an antenna in the general sense, it can nevertheless be a reasonably efficient mechanism by which radiated fields are transformed into electric currents. The amount of interference signal that might appear at some point along the cable is dependent on the standing-waves induced on the cable; this in turn is a function of the frequency and level of the environmental signal, the orientation of the cable relative to that signal, the way in which the cable is terminated, and other factors.

Although ground-support equipment was not directly involved, a good example of the dangers of high-level environmental energy coupling into a system occurred in a recent test of an A-7E aircraft in a simulated flight-deck environment. While the A-7E was running on auxiliary power (see Figure 3-1), a fire broke out in the unit that switched the plane from external to internal power. Figure 3-2 shows the extent of fire damage to the switch. A subsequent analysis of the cause of the fire indicated that environmental energy was being coupled onto the auxiliary power cable to the aircraft, and the extraneous energy caused the relay to chatter and overheat.

Figure 3-1
A-7E Aircraft on External Power, Simulated Flight Deck Environment

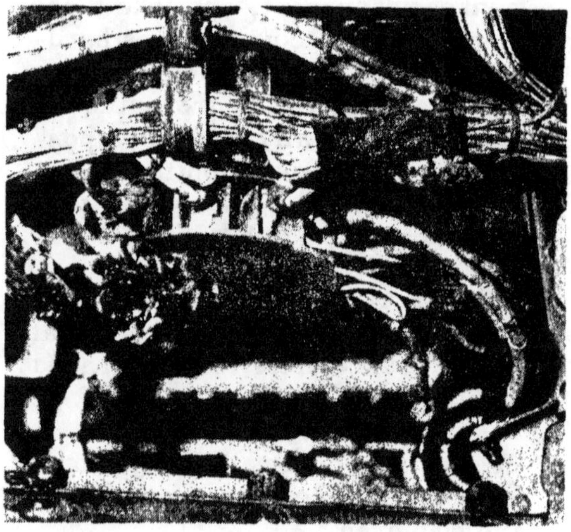

Figure 3-2
Wiring Harness Damage on Aircraft Caused by High-Level RF Radiated Energy

Environmental magnetic fields can couple to any electrically conductive configuration that forms a closed loop. Such loops might include metal ferrules around plastic knobs, cable connector rings that are not screwed onto connectors, and many other. If the field

---

*Discussions of hazards to Ordnance systems is not within the scope of this design guide. The reader is referred to NAVWEPS OD 30393, Design Principles and Practices for Controlling Hazards of Electromagnetic Radiation to Ordnance (HERO Design Guide), for information on this subject.

intensity is high enough, and the loop is oriented appropriately, a high enough current can be generated in the loop to cause it to get hot.

The transfer of environmental energy is not only affected by hardware deployed on the flight deck, but by personnel as well. The human body also displays the characteristics of an antenna, and personnel location on the deck can influence the efficiency of the transfer path of rf energy to the susceptible portions of the GSE or flight system.

3.2.2 Safety Procedures and Design Criteria

It is extremely important that proper procedures be followed in both the design and use of GSE to counteract the potential safety hazards indicated above. These procedures are as follows.

- Prior to making any connections to an aircraft or weapon, it is important that recommended avionics maintenance procedures be followed that relate to grounding the unit to the carrier flight or hangar deck.
- When a test or umbilical cable is to be connected to a GSE, to a weapon launcher, or to other components of a weapons system, the GSE should if possible, first be electrically bonded directly to such launcher or other component by contacting surfaces. A short ground strap may be employed for this purpose. Weapon-launcher systems should be designed in such a way as to provide convenient connection of the test or umbilical cables between GSE, aircraft, weapon, and launcher, after good electrical grounding of the units involved is achieved.
- When a GSE and weapon must be mated before the weapon is racked in the launcher, a potential of rf arcing also exists. Mating of the GSE and weapon skin should be accomplished prior to connection of any cabling, using a separate ground strap if necessary.
- It may be advantageous to know the potential difference between a test equipment case and the skin of an aircraft or the frame of a launcher prior to making any electrical connections between the two units. This can be accomplished using a high impedance input vacuum-tube voltmeter or solid-state VOM (at least 10 Megohms input impedance). If the potential difference is due to static charge accumulation, the voltmeter will probably dissipate the charge in the process.
- A proper filter or rf suppression device should be considered for all lines and hard contacts that must mate with aircraft or environment deck connectors. This is particularly important when the external wiring can act as a receiving antenna and couple efficiently to the radiated energy. Also, the filters must suppress the effects of arc generation (including component burnout), and the generation of any signal that could act as a triggering or ordnance firing signal.
- Uniform twisting of test circuit leads reduces the effective area of the pickup loop created by the leads. In addition, the twisted wires cause cancellation of any current that may be induced in the leads. To be effective, this balancing condition must be maintained throughout the system wiring.
- When required, static discharge resistors should be placed between each line and system ground in the

test circuit. These resistors should be matched and they should not be smaller than 100,000 ohms. In addition to providing a path for static potentials to be dissipated, the resistors will offer a high impedance between each line and ground, and help maintain balance to rf energy.

- Weapons test and firing circuits are almost always balanced to ground, and isolated from case ground so that no direct path for rf energy exists during the handling or loading of weapons. When GSE is developed to test this circuitry, the original firing circuit design philosophy must be maintained. It is important that the GSE contractor become aware of the specific precautions that controlled design, so that they are not violated when the test equipment is connected.

- Test and umbilical cables and connectors between GSE and other equipment must be completely shielded, with the shield properly bonding the GSE, the weapon, and/or the launcher. The connector must be designed with the female portion on the aircraft or launcher and the male portion on the GSE. All connector pins must be recessed, and designed in such a way that the shield mates before the connector pins mate, and the pins unmate before the shield separates.

- Multiple and long test cables create a potential hazard, because of an increased possibility of levels of ground loop currents that can cause weapon system malfunctions. The number of cables should be kept to a minimum, and the cables should always be as short as possible. Flexible cable shielding must be given particular attention; see the Shielding chapter of this design guide for further guidance.

Access doors on the GSE that might have to be opened in the high-level environment (such as battery compartments, fuse panels, etc.) should be kept to a minimum. When it is necessary to have such an access door, any cables that are exposed when the door is open must be shielded. No compartments or circuits that may be susceptible to the environment should be accessable through the access door.

- When external sections of the system are non-metallic, all circuitry behind these sections must be properly shielded and filtered. Non-metallic housings, such as fiberglass and plastic, do not afford any shielding protection, and attempts to coat these materials with a conductive finish to obtain adequate shielding have not been successful.

- As indicated elsewhere in the design guide, cable interfaces are particularly critical from the shielding effectiveness viewpoint. Special attention should be given to the uniformity of pressure applied to the shield around the connector periphery.

- Avoid using metal configurations that can act as shorted turns and thus generate high circulating currents. For - example, avoid "chicken banding" of cables, and use plastic labelling methods instead.

The above guidance is obviously somewhat general, and the GSE designer must tailor his specific requirements and approaches to meet the particular situation he encounters. His major objectives in the safety area must be (a) to assure there is no possibility of an operator shock hazard, (b) to be sure that the test equipment is not damaged by the operational environment, and (c) to assure that neither the GSE nor the equipment being tested is damaged or inadvertently triggered as a result of making or breaking connections.

As one example of options the designer has available, consider the question of grounding a piece of GSE to the flight deck before initiating tests. The discussion thus far has emphasized performing this grounding operation so that a common ground is established for all equipments and the operator.

An alternative philosophy has occasionally been used. During the development of the AN/AWM-38 Armament Wiring Test Set, it was decided that it would be advantageous to isolate the test set from the flight deck, to avoid discharging the aircraft through the test equipment. This was accomplished by using non-conducting operator controls and rubber sleeves over switches and connectors, and by mounting the case on non-conducting legs. To isolate the test device from ground, the exterior surface of the equipment was sprayed with a special high-dielectric strength paint.

3.3 FACTORS INFLUENCING GSE EMC

The achievement of EMC in avionics GSE involves a number of practical factors in both the planning and technical areas. These factors are summarized next.

First, the wide range of GSE functional objectives and test equipment complexities is obvious. Devices that are classified as GSE range from those that perform simple analog and digital checks, to others that include very complex signal processing and simulation tasks. The result of this diversification of objectives and technical approaches is that EMI reduction techniques must be tailored to the particular system being designed. In order to do this, the responsible engineer must not only be aware of the technical alternatives he has available (a major purpose of this design guide), but he must also be able to assess the extent to which operational GSE environments might affect performance of particular equipment.

Second, because of the particular environments in which GSE may operate, and because of EMI reduction techniques and devices that have been incorporated into existing GSE, some assessment of the potential areas of interference to this type of equipment can be made. The more prominent aspects of this assessment are summarized as follows:

- In general, GSE designers of weather deck and hangar deck operated equipment should concentrate on protecting the test equipment against high-power, narrowband radiated electric fields. These fields will be greatest in the HF and microwave ranges. Protection should center around maintaining the equipment case-integrity, and adequately shielding and filtering all external cables. Cable shielding is a particularly critical area and, as will be noted in Chapter 5, special consideration may have to be given to the use of well-shielded cable and to good cable-shield-to-connector electrical contact.

- In general, GSE designers of other below-decks and shore-based equipment should concentrate on protecting the test equipment against interference via the power line as well as against broadband, radiated electric and magnetic fields. Signals below 100 MHz are predominantly of concern. Narrowband radiated electric fields may also be potential interference sources to some shore-based equipment. Protection should center around power line and signal/control cable filtering, including transient suppression. Case and good grounding and bonding practices should be incorporated.

- Many GSE need to operate within only a very restricted frequency range, and the application of filtering techniques outside of this range can solve most anticipated EMI problems. For example, equipment that performs continuity checks, HI-POT tests, signal level checks and similar functions can often operate in a narrow frequency band ranging from dc to a frequency dictated by the sampling requirements of the system. Heavy filtering around the actual fault indicator device above this maximum frequency could preclude the need for extensive shielding. This is one of the reasons some GSE operate successfully on carrier flight decks without maintaining tight shielding integrity (for example, see the comments in Section 3.4 on the AN/ASM-373 Airborne Torpedo Presetter Test Set and the AN/DRM-29 Transponder Set Test Set).

- The use of GSE often requires a change to the configuration of the equipment or system under test, with the result that the performance characteristics of the unit being tested may be different from those representative of normal operation. For example, access panels on aircraft often have to be open so that GSE cables can be connected to the plane. This may expose aircraft components and wiring, including the circuitry under test, to the external EM environment and allow radiated energy to couple into either the weapon system electrical system, the GSE, or both. As a result, GSE designers must take into account the impact of mitigating the shielding or circuit integrity of the system being tested as part of the test equipment development program.

- Production quantities of GSE are almost always fairly small. In general, quantities above one or two hundred are rare. For example, production of specific GSE as of this writing was as follows:

Nomenclature	Quantity
Telemetering System Test Set AN/SKM-10	5
Guided Weapon Test Set AN/DSM-139	25
Airborne Torpedo Presetter Test Set AN/ASM-373A	18
Elective Bomb Fuzing Circuit Test Set A/E 24T-127(XAN-1)	20
Armament Wiring Test Set AN/AWM-38(V)	100
VAST AN/USM-335(V)	7
Aircraft Firing Circuit Test Set AN/AWM-54	1200

The major implication of these production volumes as far as EMC design is concerned, is that it may be more cost-effective to overdesign the system suppression requirements, if this will lead to lower engineering costs. In other words, if there is a total cost of \$15 per unit to add a specific set of RF gasketing to a GSE enclosure, and 20 units will be built, it may not pay to spend design and test engineering time to evaluate whether the gasketing could be reduced in effectiveness or eliminated entirely. On the other hand, a different conclusion might result if a 1200 unit production run was involved. The specific circumstances involved must dictate whether or not EMC overdesign is the more reasonable approach.

- The only methods currently available to assure that flight or hangar-deck GSE design is adequate from the operational environment standpoint are to actually test prototype designs aboard the ship classes of concern, or to use facilities that simulate the flight deck environment. A test bed is available at the Naval Surface Weapons Center, Dahlgren, Virginia, for simulating the flight deck EM characteristics, by illuminating a ground-plane using representative shipboard radiators. Although military EMC specifications and standards such as MIL-STD-461/462 are usually imposed on GSE designs, the direct application of such a spec does not necessarily result in operational equipment compatibility.

- While not normally categorized as EMC, another factor that should not be overlooked by GSE designers is hardware and personnel safety. The different electric potentials that may be built up on the ship decks and superstructure, on aircraft, and on shipboard maintenance personnel can result in effects ranging from annoying transients when cables are connected or disconnected, to component burnout, to shocks to GSE operators. Chapter 3.2 of this guide provides design guidance in the safety area.

3.3.1 EMC Maintenance Considerations

Since the EMC requirements associated with a piece of GSE are expected to be met over the equipment lifetime, some comments are considered appropriate relative to achieving these requirements.

High quality components should be selected for EMC applications, perhaps of even higher quality in some cases than might be used if that component were employed in a non-interference related application. The main reason for this is that the time between failure of an EMC component and its replacement may be very long compared with the equivalent time for a component used for other purposes.

Consider a filter by-pass capacitor that fails by opening. If the capacitor is a part of the equipment power supply filtering circuit, its failure will immediately be evident in improper equipment operation. If it is a part of a filter to keep interference from the power line from entering the GSE, it may be a relatively long time before evidence indicating the capacitor has opened is identified. The designer should pay particular attention in component selection to those elements of the system whose breakdown would not be immediately observable.

Many EMI suppression components are perishable, that is, they lose their effectiveness with use. For example, metallic gaskets may lose resilience after being compressed many times. The mesh in shielded rubber boots may break after continued use. Spring fingers may corrode or become blunt over a period of time. Even toothed lockwashers that have to bite into metal may become ineffective after continued applications.

GSE design engineers must recognize these limitations and, where necessary and possible, provide feedback to those individuals on the program that are responsible for establishing equipment maintenance policies. Such inputs might range from controls on torque applied to gasketed seals, to keeping a log of gasketed compartment openings and closings, to periodic replacement of suppression components in accordance with manufacturer recommendations.

3.3.2 Cost Benefit Considerations

Optimization of EMC design may not be

practical or desirable when considering the benefits to be gained against the engineering costs involved. As pointed out earlier, the typical low GSE production volumes involved, and the frequent model or modification changes, would indicate that, in general, the greater benefit to be derived is from over-design of GSE from the EMC viewpoint rather than attempting to reduce EMI protective measures to a minimum consistent with estimates of expected EM environmental levels.

In many cases, present day operational EM environmental energy levels cannot be estimated accurately, and estimates of future conditions are subject to even greater uncertainity. For these reasons, in addition to those cited above, it is more often a better choice to incorporate added filtering than to consume additional engineering time and cost in attempting to reduce EMC hardware requirements to an optimum level.

This is not to suggest to the GSE designer that he apply every conceivable EMI fix he can think of in the development of a piece of equipment. To the contrary, with a good understanding of how his system must work, the way it must interface with the equipment under test, and the type of EM environment in which it must operate, the designer will be in a position of deciding whether or not various types of interference suppression should even be considered and which reduction approaches are appropriate to employ.

3.3.3 Microprocessors and Digital Systems

Microprocessors and digital systems include all equipments that incorporate linear or digital integrated circuits. EMC problems that might involve these equipments include performance degradation due to external circuit operation. EMC risk is predominantly in the performance degradation area, with broadband emissions a possible but unlikely possibility.

Two basic types of performance degradation are possible. One type is called "upset" and includes such effects as reduced noise margins, logic state changes, and changes in output drive levels of analog and linear circuits. The second type of performance degradation is called "failure" and includes such effects as junction shorting, burnout, and metal failure.

Upset and failure occur at absorbed power levels of 10uW - 10mW and 10mW - 10W respectively. External power levels are highest at their source, and decrease exponentially with distance from the source. Metallic shielding can reduce radiated field levels by factors of 60 dB or more. Pin filters and transient suppressors can reduce conducted signal levels by an equal amount.

The development of integrated circuit technology has led to increased use of microprocessors as integral parts of avionics systems. Microprocessors are sometimes defined as one-chip processors although most are small multi-chip processors. Microprocessors usually include several hardware modules, including a processor module (CPU or MPU), one or more memory modules (RAM or ROM), and various input/output (I/O) modules. Overall, microprocessors can be characterized by their power requirements, noise immunity, impedance, and signalling or logic levels.

Microprocessor and digital circuit technology is extensive and rapidly changing. Currently, there are five basic logic types (ECL, DTL, TTL, MOS and I^2L) plus two basic linear circuit types (bipolar and bifet) that constitute various medium scale and large scale integrations (MSI and LSI). Integrated circuits have traditionally

Table 3-1

INTEGRATED CIRCUIT CHARACTERISTICS

CIRCUIT TYPE	SWITCHING SPEED	SUPPLY VOLTAGE	NOISE MARGIN	INTERF. RISK	SUSCEP. RISK
ECL	Very Fast	5	0.2	Low	High
DTL	Slow	5	1.0	Moderate	Moderate
TTL	Fast	5	1.0	High	Moderate
CMOS	Slow	16	7.0	High	Low

employed silicon technology, but in the future will also employ gallium arsenide technology to gain increased switching speeds. Very large scale integrations (VLSI) will also appear in the future. Physical limits such as gate length (0.2μm) and number of gates per chip (100,000 - 1,000,000) may be reached soon.

Increased densities of microcircuits increases the coupling between individual circuits within a module and increases the risk of undesired broadband emissions on I/O signal leads. Increased density also increases the risk of performance degradation by coupling external signals picked up by I/O leads to internal circuits. Since 5V is the standard logic supply voltage, internally generated noise will be relatively low so that interference risk is low. However, the susceptibility risk is high, especially in high ambient environments such as those created by shipboard radar, and in avionics maintenance areas. For this reason, shielding, pin filtering and good cable-to-connector bonding are important in digital systems design.

Integrated circuit characteristics are given in Table 3-1. Included are switching speed, supply voltage and noise margin, as well as interference and susceptibility risks. In general, the noise generated by the circuits increases with switching speed. ECL circuits are an exception because they have a balanced configuration in which supply current to the gate is the same in the ON and Off states, so that there is no change in current when the gate switches. Although slower, TTL circuitry is about 10 times as noisy as ECL circuitry.

Performance degradation and hence susceptibility risk is inversely proportional to noise margin. This is the input noise voltage, riding on the worst-case logic level, that will cause the gate to trigger and cause latching or data loss. Susceptibility risk increases with switching speed because faster circuits can respond to narrower noise pulses. For this reason, and also because its noise margin is low, ECL circuitry has high susceptibility risk.

Digital computers currently in use in Naval avionic systems include the AN/AYK-14. This computer includes a CPU, memory, and various I/O modules, and is utilized in the F-18 aircraft and LAMPS helicopter.

There are several different types of upsets that can occur in microprocessors and digital systems. These include:

Incorrect External Interrupt

Max Bus Interrupt

I/O Control Lost

ALU Malfunction

Memory Erase

Incorrect Data Transfer

Memory Alteration

In addition to the above, external signals can cause circuit failures from which there is no recovery.

Two types of electrical transients can cause performance degradation in microcircuits. One is electrostatic discharge transients due to handling during maintenance or operation. The other is system transients generated within the digital system or its operating environment. Static discharge transients are usually high level, short duration transients with high source impedances (500V, 150μs, 1500Ω typically). System transients are usually lower level, longer duration transients with lower source impedances (10V, 10μs, 100Ω typically). Studies of microcircuit overstress tolerance have shown that microcircuits can be separated into sensitive vs. nonsensitive categories on the basis of being able to withstand 1000V static discharge transients, and 50V system transients, with durations and impedances as given above. In a test of 40 microcircuit types, representing all common technologies, very few circuits failed the discharge transient test at voltages up to 1000V; however, several circuits failed the system transient test at 50V. The circuits that failed the system transient test included high-speed TTL, first generation I^2L, certain NMOS devices, and certain analog devices. The screening procedure has been proposed for incorporation into MIL-STD-883, Test Methods and Procedures for Microelectronics.

Electromagnetic Vulnerability (EMV) design requirements for avionics systems must reflect the avionics environment, digital system susceptibilities, and potential coupling mechanisms. The environment will include high-level radar and communication signals from co-located or nearby transmitters. Coupling will be direct to I/O lines, or indirect via intercable coupling or possibly common mode coupling. Platforms will provide shielding in some cases, but cannot be depended upon to do so. This is especially true in a maintenance environment where equipments are opened and then operated. Integrated circuit susceptibilities are a function of circuit type, but generally are not low enough to withstand external interference without protection. Integrated circuits and their interconnecting leads should be shielded, and shields should be circumferentially bonded to chassis. In some cases pin filters should be installed at the entry point to the chassis. High-level radar signals are particularly troublesome and should be

Table 3-2

Radiated Susceptibility Test Circuits for Digital Systems

Frequency (MHz)	Power Density (mW/CM^2) Peak	Average
.014-30	.106102	.0265
30-200	.106102	.000265
200-225	20	.09
225-400	1.12	.000265
400-450	1.12	.01
450-1215	.106102	.00265
1215-1365	70	.05
1365-2900	2.65	.000265
2900-3110	130	.09
3110-3235	100	.09
3235-3400	2.65	.000265
3400-3600	40	.09
3600-3700	90	.09
3700-5400	2.65	.000265
5400-5900	250	.09
5900-6000	2.65	.09
6000-7000	2.65	.000265
7000-7900	2.65	.000265
7900-8400	2.65	.09
8400-8500	2.65	.000265
8500-9600	140	.05
9600-12000	2.65	.000265
12000-40000	.00106	.00106

protected against as a matter of course.

A possible way to reduce microprocessor and digital system susceptibility to externally generated interference is to include digital information processing techniques such as forward error correcting codes in the data bus and I/O design. Error-correcting techniques are used effectively in data communications, and would provide similar benefits if used for internal data transfers within digital systems. This is an area that is yet to be explored for protection of avionic systems in an EMV environment.

EMV environment levels are defined in MIL-HDBK-235, and are applied to an avionic system at the discretion of the procuring activity. Contractors do not have a great deal of experience in meeting EMV requirements and thus must use an iterative process to generate a design. In general, the EMV design problem is solved by systematically isolating sensitive circuits from the high level EM environments. Because of weight penalties, shielding requirements need to be determined precisely; however this is difficult to do because of the wide variation (60 dB and 40 dB respectively) in component susceptibility levels and coupling factors. Since analysis techniques are not adequate, development tests must be performed.

In general digital circuits are less susceptible to EMV environments than analog circuits. Digital circuits are more sensitive than analog circuits to pulsed signals, while analog circuits are more sensitive than digital circuits to steady (CW) signals. Since pulsed signals are typically 30 dB higher than CW signals, there are many choices to be made in selecting components. Environment component susceptibility, and coupling information are all very important to successful EMV design.

EMV susceptibility tests on microprocessors and digital systems are based on viewing these devices as components. The modified MIL-STD-461 susceptibility tests are usually specified. A typical change to the RS03 test parameters is contained in Table 2.

In addition to the requirements of MIL-STD-461 one additional RF susceptibility test may be conducted. No malfunction, undesirable response, or change in indication beyond the tolerance given in the equipment specification or approved test plan shall be produced in any equipment when 1000 mV rms RF signals in the 2-30 MHz band are coupled into the interconnecting wiring.

The RF signal is modulated 80 percent with 400 Hz and with the frequency and/or type of modulation to which the test sample is most susceptible. The RF signals are applied to all interconnecting wires or cables between component parts of subsystems, and to wires and cables connected to other equipment.

The RF signals are measured in a one-turn loop with an RF voltmeter as shown in Figure 3-2A. The signal generator has an output impedance of 50 ohms or less and a maximum open-circuit voltage of at least 3 V rms. The test procedure is as follows:

a. With the current probe clamped around the wire bundles under test, set the signal generator for maximum output and slowly scan from 2 MHz to 30 MHz.

b. At those frequencies where a response is noted, move the current probe along the wire for maximum indication. Reduce the signal generator output and determine the threshold of susceptibility of the wire bundle, probe each wire in the bundle separately to determine the susceptible wire(s).

c. Record and include in the test report the threshold of susceptibility data (frequency, amplitude, and description of response) obtained in (b) above. Also include in the test report the corrective action required for each wire that does not meet the required level.

Wire bundles may be tested in sections if the current probe will not fit around the complete wire bundles. Chassis or circuit grounds not normally routed with the interconnecting wiring are measured individually and are not included in the wire bundle under test. Power ground return leads are not placed in the probe.

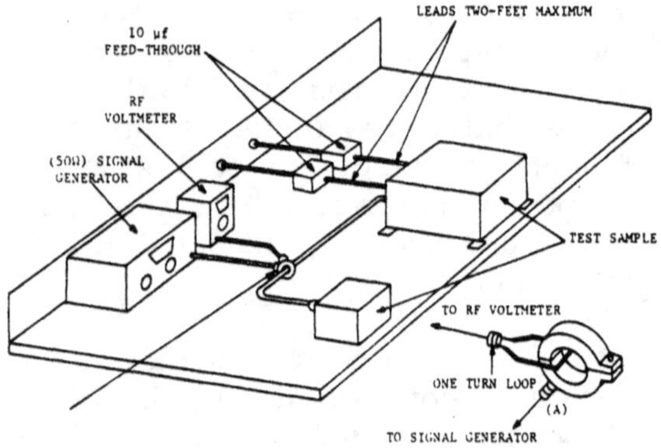

Figure 3-2A

RF Signal Measurement

3.4 EXAMPLES OF AVIONIC AND GSE DESIGNS*

GSE, in order to perform their diverse checkout-associated missions, will have to operate in a variety of different physical environments. The GSE support functions may place individual equipments on aircraft or helicopter flight decks, on hangar decks, in avionics shops, or in a number of other shipboard spaces. Specific units may also be required to perform operations on base flightlines, at base shops, at test and rework facilities, or at other ground-based locations.

Chapter 2 discussed the EM environments which might exist in some of these locations, and intended to provide some understanding of several of the factors associated with the creation of these environments.

This chapter has presented the basic design guidelines applicable to avionics and GSE. Next are provided specific examples of such designs.

3.4.1 Flight or Hangar Deck Operation

GSE designed for flight or hangar deck operation, can be grossly categorized in relation to the system or systems to which they provide support. These systems include:

- Airborne guided weapons
- Weapons delivery systems
- Aircraft

Guided missile test sets provide checks of missile operating status or capabilities by direct measurement of selected missile parameters, or by simulations of a target and noting the missile response to that target. One such device is the AN/DSM-92A Guided Weapon Test Set shown in Figure 3-3.

This test set operates and monitors the WALLEYE weapon to provide a GO/NO-GO verification check of the weapon guidance section. The test set supplies all a-c and d-c power required to operate the weapon with the exception

*Much of this material is based on the Naval Avionics Facility, Indianapolis 1971 Annual Engineering Department report

of the ram air turbine. A video indicator monitors the weapon video outputs; a GO/NO-GO type quality meter provides visual checks of the circuits under test; and various indicator lamps and switches duplicate the functions of the aircraft weapon controls. The test set has built-in self-test capabilities for checking the voltage and phase rotation of the input power and for checking the test set power supplies. An adapter cable is provided to check the video indicator using Aircraft Weapon Control Test Set AN/ASM-184. A power adapter cable is also provided to obtain three-phase power from the aircraft.

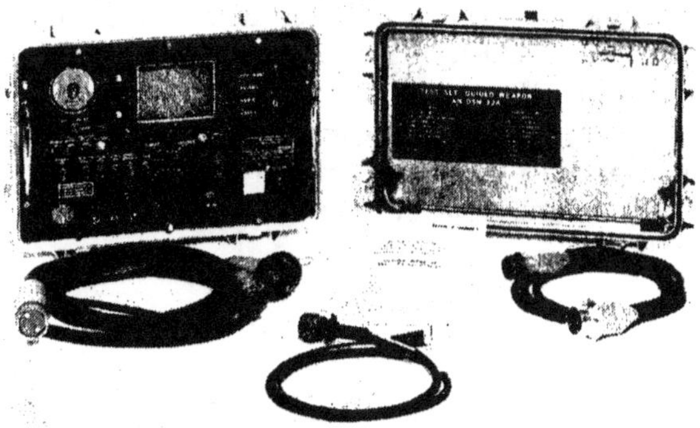

Figure 3-3
Guided Weapon
Test Set AN/DSM-92A

After initial versions of this test set were produced, it was found necessary to reduce the susceptibility of the test set to the radiated environment. This was accomplished by replacing the original test cables with shielded cables. In addition, it was found necessary to develop a modification kit that would reduce EM emissions from the AN/DSM-92 to meet MIL-STD-461.

Another example of a missile test set that is operated on flight and hangar decks is the AN/ASM-373 Airborne Torpedo Presetter Test Set. The equipment is designed to check the C-7821/A and C-8990/A torpedo depth controls installed in SH-3D and S-2G aircraft. The test set also checks the gyro cage and schackle release signals and the delay between them. The basis for test set operation is a stepper switch that sequences dc test voltages in 20 millisecond step intervals.

The major EMI suppression required on the AN/ASM-373 to assure its satisfactory performance in the operational environment included line filters on both the input and output connectors, and diodes across the stepper switch coil to reduce transients. Significantly, it was not necessary to maintain shielding integrity of the test set enclosure, since test equipment degradation occurred only due to environmental signals that were coupled onto its input and output cables. The front panel gasket is made of nonconductive rubber, front panel indicator lamps have plastic covers, and the front panel meter is not shielded.

Weapon delivery test sets provide checks of the status and capabilities of weapon launchers and ancillary devices. Several examples of such testers can be provided.

The Armament Wiring Test Set AN/AWM-38(V) is a portable field test set for performing high potential (HI-POT) and circuit continuity tests on the Aircraft Armament Monitor and Control (AMAC) Systems. It is used in checking out the AMAC system prior to loading nuclear weapons aboard Navy aircraft. The check assures the aircraft readiness as a preflight test of the AMAC system. The test set is shown in Figure 3-4.

In order for this unit to operate in a flight-deck environment without burning out transistors, it was necessary to use very tightly-braided shielded cabling, and to incorporate a "Magnaforming" technique for assuring shielding integrity between the cable and its connectors. Magnaforming will be discussed in the Shielding chapter of this design guide. In addition, filters were required to prevent damage to the test set due to transients introduced onto the cable when connecting to the AMAC systems.

Figure 3-4
Armament Wiring
Test Set AN/AWM-38(V)

The Aircraft Firing Circuit Tests Sets (AFCTS) AN/AWM-53 and AN/AWM-54 are portable, battery operated flight line test sets used to perform stray voltage and firing circuit tests. The AFCTS sets provide a visual indication of unsafe low-level voltages and of the presence or absence of firing current. The indications are furnished by a red NO-GO and a green GO indicator visible under all lighting conditions including red light. The AWM-53 is a hand-held unit, while the AWM-54 is waist-mounted (see Figure 3-5).

During the stray voltage test, the AN/AWM-53 test set monitors the aircraft's firing circuit to determine that the ambient voltage of the firing circuit is not sufficient to detonate electro-explosive devices. The test set monitors voltages with frequencies from d-c to 1 MHz.

During the firing circuit test, the test set monitors the aircraft's firing circuit to determine firing capability. The aircraft must be capable of producing a minimum firing current for a specified period of time before the test set will produce GO indication.

During self test, the test set checks the threshold response of the stray voltage circuit; the function of the lamp and logic circuits; the charge status of the battery pack; and the continuity of the fuse, input circuitry, and the aircraft adapter cable.

The AN/AWM-54 differs from the AFCTS AN/AWM-53 because it is designed to the requirements of MIL-T-21200, utilizes a stray energy sense circuit in lieu of a stray voltage circuit, and employs a resistance sensitive circuit for firing circuit testing in lieu of a current sensitive circuit.

Figure 3-5
Aircraft Firing Circuit
Test Set AN/AWM-54

Extensive EMI reduction techniques have been applied to these units. These include separately shielded battery packs, filtering of battery output leads, filters in signal leads, shielding of cables and good bonding of cable shields to connector shells, conductive gaskets around front panels and other covers, use of conductive finishes, incorporation of waveguide below cutoff recesses for lights, and many others. This is particularly significant, since a relatively large quantity of AN/AWM-54 test sets (about 1500) are being produced.

Aircraft test sets provide the means for evaluating a wide variety of aircraft-associated components, equipments, and subsystems prior to takeoff. Equipment checkouts might range from responses from IFF transponders to turbine temperature testing, or from radar power output measurements to engine trim box operation. Two examples of these types of units follow.

The Transponder Set Test Set AN/DRM-29 (See Figure 3-6) is a portable flight-line test set to be used to test the Transponder Set AN/DRQ-4 (V) and its associated wiring and antennas when installed in a target aircraft or drone. The test set is a portable hand-held battery-powered unit with a built-in battery charger.

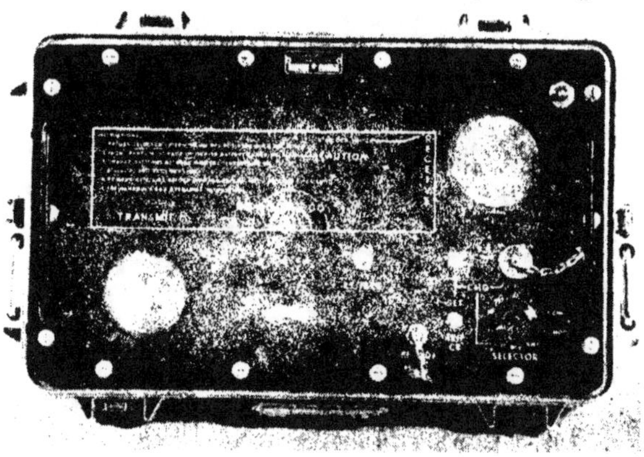

Figure 3-6
Transponder Set Test Set AN/DRM-29

A GO/NO-GO indication is given for the transponder set condition, and for all self-checks for the test set.

The test set transmits to the transponder set an FM signal in the 1.76 to 185 GHz band. A FM signal is then received from the transponder set in the 2.20 to 2.29 GHz band. The received signal is analyzed and a meter gives a GO/NO-GO decision on the transponder set performance.

A microstrip board containing spiral transmitting and receiving antennas, 1760 MHz bandpass filter, delay lines, coupler, and 180 degree hybrid discriminator is used. Another board contains amplifier, oscillator, voltage regulators, and battery charger circuits.

The test set is completely self-contained, and this feature eliminates the need to consider power and signal cable filtering. It was found that the set was relatively insensitive to the flight deck radiated environment, thus no rf gasketing was employed, and case assembly using screw fastners was considered adequate. However, portions of internal shields (particularily those on the front surface of the unit) have been solder-seamed, and microwave absorbent material has been used to cut down on internal reflections.

The TF34-2 Engine Trim Test Set is a solid-state device for trimming installed engines in the S3A aircraft. The unit provides monitoring of temperatures, speed and inlet guide vane angles during engine flight checkout. Particular attention has been given in design of the test set to interference from radiated electromagnetic sources, and an environmentally sound case and shielded cables are employed.

RF gaskets are used extensively to provide a good electrical seal between the front panel of the test set and the box.

Push button and toggle switches incorporate neoprene-filled conductive gaskets, and rotary switches include EMI shaft seals. Indicating lamps are EMI-protected and backed with neoprene-filled conductive gaskets, and LED displays are behind a wire screen. A thumbwheel switch for presetting parameters is located behind a door whose edges are sealed with a neoprene-filled conductive gasket. Cable shields are required to have 95% coverage.

The Multiapplication Engine Vibration Test Set is used to amplify and display signals produced by self-generating Piezoelectric accelerometer or velocity type transducers attached to gas turbine engines. It incorporates considerable EMI suppression, including weather seal/conductive gaskets, conductive boots on toggle switches, and bulkhead-mounted power and signal line filters.

Conductive mylar is used to shield the LED tachometer display window, while other indicating lights and meter dials are protected with wire mesh screens. All test set cabling is shielded.

3.4.2 Shop Testing

GSE designed for shop use, or for use at test and rework facilities, typically involve more comprehensive equipment or system checkout capabilities than is normally possible on hangar or weather decks because of space and time limitations. These units are normally ac powered, frequently non-portable, and sometimes contain or are associated with mechanical test fixtures. Compliance with MIL-STD-461A will generally insure the electromagnetic compatibility of this type of equipment.

4.0 CONTROL & TEST PLANNING

4.1 GENERAL

The success in controlling EMI effects during GSE design is very often dependent on the extent to which an electromagnetic compatibility management program is established and implemented. This program is necessary whether the equipment under development is a relatively simple and self-contained HI-POT and continuity tester, or a roomful of missile test and simulation gear - although the type of management effort required in these two instances will obviously be distinctly different.

This chapter of the design handbook provides guidance to the GSE developer in organizing and implementing efficient interference regulation. Specifically, it discusses two of the major documents that will be needed to achieve a high level of EMI control. These documents, which are normally prepared by the design activity, are:

- The EMC Control Plan
- The EMC Test Plan

4.2 THE EMC CONTROL PLAN

4.2.1 The Role of the Control Plan

The Interference Control Plan is the formal document that will place into effect the control procedures and activities to be employed during an equipment design/development cycle. In the case of a hardware contractor, the plan will respond directly to commitments of a bid proposal or contract. For government in-house developers, the plan will reply to directives that define or interpret the design and operational objectives of the GSE.

The Interference Control Plan is the document that should communicate to each engineer, each department head, each subcontractor, and each representative of the procuring agency the EMC work effort, emphasis and guidance to be imposed during the project. It is the mechanism for establishing EMC engineering and management milestones, including points in time at which the effectiveness of the program can be judged, and is the device in which individual and group responsibilities are defined.

The importance of the EMC Control Plan cannot be overestimated. Since it will outline the entire program by which subsequent EMC documentation, requirements, specifications, test methods, and criteria are controlled, the manner in which the plan is developed can be seen to have an important bearing on successful achievement of electromagnetic compatibility.

Even when not called for separately, an EMC Control Plan is automatically imposed on a development when compliance with certain interference specifications is required. For example, both MIL-STD-461A and MIL-E-6051D identify the need for such a plan and provide outlines of general plan content. The initial version of the plan is usually prepared by the equipment developer and distributed within 90 days after program initiation or contract award. However, periodic review and update is encouraged (typically on a six-month basis) to be sure that the document continually reflects the current objectives and needs of the GSE development effort.

4.2.2 The Contents of the Control Plan

The basic outline of a representative EMC Control Plan is as follows:

1. Introduction
2. Application Interpretation
3. Specification Interpretation
4. Program Management Requirements
5. Design Requirements
6. Test Requirements
7. Quality Assurance Tests
8. Definition of Terms

A discussion of the possible contents of the various sections of this outline should provide a better indication of the role the plan can play in support of GSE systems engineering.

4.2.2.1 Introduction

This section presents the scope and objectives of the control plan. It should introduce the system to which the plan is applied in enough detail to acquaint the reader with the functional characteristics of the equipment involved and the EM conditions it is required to meet. Plan review and up-dating procedures should be specified; as indicated previously, a review cycle of no more than six months is recommended.

4.2.2.2 Applicable Documents

All pertinent standards, specifications, drawings, plans and reports related to EMI control are listed in this section. The EMC documents developed for the system by the military, the contractor, and subcontractors and vendors should be included. The effective dates of all documents and any restrictions or limitations on their use should be stated.

4.2.2.3 Specification Interpretation

Many EMC specifications and standards are directed toward establishing limits on individual components or units and, by themselves, do not guarantee any specific degree of compatibility when the units are assembled as part of a larger system. A particular limit may be set too low, thereby causing devices to be very susceptible to their EM environment, or be established too high, resulting in an unnecessary expenditure of suppression dollars. For this reason, and because two specifications and standards may conflict in some areas, it is necessary to analyze and modify specification requirements so that they more realistically represent the limits of expected operating conditions.

It should be noted that the need to tailor specifications to an anticipated EM environment applies to small and large GSE alike. The broad scope of a specification such as MIL-STD-461 may require modification even in the case of a battery powered, hand-held test set, when that test set must operate in a flight deck environment, or in direct contact with a Sidewinder missile launcher.

This section of the EMC Control Plan should outline the approach the developer will take to interpret designated specifications, establish their applicability, and propose modified or additional requirements. Many of the analysis techniques to do this evaluation are contained in this design guide.

4.2.2.4 Program Management Requirements

One of the most important functions of the EMC Control Plan is that of documenting the responsibility and authority of the individuals who will direct and implement the Electromagnetic Compatibility Program of the equipment developer. The success of the EMC Program will rest upon its effective administration.

The management portion of the plan should establish the organization and responsibilities of the EMC group, stating its functions and objectives. In general, this group participates in periodic design reviews, acts in an advisory capacity on EMC matters, performs EMC testing, and generates

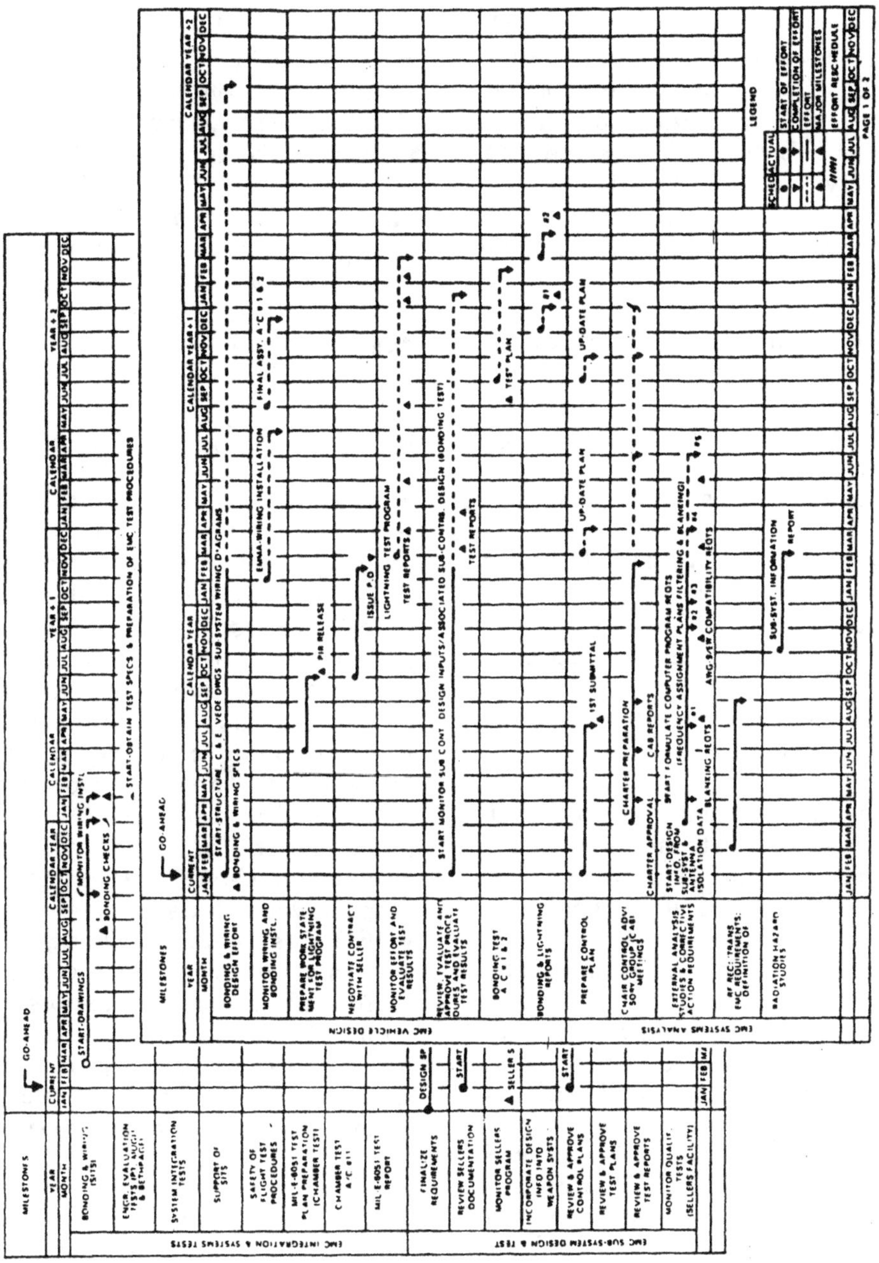

Figure 4-1 Typical EMC Program Milestones

recommendations for the solution or further definition of EMC problems uncovered. Management of electromagnetic compatibility extends to the standards and specifications imposed contractually on vendors and suppliers of equipments and components.

The management portion of the EMC Control Plan should also specify all EMC documents and reports to be submitted during the course of equipment development. Target dates and milestones for the EMC effort should be included. A typical EMC program milestone chart is shown in Figure 4-1[1]. At a minimum, the milestones should include:

a. Preparation and submission of the EMC Control Plan
b. Qualification testing of equipment to interference and susceptibility requirements
c. Preparation of the system EMC test plan
d. Preparation of the system EMC test procedures
e. Initiation of the EMC test effort
f. Initiation of the general acceptance test
g. Any other milestones or target dates that are considered pertinent.

Another key management action for GSE's of even moderate size and complexity should be the establishment of an EMC Advisory Board, Group, or Committee (EMCAB, EMCAG, EMCAC). This activity would act in an advisory capacity to the EMC program, to assist in resolving problems that may arise, expediting solutions, and establishing channels of coordination. NAVAIR AR-43, "Electromagnetic Compatibility Advisory Board, Requirements for," provides guidance for the formation and operation of an EMCAB and is written for use on EMC programs organized for NAVAIR contracts. EMCAB activities include EMC program review, discussion and evaluation of potential problem areas, definition of specific problems, determination of possible problem solutions, recommendation of best solutions to the responsible activity, and review of results of the implementation of the solutions.

Members of the board are typically representatives of the developer/contractor, subcontractors, vendors, and the procuring activity. The charter for the board is usually included in the EMC Control Plan. Typically, the charter details membership, defines the normal interval between regular meetings and the meeting locations, and identifies the schedule for EMC design reviews.

4.2.2.5 Design Requirements

The means of assuring satisfactory EMI/EMC design control should be presented in the EMC Control Plan. This includes a statement of the design guidelines that are to be used, how the the design guidelines are to be presented to the design organizations, the controls to assure that the design guidelines are being followed, and the means by which the design organizations are to be informed of the EMI/EMC philosophies of the EMC Control Plan. The guidelines should be written to apply to management as well as to the individual designers. Design organizations should be told of these guidelines in formal training sessions, through publications prepared and distributed by the responsible organization, and in conferences with representatives from all affected organizations.

As a basis for establishing GSE design guidelines, it is necessary to identify degradation or performance threshold criteria that must be met at various locations within a system, and the consequence of failure to meet these criteria. One way in which the impact of interference may be categorized is as follows:

Category I: Subsystems or equipments where interference could result in loss of life or vehicle, aborted mission, costly delays, and unacceptable reduction in systems effectiveness.

Category II: Subsystems or equipments where interference could result in injury, damage to vehicle, or a reduction of systems effectiveness.

Category III: Subsystems or equipments where interference will cause only annoyance, minor discomfort, or loss of performance that does not decrease desired system performance.

For equipment in Categories I or II it may be advisable to develop additional degradation criteria beyond the limits specified in standards.

In addition, it may also be desirable to specify interference safety margins as a function of these categories, to assure that more critical operations receive a higher level of protection (e.g., 6 dB margins for Categories II and III, 20 dB margin for Category I). This problem should be addressed in the control plan.

The purpose of this design requirements task is to assure that every essential EMI/EMC control measure is incorporated as early as possible into equipment design. Otherwise, costly time delays, major redesigns, or serious compromises of equipment operation will result.

In most cases, a subsystem from one organization will ultimately be joined with a subsystem from another organization. The design portion of EMC Control Plans submitted by the two different organizations must present EMC design policy in enough detail to assure the procuring activity that the design of one system will not be violated by the design of an associated system.

Some of the areas for which EMC GSE design specifications must be prepared are:

a. Electrical bonding and grounding: Equipment - to - ground reference plane, electrical interface with other equipments or systems and subsystems, electrical power returns, and conductor shields.

b. Shielding: Equipment and subsystem case shielding, shielding provisions of the system structure, shielding and twisting of conductors, and anticipated electromagnetic and electrostatic environments.

c. Transient control: Suppression of transients from inductive sources, suppression of contractor transients, and surge limiting within the system power profile.

d. Radiated signal control: Spurious signals from intentional radiation sources, unintentional radiation sources, control of response to radiated signals, and projected intentional radiation environment created by the system.

e. Interference and susceptibility prediction: Frequency allocations, antenna locations, antenna patterns, and signal levels, when applicable.

f. Cable and/or conductor routine consideration: Conductor separation and isolation, and location of equipments and subsystems.

g. Electromagnetic radiation hazard (RADHAZ) and hazards of electromagnetic radiation to ordnance

(HERO): Design policy and test procedures to prevent electromagnetic radiation hazard to personnel or harmful operation of weapon systems and electro-explosive devices.

h. Any special EMC considerations: Conditions peculiar to the type of GSE involved, or to the nature of its intended mission may call for more stringent EMC control or variations to the cited specifications.

4.2.2.6 Test Requirements

The test requirements of the EMC Control Plan should provide for:

a. Preparation of an EMC Test Plan that defines equipment, subsystem, and system level testing to be performed, and that, in the case of a contracted program, is a requirement over which the procuring activity has approval authority. The next section of this design guide (Section 4.3) will discuss test plan preparation in some detail.

b. A statement of criteria used in determining which equipments and subsystems shall be tested.

c. Identification, by nomenclature and serial numbers if possible, of equipments and subsystems to be

TABLE 4-1 EMC CONTROL PLAN CHECKLIST

1. Introduction

1) Is the purpose of the document specified?

2) Is the document applicable to all subsystems and components?

3) Is the document applicable to all subcontractors and to all off-the-shelf purchased parts and components?

2. Management Section

1) Is there a positive statement concerning program management's intent to support fully an effective EMC control program?

2) Is there a sufficient organizational presentation to assure ready implementation of the EMC control program?

3) Have all critical organizational elements, including program management, EMC engineering, testing, quality assurance, materials, publications, and system design been identified in the organization?

4) Does the control plan specify all applicable EMC documentation to be provided?

5) Has the responsibility for preparation of critical EMC documentation including control plans, procedures, test plans, and other documentation been made?

6) Has responsibility for testing at subsystem and component levels as well as system levels been made?

7) Have appropriate distinctions been made between qualification acceptance and compliance testing?

8) Is an individual or group of individuals designated to provide representation on behalf of EMC during critical design reviews and preliminary design reviews?

9) If the representation provided above is made by a team of individuals, is at least one of the individuals an EMI engineer?

10) Is the organization responsible for preparing, approving and submitting deviation requests and/or waivers against EMC requirements, limits, or specifications been identified?

11) Does the document provide for related status of EMC efforts in relation to critical milestones of the entire program?

12) Does document show program milestone for EMC?

13) Does the document provide for a duly constituted EMC board that will merge management and EMC engineering organizations?

14) Is it clear how design testing and management organizations interface together?

15) Does descriptive information on the EMC organization make clear how liason among the elements are to be carried out?

16) Is there provision for EMC presentations to be made concurrent with preliminary design reviews and critical design reviews?

3. Applicable Documents

1) Have all applicable systems docu-

tested.

d. A statement of criteria used in determining the extent of EMC tests to be performed.

e. A design review to assure the equipments and subsystems have complied with the applicable design requirements of the EMC Control Plan.

In large GSE developments, many different organizations may be involved in testing equipments and subsystems to EMC specifications and standards. When this is the case, the EMC Control Plan should indicate the particular tests to be performed by each of the organizations and specify how the results of these tests are to be coordinated from a systems point of view. Active liaison is especially important between testing organizations performing subsystem and equipment tests to MIL-STD-461A and MIL-STD-462 standards, and those performing tests to assure overall system performance.

This is true because the results of the subsystems and equipment level tests are used both to indicate points to be monitored and to establish tolerable limits for the systems level testing.

TABLE 4-1 EMC CONTROL PLAN CHECKLIST (Continued)

ments for systems and subsystems been listed?

2) Have the latest or applicable version of EMC specifications been listed?

3) Have all applicable documents relating to reliability and quality assurance been listed?

4) Have all applicable specifications regarding materials and processes been indicated including those for metal treatment, finishes, chemical films, etc.

5) Have all ancillary drawings and specifications regarding grounding planes, harnessing schemes, cable routing, etc., been listed?

6) Where more than one specification applies, is the order of precedence indicated?

7) Is it clear which way the control plan interfaces with applicable directives and all other specifications regarding EMC?

8) If separate individual specifications have been developed for system level testing and component and subsystem level testing, have these been listed and in the correct order of precedence?

9) Have all special provisions for deviations from specification limits, frequency ranges, test procedures, etc., been listed?

4. Analysis and Design

1) Have key operational parameters of the overall system been adequately and concisely described?

2) Have the operating frequency and power output levels of all emitters and susceptible devices been delineated?

3) Has a general presentation been provided on potential problem areas?

4) Have critical potential victim elements been defined?

5) Has an analysis been made of the power system utilization and potential susceptibilities via the power system, including grounds?

6) Have harmonics been considered when providing a table for predicting interference problems?

7) Have potential nonlinear elements in the system been identified which could generate spurious responses?

8) Have basic and specific guidance been provided toward the following areas?

a) Selection of interference-free parts and components?

b) Bonding and grounding.

c) Wiring and cabling, including connectors, grouping, bundling, and routing.

d) Shielding and filtering of sensitive circuits.

e) Printed circuit coupling.

f) Diode and transistor logic levels.

g) Filters.

4.2.2.7 Other Information

Additional information is normally provided in the EMC Control Plan, including definitions of terms used; methods to be used in the education of design, management, and test personnel to EMC awareness, system quality assurance and production control; EMC maintenance procedures for operational equipment; or any studies that will be initiated for improvement of the EMC control program.

4.2.3 Control Plan Checklist

The checklist contained in Table 4-1[2], in addition to the design guidelines sections of several chapters of this publication, provides good initial guidance on the features to incorporate in an EMC Control Plan. This must be supplemented with any requirements unique to the equipment or system under development.

4.3 THE EMC TEST PLAN

The purpose of an EMC test plan is to provide the details necessary to implement the collection of equipment or systems performance characteristics pertinent to EMI evaluation. It should precede any major testing effort as a matter of good engineering practice, even if it does not represent a requirement imposed by the Navy.

Very often, the requirement to prepare the EMC test plan is levied by a specification or standard. For example, MIL-STD-461A requires such a plan be submitted and approved prior to formal compliance testing. A 90-day review of a plan by the cognizant organization is typical.

A suitable GSE test plan should contain but not be limited to the following:

- Justification for running each test, including how the test will support any GSE performance evaluation, and the application of the test results in identifying and resolving real or potential problem areas.
- A description of each test to be performed, including the test instrumentation to be used, the test configurations to be employed, the status of all controls and switches that can affect the test, the test modulations to be employed, the specific location (by terminal or pin identification) of test points, the performance criteria to be used, and the output format for the results of each test.
- Detailed step-by-step test procedures.
- A description of how the required characteristics of any simulated signals will be established, and how these signals will be obtained and synchronized.
- A detailed description (including characteristics, operating procedures, calibration procedures, and circuit diagrams) of any specialized or unique test equipment or test component not normally employed for the particular test in question.
- A description of how the uncontrolled ambient electromagnetic environment will be measured, and the action to be taken to prevent the uncontrolled ambient level from introducing errors into the test measurements.
- A description of procedures to be employed to assure proper operation of the systems or equipments under test prior to and during the testing.
- A description of test instrumentation calibration procedures to be employed.

An outline of a typical test plan is shown in Table 4-2.

When preparing the test plan, care must be taken to be sure that conditions that result in maximum interference impact are tested. For example, emitter/repeater pairs should be studied to be

TABLE 4-2 OUTLINE OF TYPICAL TEST PROCEDURE

<u>Contents</u>
    Includes major paragraphs and appendices

<u>Administrative Data</u>
    Include the following headings:
        Purpose of Test
        Applicable Documents - Specifications
        Standards, Drawings
        Test Sample Identification
        Manufacturer
        Security Classification
        Test to be Performed by
        Test Location
        Date of Test
        Disposition of Test Sample

<u>Scope</u>
    Define the scope of the EMC tests to be performed, and identify the applicable EMC specification

<u>Description of Test Sample</u>
    Functional and Physical Description
    Connectors Associated with the Test Sample
    External Controls

<u>EMI Tests To Be Performed</u>
    Interference Tests
    Susceptibility Tests
    Tests Not Applicable
    Exceptions to EMI Control Specification

<u>Anticipated Interference</u>
<u>Test Equipment</u>
    Test Equipment Required
        List all ground support equipment(GSE), EMC test equipment, hardware, special test interfaces, include part no., S/N and calibration dates
    Test Equipment Calibration
    Line Impedance Stabilization Networks (LISN) or Feed-Through Filters
    Test Equipment Bonding

<u>Test Conditions</u>
    Test Location
    Test Arrangement
    Test Sample Bonding
    External Loads
        Describe electrical and/or mechanical loads to be used
    Lead Lengths and Position with respect to Ground Plane

sure that possible tuning arrangements in which fundamental, harmonic, subharmonic, or spurious frequency overlaps can occur are investigated. In a similar vein, extensive probing should be specified for radiated tests to be sure that the most serious case leakage paths are located.

TABLE 4-2 OUTLINE OF TYPICAL TEST PROCEDURE (Continued)

1. Primary Power Leads
2. Interconnecting Leads
3. Load Leads

Shielded Leads
Coaxial Cables
Test Sample Operation

1. Modes of Operation
2. Steady State Interference Test Mode
3. Transient Interference Test Mode
4. Susceptibility Test Mode
5. Functional Tests
6. Control Settings

Accuracy of Test Equipment
Define recalibration conditions, etc.

<u>Detailed Test Procedure</u>

Operation of Interference Measuring Instruments
(State frequency range, detector function and calibration)
Selection of Test Frequencies
Ambient Interference Measurements
Test Limits
Describe design parameters and specify acceptable variations
Future Criteria
List conditions that constitute failure due to susceptibility, i.e., where changes from operational test values beyond allowable limits or degradation of performance occurs
Specific EMI test procedures for each EMI test to be performed
Test #1, Conducted emissions from GSE power leads (30 Hz to 20 KHz)
Test #2, Conducted emissions from GSE output leads (30 Hz to 20 KHz)
(Continue list as required)

<u>Test Report</u>

Acknowledge that a test report will be prepared

<u>Test Personnel</u>

Acknowledge that personnel performing test will be identified in test report

<u>Appendix I</u>

Includes test arrangement diagram, electrical loads, test equipment list

<u>Appendix II</u>

Sample data sheets and graphs

The requirement for testing in a low ambient environment usually dictates use of a shielded enclosure (See Section 9.2). The test plan must be particularly detailed regarding precautions to be taken when using an enclosure, to be sure that the test results obtained are meaningful. The plan should define the physical and electrical requirements of the enclosure and ancillary equipment within the unit (such as its inside dimensions, the size and type of ground plane, the method to use to ground or bond the

		SPECIFICATION:MIL-STD-461			TEST METHOD: RE02		
FREQUENCY RANGE	ITEM	MANUFACTURER	MODEL	S/N	CAL. PERIOD	DATE OF LAST CAL.	ACCURACY
.014-1000 MHz	RIFI Meter	Singer Metrics	NF-105	1824	6 months	12-15-72	freq ±3% amp ± 2dB
.014-.14 MHz	Tuning Unit	Singer Metrics	TX/NF-105	3106	6 months	1-12-73	freq ±3% amp ± 2dB
.15-30 MHz	Tuning Unit	Singer Metrics	TA/NF-106	3057	6 months	12-20-72	freq ±3% amp ± 2dB
20-200 MHz	Tuning Unit	Singer Metrics	T1/NF-105	1817	6 months	10-13-72	freq ±3% amp ± 2dB
200-400 MHz	Tuning Unit	Singer Metrics	T2/NF-106	1817	6 months	1-15-73	freq ±3% amp ± 2dB
400-1000 MHz	Tuning Unit	Singer Metrics	T3/NF-105	1817	6 months	2-23-73	freq ±3% amp ± 2dB
.014-.15 MHz	41" Rod Antenna	Singer Metrics	VR1-105	L-428	12 months	10-21-72	± 2dB
.15-30 MHz	41" Rod Antenna	Singer Metrics	VA-105	1817	12 months	10-12-72	± 2dB
20-200 MHz	Bi Conical Antenna	NSL					
200-1000 MHz	Log Spiral Antenna	EMCO					

TABLE 4-3
SAMPLE TEST EQUIPMENT LIST

TABLE 4-4 CHECKLIST FOR PREPARATION OF TEST PLAN

1. <u>Introduction</u>

1) Is the purpose of the test indicated?

2) Are applicable drawings, documents, specifications, and standards listed? Is the listing complete?

3) Is the test sample or system completely identified?

4) Is it clear who would perform the test and at what location?

2. <u>Description of Test Sample</u>

1) Is the test sample or system under test functionally and physically identified?

2) Is it clear whether the system under test is a production sample, prototype, or engineering model?

3) Are external controls, ground support equipment, or ancillary systems required for testing identified?

4) Is the operational configuration of the system or sample during test clearly specified?

3. <u>Test Conditions</u>

1) Is the test facility and test location completely identified and described?

2) Are the facilities for test including shielded enclosures, power line filtering, power line load distribution completely specified?

3) Are the ambient conditions clearly identified?

4) Is the arrangement of the test system during test clearly specified?

5) Do the test setup drawings and specifications completely identify the power distribution to the system under test, location and orientation of exercising equipment and placement location of breakout boxes monitoring circuitry and equipment locations?

6) Are external loads clearly labeled?

7) Is the test sample bonding in accordance with MIL-STD-461?

8) Are test and monitoring points appropriately listed?

9) Are the system operating modes clearly delineated?

10) Are separate tests to be performed in terms of steady state interference and susceptibility and transient performance?

11) If so, are these completely delineated?

12) Are susceptibility test modes responsive to stimuli requirements of the applicable specifications?

13) Are all possible susceptible parameters listed and specified?

14) Are the initial system or sample operating conditions and control settings listed for susceptibility and interference testing?

15) Are critical cables between support equipment and bench test equipment shielded in order to eliminate interaction with the test setup?

16) Has care been taken to ensure that the test monitoring equipment themselves are not susceptible or interfering sources?

4. <u>EMC Test to be Performed</u>

1) Are all tests in consonance with

ground plane to the enclosure and to the equipment under test, the room attenuation characteristics, etc.). It should also identify the locations of the test equipment and the GSE within the enclosure, and the interconnecting lead dress, to minimize reflection and undesired coupling effects.

The test procedures themselves should be sufficiently detailed to enable repeating the same test at a later date if desired. The procedures should include the output criteria to be employed or monitored, the specific test points at which signals are to be injected or removed, the matching techniques to be

TABLE 4-4 CHECKLIST FOR PREPARATION OF TEST PLAN (Continued)

applicable specifications?

2) Are deviations and special exceptions clearly documented?

3) Is the listing of each test to be performed complete?

4) Are all legitimate operating modes of the systems considered for these tests?

5) Is the test listing compatible with performance requirements?

6) Are there any tests that are not required by reason of prior qualification? If so, has the prior qualification been evaluated with regard to the applicable specifications?

7) Are exceptions being made to the EMC control plan?

8) Are the EMC tests to be performed in complete agreement with the EMC control plan?

9) Are there any deviations from the EMC control plan test requirements?

10) Are test limits thoroughly described for interference?

11) Have special tests been identified in relation to unique system requirements?

12) Have susceptibility threshold limits been carefully interpreted?

5. Test Equipment

1) Has all of the required EMC test instrumentation been listed?

2) Are the support equipments for monitoring and exercising the system or sample been made complete?

3) Have special test interfaces of breakout boxes been identified?

4) Is the equipment list complete?

5) Does the equipment list include all applicable calibration dates?

6) Does the calibration conform to MIL-C-45662?

7) Have all special-purpose devices which have been identified and described, including feedthrough filters, LISN's and voltage probes?

8) Does the list of equipment include all applicable operational parameters including frequency range sensitivity, etc., as applicable to the parameters to be investigated?

9) Is the accuracy specified?

6. Detailed Test Procedures

1) Do the test procedures conform to MIL-STD-461, MIL-STD-462, MIL-E-6051D, or specific specification?

2) Have test frequencies been selected on the basis of real system operational requirements?

3) Have deviations necessitated by system operational parameters of from commonly used specifications been clearly delineated?

4) Have the test limits been accurately identified?

5) Have the failure criteria been interpreted into meaningful test procedures?

6) Is there a specific EMI test procedure and detail including the operation of each monitoring and EMC test instrument for each test procedure?

7. General Requirements

1) Does the test plan acknowledge that the test report will be prepared?

2) Is there a sample test data sheet, including provisions for all critical system measurement equipment parameters, antenna correction factors, cable losses, etc.?

3) Are the test personnel to be performing the tests identified?

4) Are all important test arrangement diagrams, electrical loads, test equipment lists, etc., included?

5) Does the documentation methods proposed conform to MIL-STD-831 or the applicable documentation standard?

used at these points (including detailed descriptions of couplers, matching networks, etc.), test modulations, positions of all adjustable controls, test setup block diagrams, and any other factors that have to be defined before a test can be unambiguously performed.

A tabulation of planned test equipment should accompany the test plan. This tabulation should at least include general calibration and accuracy data (See Table 4-2), but should be expanded considerably for unique or nonstandard devices. In many cases (such as when testing in accordance with MIL-STD-461), approval must be obtained on instrumentation before it can be used.

It must be appreciated that an EMC Test Plan is an evolving document. The initial draft of the plan can be expected to emphasize the types of tests that should be performed and the general philosophy to be implemented in the testing program. As the GSE design proceeds, the plan should be updated and expanded so that, when detailed test procedures are needed, they will be available. The checklist of Table 4-4 offers a good start in identifying items to be considered in test plan preparation.

BIBLIOGRAPHY

1. NAVAIR 5335, Naval Air Systems Command, Electromagnetic Compatibility Manual

2. P. J. Mondin, "Control and Test Plans in EMC Programs," Seminar on Electromagnetic Compatibility, May 1973.

5.0 SHIELDING

5.1 GENERAL SHIELD DESIGN CONSIDERATIONS

Shielding has two main purposes: (a) to keep radiated EM energy confined within a specific region, and (b) to prevent radiated EM energy from entering a specific region. Thus, shielding is essentially a decoupling mechanism used to reduce radiated interactions between equipments, or between portions of a given equipment.

Shielding requirements, as far as representative GSE design is concerned, are of greatest importance with regard to shipboard topside units. As indicated previously, a major consideration in the development of flight and hangar deck test equipment is the maintenance of case shielding integrity, and the protection of associated cables from the expected high-level EM environment.

The shielding effectiveness of an equipment or subassembly enclosure is a complex function of a number of parameters, the most notable of these being the frequency and impedance of the impinging wave, intrinsic characteristics of the shield materials, and the numbers and shapes of shield discontinuities. The equipment design process consists of establishing undesired signal levels on one side of a proposed shielding barrier, estimating tolerable signal levels on the other side, and trading off shield design options to achieve the necessary effectiveness level.

This chapter will identify many of these shield design options. It is subdivided into discussions of solid shields (including multiple wall and thin-film shields), non-solid shields, cable shielding and the maintenance of shielding integrity through a connector, and shield discontinuities (including seams, gasket requirements, the use of waveguide attenuators, conducted glass, and other aspects).

Throughout the discussion, it should be recognized that shielding, although an important technique for reducing EM interference effects, is not the only technique available for this purpose. Application of shielding techniques should not be made without due regard to the roles that filtering, grounding, and bonding will play in the suppression program.

5.2 SOLID SHIELDING MATERIALS

5.2.1 Shielding Analysis

Interference signal attenuation by a solid shield is due to two distinct effects: (a) reflection of the interference wave at the air-metal boundary as the wave strikes the metal surface, and reflection at the metal-air boundary as the interference wave emerges from the metal shield and (b) absorption of the interference wave in passing through the metal shield between the two boundaries. The first loss is generally called Reflection Loss and the second is called Penetration or Absorption Loss. The combined loss due to these two effects is considered the shielding effectiveness of the shield.

It is convenient to separate the initial reflections from both surfaces of the shield from subsequent reflections that may take place at these surfaces. These effects are called the Single Reflection Loss and Multiple Reflection Correction Term, respectively. Under circumstances when the absorption loss is greater than about 15 dB, the multiple reflection term can be ignored.

Using transmission line theory, the Shielding Effectiveness, S, of solid shielding materials is defined as[1, 2, 3]

$$S = A + R + B \qquad (5.1)$$

where

A = Penetration or Absorption Loss by the material, in dB

R = Single Reflection Loss from both surfaces of the sheet, in dB

B = Multiple Reflection correction term, in dB.

Magnetic shielding depends primarily on absorption losses, since reflection losses for magnetic fields are small for most materials. Electric fields are readily stopped by metal shields because large reflection losses are easily obtained. The penetration (absorption) loss, which is essentially independent of wave impedance, is the same for electric and magnetic fields.

The absorption loss term, A, can be described by the relationship

$$A\ (\text{in dB}) = 3.34\sqrt{ufg}\ d \qquad (5.2)$$

where

d = shield thickness, in inches

g = material conductivity, relative to copper

u = material permeability, relative to copper

f = frequency, in Hertz

For a given material, absorption loss in dB at a specific frequency is a linear function of material thickness. The characteristics of the material that influence this loss are its conductivity and permeability. Table 5-1 lists values of conductivity and permeability relative to copper for a variety of useful shielding materials, and indicates calculated values of absorption loss at 150 kHz.

The <u>single reflection loss</u> term, R, is wave-impedance dependent, and can be described by the relationships:

R(low impedance source) =

$$20 \log\left[\frac{0.462}{\sqrt{\frac{f_g}{u}}\ r} + 0.136\sqrt{\frac{f_g}{u}} + 0.354\right] \qquad (5.3)$$

R(high impedance source) =

$$354 - 20 \log\left(\sqrt{\frac{uf^3}{g}}\ r\right) \qquad (5.4)$$

R (plane wave source) =

$$168 - 20 \log\left(\sqrt{\frac{uf}{g}}\right) \qquad (5.5)$$

where r = source to shield distance, in inches. These equations are plotted in Figures 5-1 through 5-3.

The amplitude of a wave reflected from solid shielding material depends upon the degree of mismatch between the impedance of the air medium and the impedance of the shield. The impedance of the impinging wave is given by the ratio of its electric to magnetic field strength in space in the vicinity of the shield.

The impedance of the shield is a complex function of the shield electrical parameters, shield thickness, and frequency of interest; in general it is low for good conductivity shields, and high for

TABLE 5-1

Characteristics of Various Metals
Used for Shields

Metal	Relative Conductivity g	Relative Permeability at 150 kHz u	Absorption Loss db/mil at 150 kHz
Silver	1.05	1	1.32
Copper-Annealed	1.00	1	1.29
Copper-Hard Drawn	.97	1	1.26
Gold	.70	1	1.08
Aluminum	.61	1	1.01
Magnesium	.38	1	.79
Zinc	.29	1	.70
Brass	.26	1	.66
Cadmium	.23	1	.62
Nickel	.20	1	.58
Phosphor-Bronze	.18	1	.55
Iron	.17	1,000	16.9
Tin	.15	1	.50
Steel, SAE 1045	.10	1,000	12.9
Beryllium	.10	1	.41
Lead	.08	1	.36
Hypernick	.06	80,000	88.5*
Monel	.04	1	.26
Mu-Metal	.03	80,000	63.2*
Permalloy	.03	80,000	63.2*
Steel, Stainless	.02	1,000	5.7*

* assuming material not saturated

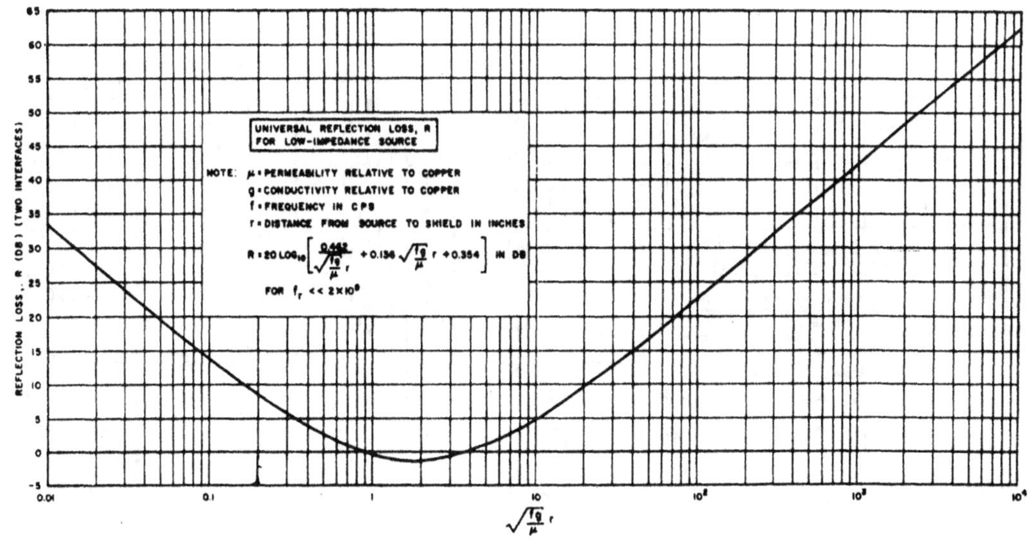

Figure 5-1

Universal Reflection Loss for Low-Impedance Source

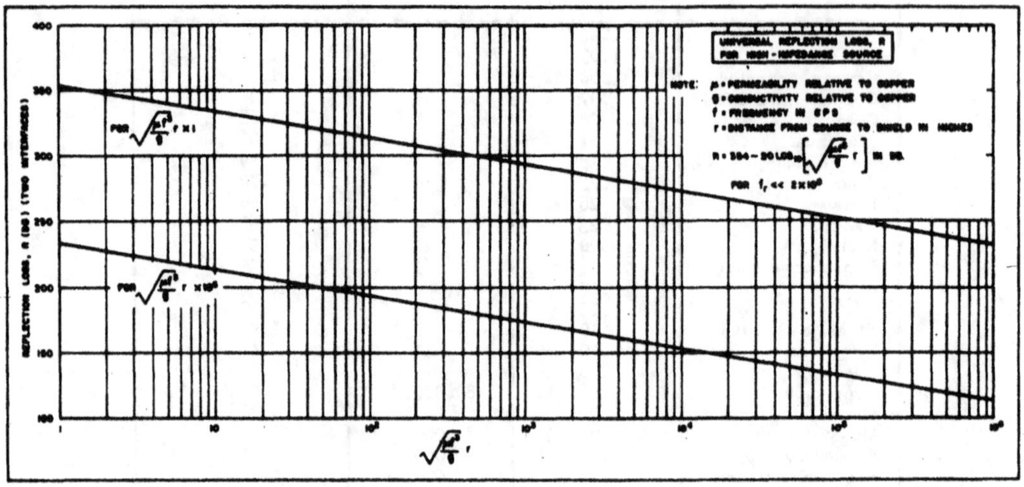

Figure 5-2

Universal Reflection Loss for High-Impedance Source

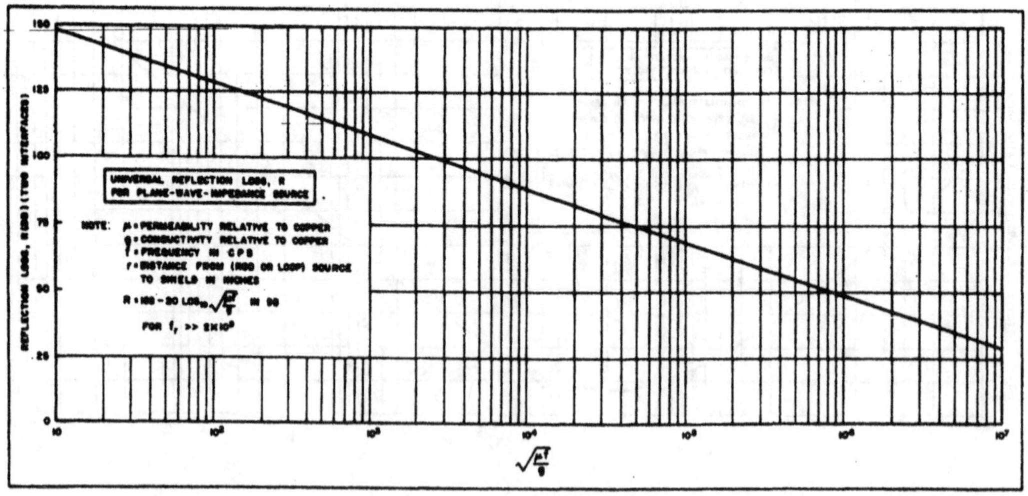

Figure 5-3

Universal Reflection Loss for Plane-Wave Impedance Source

shields that have high permeability. At radio frequencies of 100 kHz and above, up to 10 GHz or more, the thickness of the shield is generally great enough and the shield hole fraction small enough that the shield impedance can be approximated by that of the solid metallic surface of relative conductivity g and relative permeability μ. This impedance is

$$2\pi(10^{-7})\sqrt{\frac{0.4f\mu}{g}} \tag{5.6}$$

For a non-magnetic shield, $\mu \approx 1$. For most shield materials (aluminum, copper, silver, etc.) $g \approx 1$. Thus even at 10 GHz, the surface impedance is only on the order of 0.04 ohm.

In order that the reflected wave be as large as possible, or that the reflection loss be great, the shielding sheet should have an impedance that is either very much greater than the wave impedance or very much less. In shielding against plane waves, it is more practical to establish a mismatch by using shield material having a very low impedance than it is to use very high impedance material, and that is why conducting material is always employed.

In the case of high or low impedance sources, it can be seen from equations 5.3 and 5.4 that the reflection loss obtained will be a function of the separation distance between the source and the shield. The equations have been derived on the assumption that these sources are point sources (that is, small compared to the wavelength of concern). The reason the loss is distance-dependent is because the impedance of the wave impinging on the shield is a function of this separation. Figure 2-6 illustrates how the wave impedance varies for both an electric and magnetic source.[4]

The multiple reflecting correction term, B, is defined by the equation

$$B(\text{in dB}) = 20\log\left|1 - W^{-A/10}(\cos 0.23A - j\sin 0.23A)\right| \tag{5.7}$$

where

$$W = 4\frac{(1-m^2)^2 - 2m^2 - j2\sqrt{2}\,m(1-m^2)}{1 + (1+\sqrt{2}\,m)^2} \tag{5.7a}$$

$$m = 0.766\sqrt{\frac{fg}{u}}\ r \tag{5.7b}$$

As indicated previously, the multiple reflection correction term may be neglected if the absorption loss exceeds 15 dB. If the absorption loss is less than 15 dB, the correction must be determined.

The multiple reflection correction term is complicated to compute, and depends on material, dimensional, and frequency parameters. Fortunately, in almost all practical cases, W in equation (5.7a) equals unity. The only notable exception is the special case of extremely low frequency-shielding against low-impedance (chiefly magnetic) fields.

Equation (5.7) is combined with the absorption loss and is plotted in Figure 5-4 for W = 1. Note that the sign of B can be positive or negative, depending on the phase relationships between the reflections. When W is not equal to unity, its value may be obtained using Figure 5-5.

Table 5-2 illustrates a typical set of shielding effectiveness calculations for solid sheets of copper and iron.[2] Note the relative magnitudes of the reflection and penetration components. Reflection loss in the electric field cases decreases inversely with frequency while reflection loss in the magnetic field cases increases inversely with frequency.

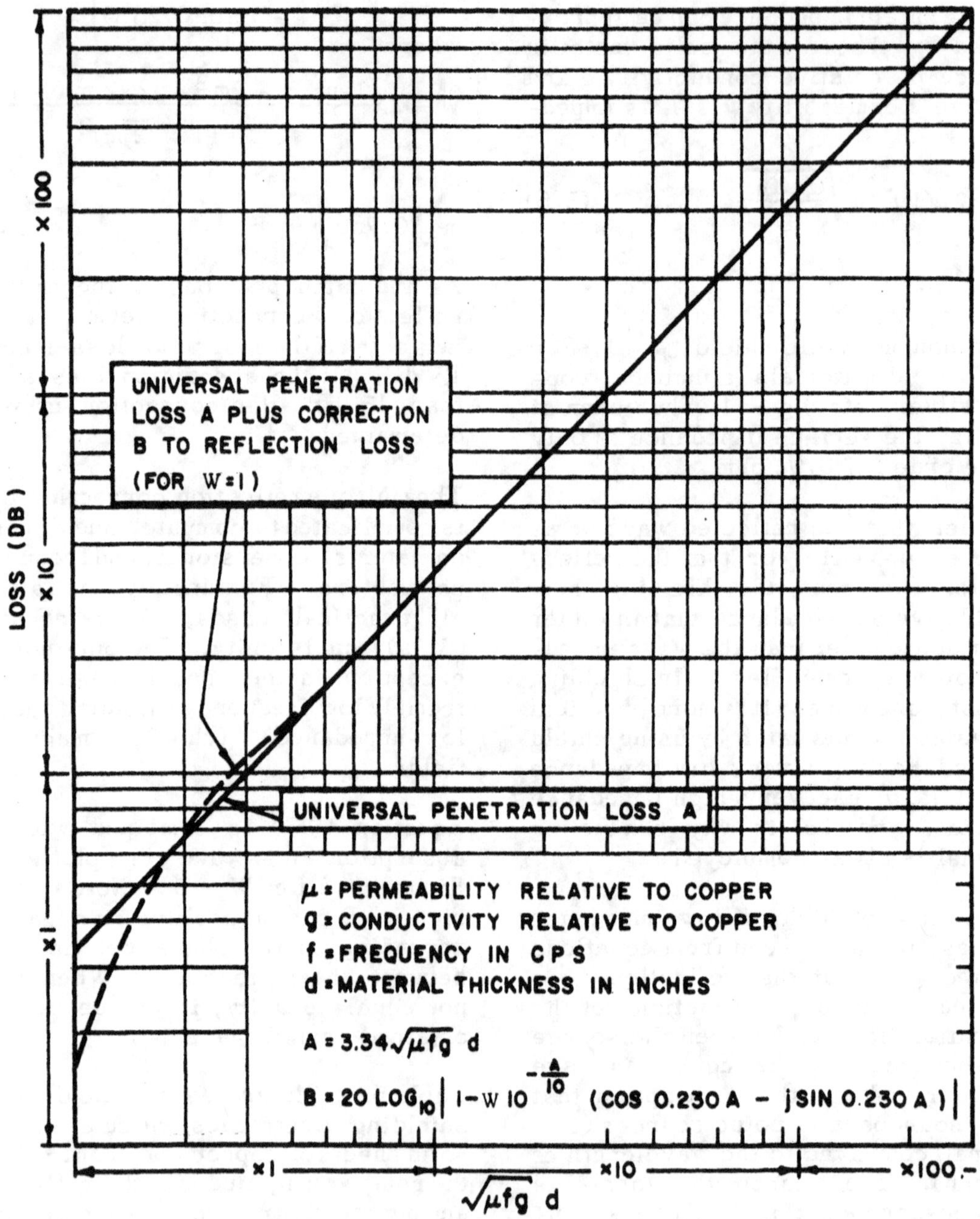

Figure 5-4

Penetration Loss and Multiple Reflection Correction Term When W = 1

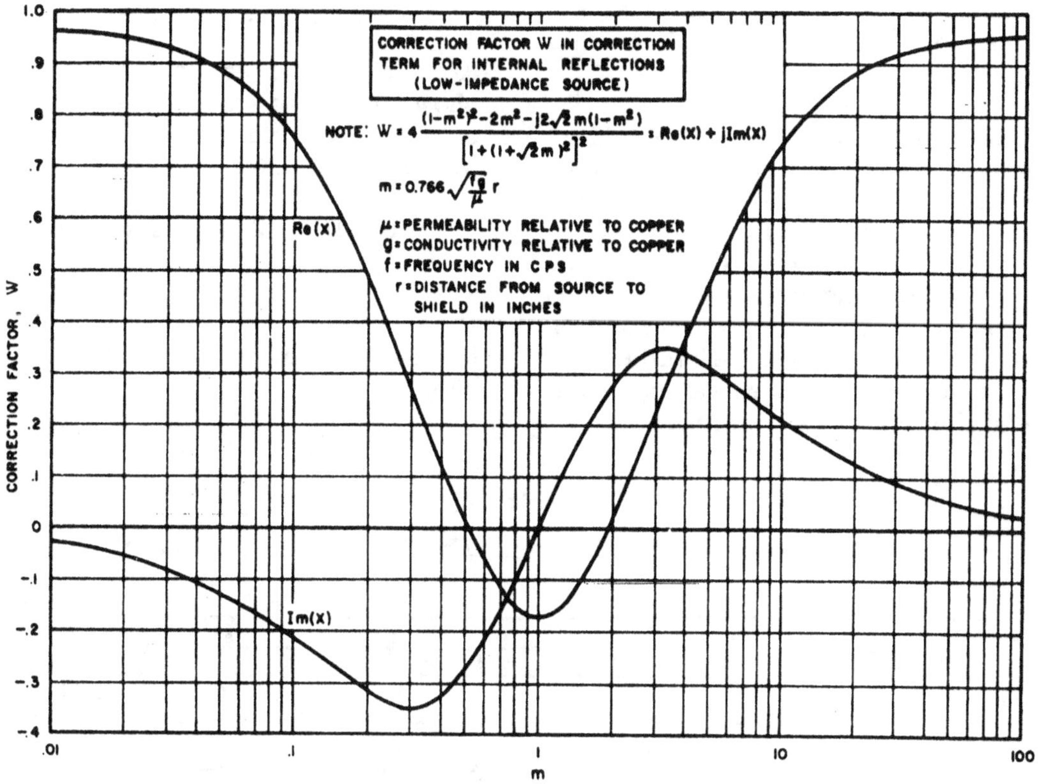

Figure 5-5

Correction Factor in Correction Term for Internal Reflections

Table 5-2

Typical Calculated Values of Shielding Effectiveness of Solid Sheets

Material	Fre-quency	Type of Field	Metal Thickness (mils)	R Reflection Loss (db)	A Penetration Loss (db)	B Correction Term (db)	Total Shielding Effectiveness S = R + A + B (db)
Copper	60 Hz	Magnetic	1	22.4	0.026	-22.2	0.23
	60 Hz	Magnetic	10	22.4	0.26	-19.2	3.46
	60 Hz	Magnetic	300	22.4	7.80	+0.32	30.52
	1 kHz	Magnetic	10	34.2	1.06	-10.37	24.89
	10 kHz	Magnetic	10	44.2	3.34	-2.62	44.92
	10 kHz	Electric	10	212.0	3.34	-2.61	212.73
	10 kHz	Plane Waves	10	128.0	3.34	-2.61	128.73
	10 kHz	Magnetic	30	44.2	10.02	+0.58	54.80
	150 kHz	Magnetic	10	56.0	12.9	+0.5	69.4
	150 kHz	Electric	10	176.8	12.9	+0.5	190.2
	150 kHz	Plane Waves	10	117.0	12.9	+0.5	130.4
	1 MHZ	Magnetic	10	64.2	33.4	0	97.6
	1 MHZ	Electric	10	152.0	33.4	0	185.4
Copper	1 MHZ	Plane Waves	10	108.2	33.4	0	141.6
Iron	60 Hz	Magnetic	1	-0.9	0.334	+0.95	0.38
	60 Hz	Magnetic	10	-0.9	3.34	+0.78	3.22
	60 Hz	Magnetic	300	-0.9	100.0	0	99.1
	1 kHz	Magnetic	10	0.9	13.70	+0.06	14.66
	10 kHz	Magnetic	10	8.0	43.5	0	51.5
	10 kHz	Electric	10	174.0	43.5	0	217.5
	10 kHz	Plane Waves	10	90.5	43.5	0	134.0
Iron	10 kHz	Magnetic	30	8.0	130.5	0	138.5

Nomographs for computing the shielding effectiveness of solid sheet materials may be found in Chapter 11 of this publication.

Representative solid sheet shielding effectiveness measurements are shown in Table 5-3. The data encompasses a variety of materials subjected to either low or plane-wave impedance energy at frequencies ranging from 100 Hz to 10 MHz. For the materials shown, the shielding effectiveness against high-impedance waves would exceed the plane-wave cases.

5.2.2 Additional Comments on Magnetic Shielding

At very low frequencies, it can be shown that the shielding effectiveness of solid materials to low impedance waves approaches

$$S \text{ (in dB)} = 20 \log \left(1 + \frac{1}{2}\frac{ud}{r}\right) \tag{5.8}$$

Measurements confirm this relationship.

Table 5-3

Measured Shielding Effectiveness Data for Solid Sheet Materials

Impinging Wave	Material	Thickness (mils)	Nominal Effectiveness 0.1 kHz	1 kHz	10 kHz	85 kHz	200 kHz	2 MHz	5 MHz	10 MHz
Plane	Cu	2.5					109	106	114	
Plane	Al	5					107	109	118	
Plane	Stainless Steel	18					97	95	99	
Plane	Steel (U_r=250)	4.5					105	99	101	
Plane	AA-Conetic Foil (U_r-10,000)	3.5					97	130		
Low Impedance	Cu	125	8	22	58					
		63				97				
		31	4	11	29	59		120		
		4.5				34		55		
Low Impedance	Al	125	5	18	50					
		63	1	16	35	78				
		31	1	10	24					
Low Impedance	Steel (u=242)	63	25	40	80					
		31	4	28	59	94		92		120
Low Impedance	Brass	31				42				
Low Impedance	Cu-Clad Steel Clad 2 Sides	31				107				
Low Impedance	Cu-Clad Steel Clad 1 Side	94				103				

Figure 5-6 illustrates representative shielding effectiveness data taken for a variety of high-permeability sheet materials. Loop sensors were located 1/8" from each sheet. The figure shows the typical leveling off in shielding effectiveness as frequency is decreased, with the breakpoint occurring in the 1 kHz range. Low frequency magnetic shielding is essentially achieved by establishing a low reluctance path in which the magnetic field is contained.

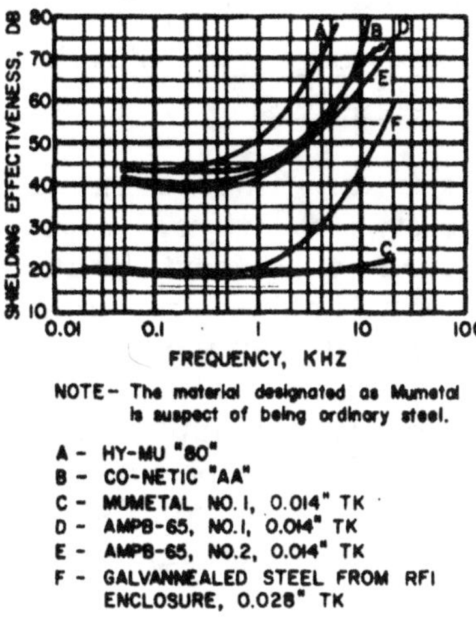

Figure 5-6

Shielding Effectiveness of High-Permeability Sheets

The variation of shielding effectiveness as a function of loop sensor separation is shown in Figure 5-7a for one of the materials plotted in the previous figure. A change in effectiveness of about 5 dB over the range of the test at a particular frequency is indicated.

A difficulty with most magnetic shielding materials is their tendency to change permeability when formed, machined, subjected to rapid or extreme temperature changes, or dropped. These processes change the orientation of the magnetic domains

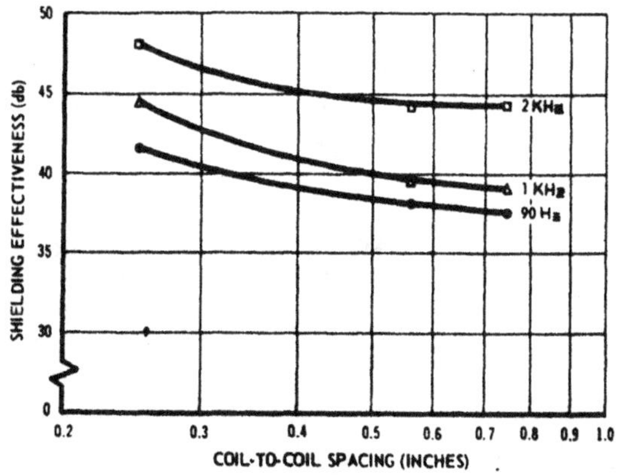

Figure 5-7a

Shielding Effectiveness of AMPB-65 as a Function of Coil-to-Coil Spacing

in the material, and it is necessary to reorient the domains by annealing to restore the initial magnetic properties. A typical annealing process involves heating the material to about 2000° F (sometimes in an inert gas environment), holding it at that temperature for a couple of hours, and letting it slowly cool to room temperature.

5.2.3 Multiple Solid Shields

There are cases when it is appropriate to consider using two or even three layers of shielding material rather than a single sheet to obtain particular total shielding characteristics. The most frequently encountered circumstances are when good protection against both electric and magnetic fields is desired, although other situations also occur.

Although Mumetal and similar types of high-permeability alloys provide good shielding for low-frequency weak magnetic fields, they tend to be less effective under the saturating effects of high-level fields. Where magnetic shielding in high level environments is necessary, it is often desirable to use a multiple shield, where the outer material (material closer to the EM source) has a lower permeability but a higher saturation level than the inner material. Such a structure might be created with materials having the following characteristics:

	INNER MATERIAL (CO-NETIC AA)	OUTER MATERIAL (NETIC S3-6)
Initial Permeability	20,000	300
Permeability at 200 Gauss	80,000	500
Saturation Inductance (Gauss)	7,500	22,000

The material thickness necessary would be dictated by the expected levels of external fields, and the desired suppression.

When much of the usefulness of shielding is due to reflection loss, two or more layers of metal, separated by dielectric materials and yielding multiple reflections, will provide greater shielding than the same amount of metal in a single sheet. The separation of the two layers of metal is necessary to provide for the additional discontinuous surfaces.

Table 5-4 compares single and double shields of copper, bronze, and galvanized sheet steel at various frequencies. Table 5-4A compares single and double shielding, both nonisolated and isolated. Table 5-4B shows how the attenuation varies with frequency for various combinations. Curves for electric field shielding effectiveness often are not drawn because of the very high values encountered, often beyond measurement equipment capabilities. However, examples appear in Fig. 5-13 and 5-14.

A similar advantage has been noted with magnetic sheet materials[5] (see Figure 5-7b.

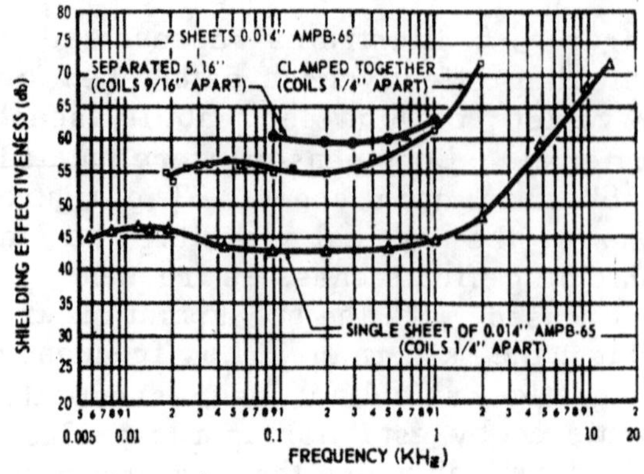

Figure 5-7b
Shielding Effectiveness of Two Sheets of AMPB-65 Compared with Single Sheet

5.2.4 Coatings and Thin-Film Shielding

Thin shielding* has been employed in a variety of ways, ranging from metallized component packaging for protection against rf fields during shipping and storing, to conductive glass, to vacuum deposited shields for microelectronics applications. Since future GSE developments might incorporate thin solid shields, come comments regarding such shields are considered appropriate.

Solid material shielding theory is applicable to thin-film shields. For shields much thinner than $\lambda/4$, the Absorption Loss is very small, but the Multiple Reflecting Correction Term, B, is fairly large and negative, thus offsetting a portion of the Single Reflection Loss. The implication of the negative term is that the various reflections have additive phase relationships, and thus reduce the effectiveness of the shield. The shield effectiveness is essentially independent of frequency.

* Thin shielding is loosely defined as shield thickness less than one quarter wavelength at the propagation velocity dictated by the material.

ATTENUATION AT 1,000 MEGAHERTZ
(1 GHz)

Type of Room	Electric Field and Plane Wave Field	
Single Shield Room	55-65	Copper Screen
Double Shield Room not Isolated	80	
Double Electrically Isolated Room	120	

Table 5-4A

ATTENUATION PROVIDED BY BARRIER TYPES
IN DOUBLE ELECTRICALLY ISOLATED ROOMS

	Magnetic Field		Electric and Plane Wave Field		
Barrier Material	60 Hz	15 KHz	15 KHz	1 GHz	10 GHz
3 oz. (per sq. ft.) Copper, both layers	2-3	64	120	120	120
24 ga. Steel, both layers	15	80	120	120	90 to 106
Combination Copper and Steel	18-25	90	120	120	120
Copper Screen, both layers, 22 x 22 x .015	2-3	68	120	120	77
Bronze Screen, both layers, 18 x 20 x .011	0	40	120	110	57
Four shields, two 24 ga. Steel, two 24 oz. Copper	30-36	120	120	120	120

Table 5-4B

Table 5-4
Comparison of Single and Double Shields[26]

Table 5-5
Calculated Values of Copper Thin-Film Shielding Effectiveness Against Plane-Wave Energy

Shield Thickness	1050 Å		12500 Å		21960 Å		219600 Å	
Frequency	1MHz	1GHz	1MHz	1GHz	1MHz	1GHz	1MHz	1GHz
Absorption Loss, A	.014	.44	.16	5.2	.29	9.2	2.9	92
Single Reflection Loss, R	109	79	109	79	109	79	109	79
Multiple Reflection Correction Term, B	-47	-17	-26	-.6	-21	.6	-3.5	0
Shielding Effectiveness, S	62	62	83	84	88	90	108	171

When the shield thickness exceeds $\lambda/4$, the Multiple Reflection Term becomes negligible, and there is no offsetting effect to the other losses. Thus the material shielding effectiveness increases, and is frequency dependent.

Table 5-5 provides representative calculations of the shielding effectiveness of thin-film copper for different thicknesses and frequencies, using equations 5.1, 5.2, 5.5 and 5.7. One-quarter wavelength in copper is approximately 32500 Å (1.3×10^{-4} inches) at 1 GHz*, and it can be seen that shield effectiveness changes significantly above this thickness.

Figures 5-8 and 5-9 contain measured shielding effectiveness data on copper, aluminum, silver, gold, and nickel materials[6]. A comparison between values computed in Table 5-5 and Figure 5-8b indicate differences between thin-film measurements and calculations of about 10 dB. One factor that may account for some of this difference is the difficulty in depositing a uniformly thick film. The dotted curves in Figures 5-8 and 5-9 are the experimenter's estimate of corrections necessary as a result of an error noted in his test setup that altered low frequency data.

Conductive coatings can be applied in a number of ways. Vacuum metallizing involves use of vacuum deposition equipment to apply a metallic film (usually aluminum) onto a surface; thicknesses of about 10^4 A are typical.

Flame spraying is a somewhat less expensive and time-consuming process than vacuum deposition and typically results in film thicknesses of 2.5×10^5

*One-quarter wavelength in meters $= \frac{.112}{\sqrt{f \mu g}}$

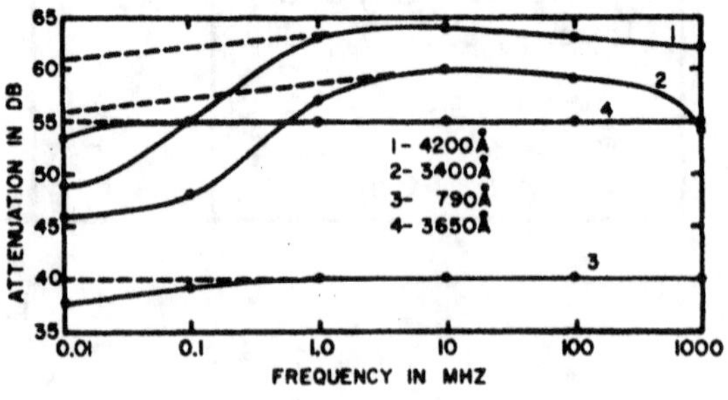

Figure 5-8a

Shielding Effectiveness Curves, Copper

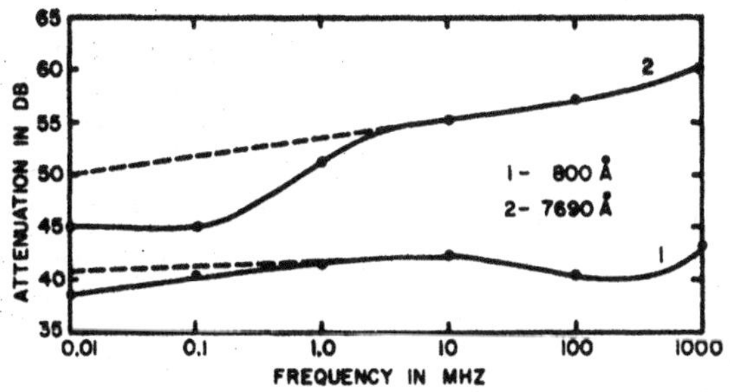

Figure 5-8b

Shielding Effectiveness Curves, Copper

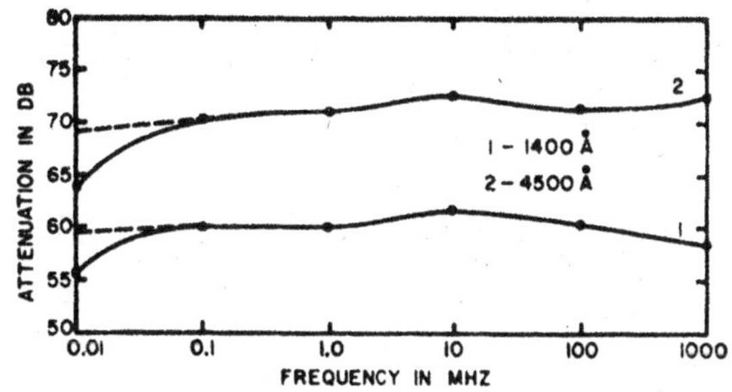

Figure 5-9a

Shielding Effectiveness Curves, Silver

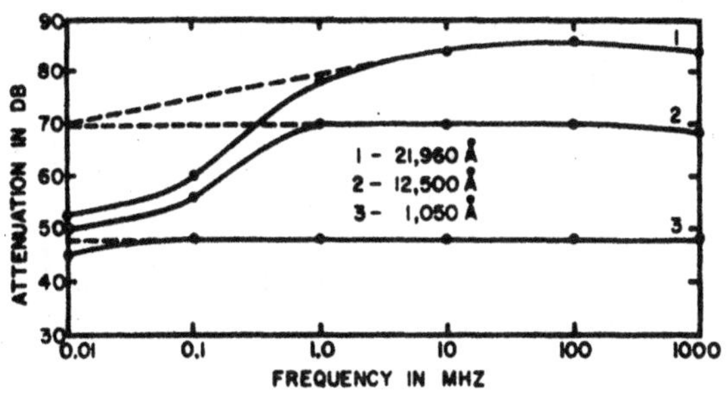

Figure 5-8c

Shielding Effectiveness Curves, Aluminum

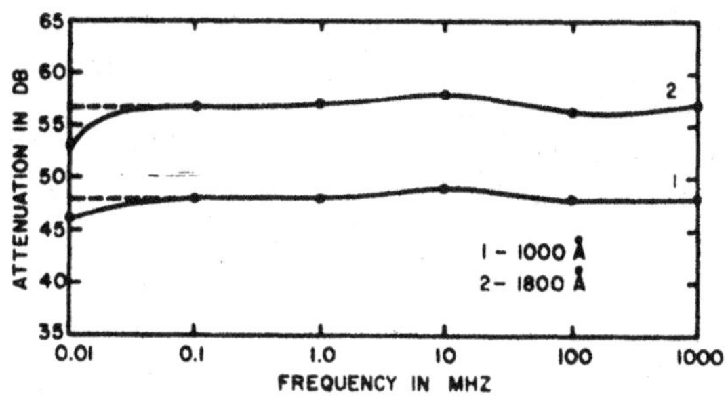

Figure 5-9b

Shielding Effectiveness Curves, Gold

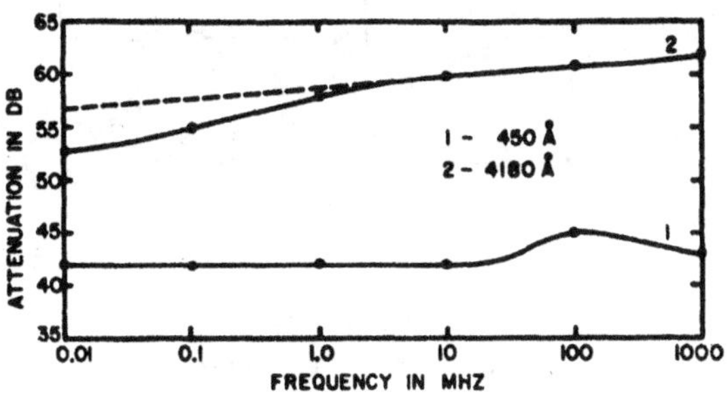

Figure 5-8d

Shielding Effectiveness Curves, Aluminum

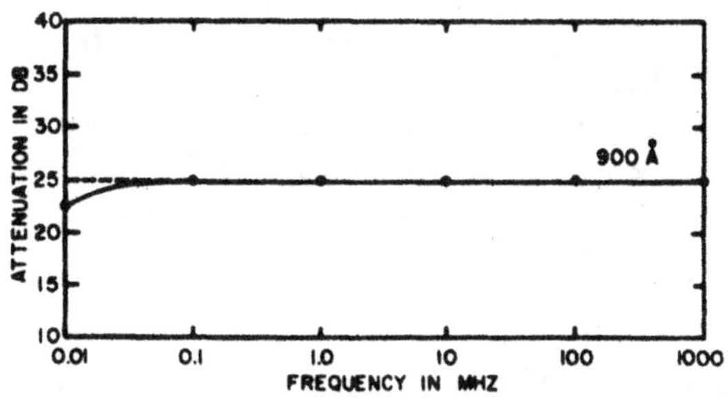

Figure 5-9c

Shielding Effectiveness Curves, Nickel

to 5×10^5 Å for acceptable conductivity. One version consists of two consumable wires that are fed through electrode tips of a spray gun, where a 600 ampere electric current is transferred to them. As the wires emerge from the gun on intersecting paths, an arc is struck. The arc melts the consumable wires, and an air nozzle behind the wires atomizes the molten metal (see Figure 5-10a).

The flame-spray process indicated above, sometimes called arc or electrospray, operates at arc temperatures of 7000°F. This high temperature provides fast melting of the consumable wires, and a corresponding high spray deposition rate. For example, the flame spray process can deposit about 25 pounds of aluminum or 50 pounds of stainless steel per hour. Good bonds are achieved with less than ideal surface preparation, and relatively dense coatings are obtained.

Another form of flame spraying, sometimes called wire spray, involves a single consumable wire that is vaporized in a fuelgas and oxygen atmosphere. Compressed air impels the molten particles to the surface being sprayed. A cutaway view of a wire spray gun is shown in Figure 5-10b. Temperatures as high as 4500°F can be obtained, and a variety of spray rates can be achieved. This approach provides relatively fine-grain coatings at somewhat lower deposition rates than is available using the flame spray process.

A third form of flame spraying uses an oxy-gas flame in which powdered metal is melted. The powdered material is held in a hopper on top of the spray gun, is gravity-fed into the gun proper, and is carried to the gun nozzle by an oxygen-acetylene or hydrogen gas mixture. Then it is melted and projected onto the prepared surface. This process is illustrated in Figure 5-10c.

The use of powdered metal offers a wide range of materials to be sprayed. The texture of the finish ranges from rough to smooth (depending on the powder used). Coating thickness can be typically 5000-8000 Å, while coverage rates are as high as 500 square feet per hour at one mil thickness.

In addition to flame-spraying, conductive coatings can be applied in a number of other ways. These include:

> Paint consisting of an acrylic-type carrier with silver-metallic or other conductive filter. Paint is sprayed on.
>
> Plating of either rigid materials or films of plastic.
>
> Conductive tape or foil, with or without adhesive backing.

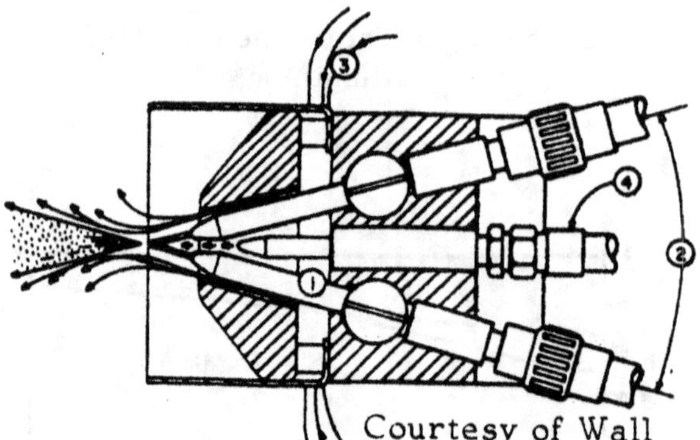

Courtesy of Wall Colmoney Co.

1. Electrode types
2. Consumable wire route
3. Cooling air circulation
4. Air nozzle

Figure 5-10a

Details of Arc Flame Spray Gun

Conductive glass (a very thin conducting surface on a clear glass substrate) is often used in applications that require electromagnetic shielding properties but must maintain certain visual requirements. A typical shield thickness in this application is 10^4 Å ($4x10^{-5}$ inches). A more detailed discussion of the applications of conductive glass will be found in Section 5.5.5.

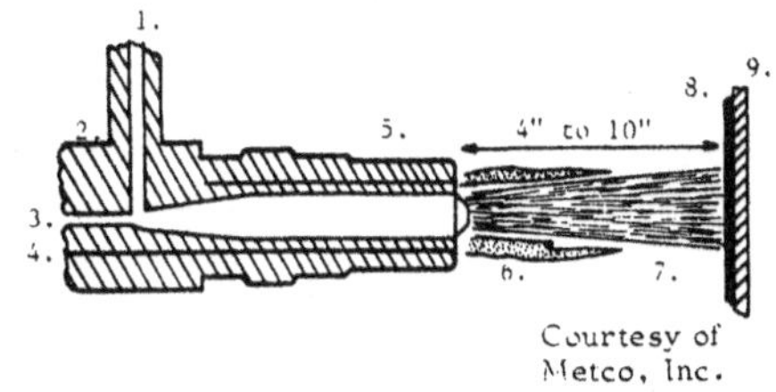

1. Powder
2. ThermoSpray Gun
3. Powder Carrying Gas
4. Fuel Gases
5. Nozzle
6. Burning Gases
7. Spray Stream
8. Sprayed Coating
9. Prepared Base Material

Figure 5-10c

Details of Powdered Metal Spray Gun

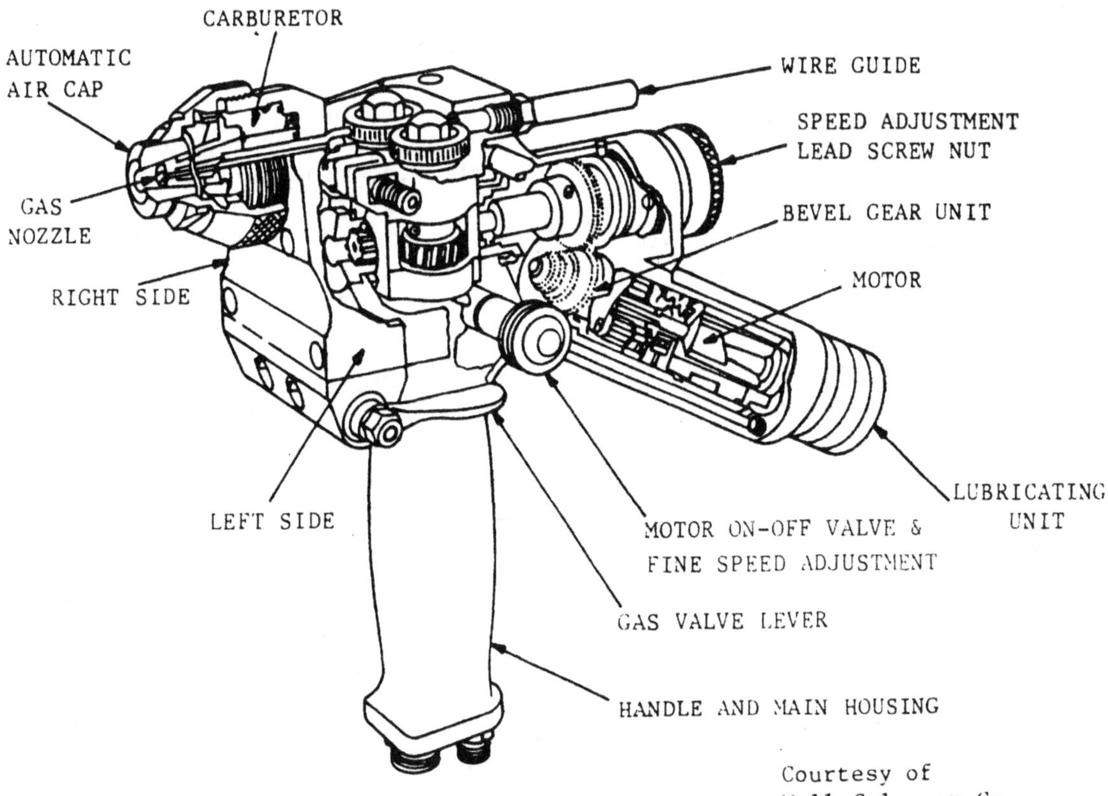

Figure 5-10b

Cutaway View of Wire Spray Gun

5.3 NON-SOLID SHIELDING MATERIALS

5.3.1 Types of Discontinuities

An ideal shielded enclosure would be one of seamless construction with no openings or discontinuities. However, personnel, power lines, control cables, and ventilation ducts must have access to any practical enclosure. The design and construction of these discontinuities become very critical in order to incorporate them without appreciably reducing the shielding effectiveness of the enclosure. Some types of discontinuities commonly encountered include:

a. Seams between two metal surfaces, with the surfaces in intimate contact (such as two sheets of material that are riveted or screwed together).

b. Seams or openings between two metal surfaces that may be joined using a metallic gasket.

c. Holes for ventilation, or for exit or entry of wire, cable, light, film, water, meter faces, etc.

d. Screened openings.

Design and construction considerations associated with such openings will be discussed in the next sections of this design guide.

5.3.2 Shielding Analysis[7,8]

Leakage through openings in metal shields has been studied using transmission line theory. Based on these studies, the Shielding Effectiveness, S, of non-solid shielding materials has been defined as

$$S = A_a + R_a + B_a + K_1 + K_2 + K_3 \quad (5.9)$$

where

A_a = The Attenuation introduced by a particular discontinuity, in dB

R_a = The Aperture Single Reflection Loss, in dB

B_a = The Multiple Reflection Correction Term, in dB

K_1 = A correction term to take into account the number of like discontinuities

K_2 = A low-frequency correction term to take into account skin depth

K_3 = A correction term to take into account coupling between adjacent holes.

The first three terms in Equation (5.9) generally correspond to the three terms of Equation (5.1), while the last three terms encompass other factors that need not be considered for solid sheets. The analysis is most appropriate for single discontinuities, or for identical and uniformly spaced apertures (such as screening or perforated sheets), but can be extended to somewhat more complex configurations as well.

The absorpiton or penetration loss term, A_a, is obtained by assuming that the impinging wave is well below its cutoff frequency.* Under these

* The cut-off frequency is that frequency below which the propagation constant of the aperture is real. It is defined by $f_c = c/\lambda_c$, where c is the velocity of light and λ_c is the cut-off wavelength, in the same units. The cut-off wavelength is equal to twice the maximum dimension of a rectangular aperture, or 3.412 times the radius of a circular aperture. The multiplying factor to be applied to equation (5.10a and 5.10b) for frequencies f not well below cut-off is $\sqrt{1 - (f/f_c)^2}$.

circumstances

$$A_a = 27.3 \frac{d}{w} \quad \text{for rectangular apertures} \quad (5.10a)$$

$$A_a = 32 \frac{d}{D} \quad \text{for circular apertures} \quad (5.10b)$$

where

d = depth of aperture, in inches

w = width of aperture perpendicular to the E-field vector, in inches

D = aperture diameter, in inches.

These equations are plotted in Figure 5-11a.

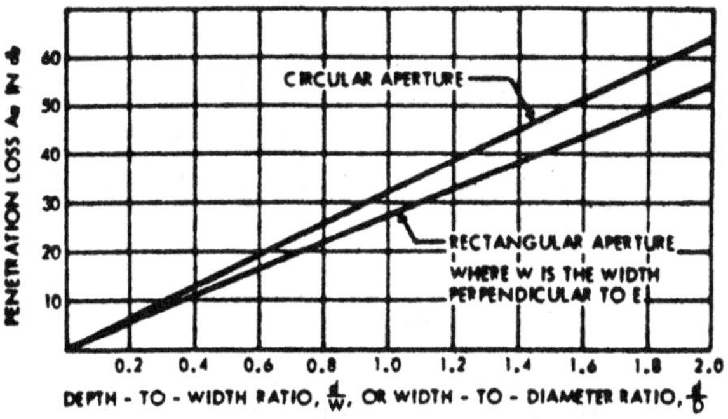

Figure 5-11a
Aperture Shielding
Absorption Loss

The aperture single reflection loss, R_a, is wave-impedance dependent and is also a function of aperture shape. It can be expressed as:

$$R_a = 20 \log \frac{1 + |K|^2}{4\,|K|} \quad (5.11)$$

where

$K = \frac{W}{\pi r}$ for low impedance fields and rectangular apertures

$K = \frac{D}{3.682r}$ for low impedance fields

$K = jfW(1.7 \times 10^{-4})$ for plane-wave fields and rectangular apertures

$K = jfd(1.47 \times 10^{-4})$ for plane-wave fields and circular apertures

f = frequency, in MHz

r = source to shield distance, in inches

$j = \sqrt{-1}$

Graphs of R_a for the low-impedance case are shown in Figure 5-11b.

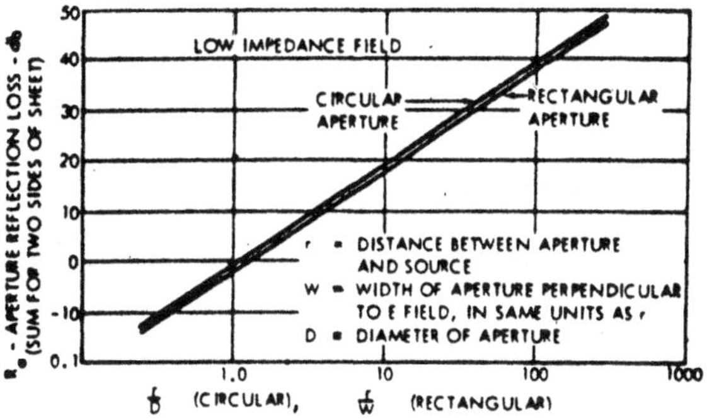

Figure 5-11b
Aperture Shielding
Reflection Loss

The multiple reflecting correction term, B_a, is defined by the equation:

$$B_a = 20 \log \left[1 - \frac{(K-1)^2}{(K+1)^2} \, 10^{\frac{-A_a}{10}} \right] \quad (5.12)$$

and is applicable only when A_a is less than 15 dB. A curve for the cases when either $W/\pi r$ or $D/3.682r$ are much less than unity is plotted in Figure 5-11c.

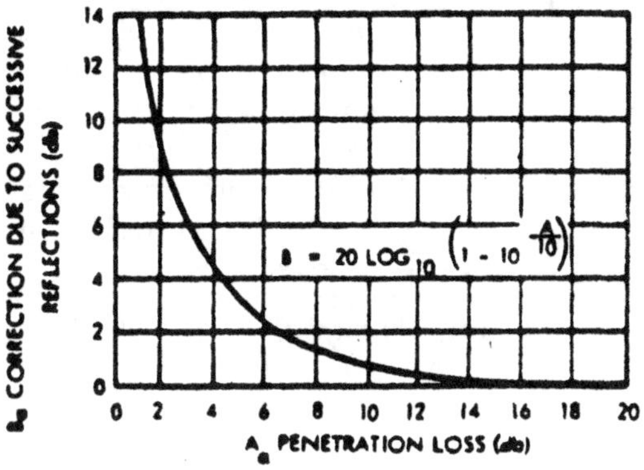

Figure 5-11c
Aperture Shielding
Correction Due to Re-reflection

The <u>number of discontinuities correction term,</u> K1, is employed when the source is located a large distance from the shield, in comparison with the aperture spacing in the shield. It is represented by

$$K_1 = -10 \log an \qquad (5.13)$$

where

a = area of each hole, in square inches

n = number of holes per square inch

The correction term can be ignored for sources close to the shield. Equation (5.13) is plotted in Figure 5-11d.

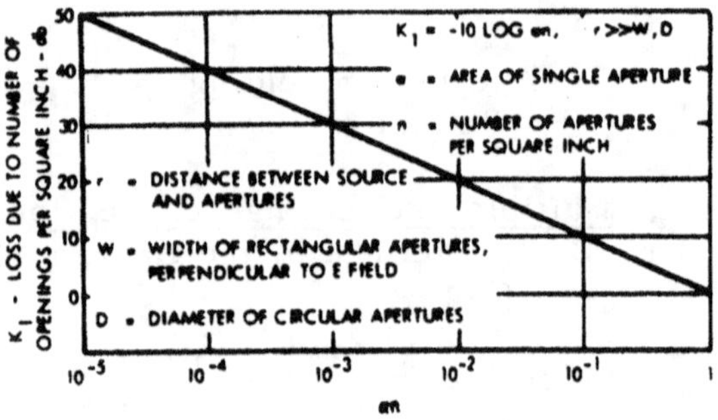

Figure 5-11d

Aperture Shielding
Loss Due to
Density of Openings

The <u>skin depth correction term</u>, K_2, recognizes that, at low frequencies, when the skin depth becomes comparable to the screening wire diameter or dimension between apertures, a reduction in shielding effectiveness can occur. An empirical relationship for this effect was developed:

$$K_2 = -20 \log\left(1 + \frac{35}{p^{2.3}}\right) \qquad (5.14)$$

where

p = the ratio of wire diameter to skin depth for screening

p = the ratio of conductor width between holes to skin depth, for perforated sheets

For convenience, the equation for skin depth in copper, in inches, is

$$SD_{cu} = \frac{2.6 \times 10^{-3}}{\sqrt{f}} \qquad (5.15)$$

Equation (5.14) is presented in Figure 5-11e.

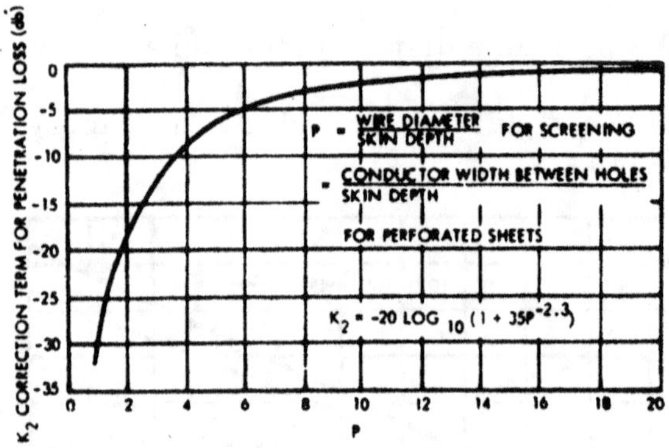

Figure 5-11e

Aperture Shielding
Correction for Skin Effect

The <u>adjacent hole coupling correction term</u>, K_3, is the result of noting that shielding efficiency was higher than expected when apertures in a shield were closely spaced and the depth of the openings were small compared to the aperture width. This is interpreted as the result of coupling between adjacent holes, and becomes important for small openings. The equation for computing K_3 is

$$K_3 = -20 \log \left(\coth \frac{A_a}{8.686}\right) \qquad (5.16)$$

This equation is plotted in Figure 5-11f.

The elements of the non-solid shielding effectiveness equation just discussed are summarized in Table 5-6 for convenience. Calculations in accordance with these equations have resulted in

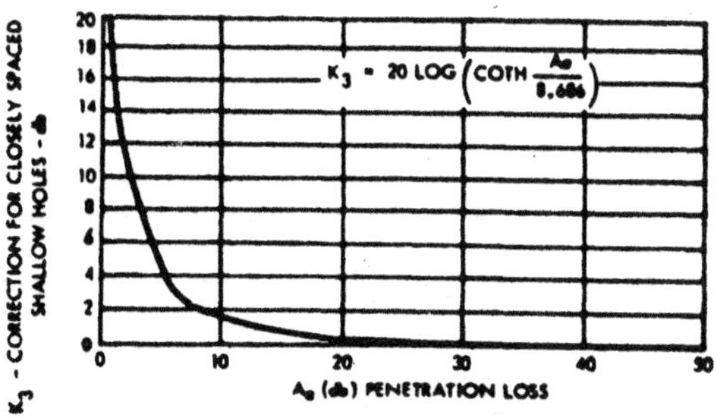

Figure 5-11f

Aperture Shielding Correction for Closely Spaced Shallow Holes

fairly good agreement with measurements, so long as the aperture pattern is uniform. Examples of this correlation are shown in Figure 5-12 for low-impedance shielding effectiveness of a variety of mesh copper screens. Both measurements and calculations used a loop-to-loop separation of 3.5 inches, and the measurements employed 3-inch diameter loops in coplanar orientation. Table 5-7 also provides similar comparative data.[7]

Figures 5-13 and 5-14 show the shielding effectiveness, respectively, of 3 oz.

Table 5-6

Terms for Aperture Shielding

Symbol	Item	Aperture: Rectangular	Aperture: Circular	Comments
A_a	Penetration Loss	$27.3\frac{d}{w}$	$32\frac{d}{D}$	d=Depth of Aperture, inches w=Width of Rectangular Aperture, Perpendicular to E field, inches D=Dia. of Circular Aperture, inches
R_a	Aperture Reflection Loss (sum for 2 sides of sheet)	$20 \log_{10} 0.785\frac{r}{w}$	$20 \log_{10} 0.92\frac{r}{D}$	<< 377 Ω
		$20 \log_{10} \frac{1+(1.7\times10^{-4}Fw)^2}{4(1.7\times10^{-4}fw)}$	$20 \log_{10} \frac{1+(1.47\times10^{-4}fw)^2}{4(1.47\times10^{-4}fw)}$	377 Ω
		----------------	----------------	>>377 Ω
B_a	Correction Term Due to Successive Reflections When A_a < 15 db	$20 \log_{10} (1-10^{-\frac{A_a}{10}})$		$\frac{w}{\pi r'}$ $\frac{D}{3.68r}$ << 1
K_1	Loss Term Due to Number of Openings Per Unit Square	$-10 \log_{10} an$, r >> ω, D zero, r << ω, D		a Area of Single Aperture, inches^2 n=No. of Holes per sq. inch
K_2	Correction Term for Penetration of Conductor at Low Frequencies	$-20 \log_{10} (1+35p^{-2.3})$		$\frac{\text{Wire Dia.}}{\text{Skin Depth}}$, Screening P= Conductor Width Between Holes/ Skin Depth Perforated Sheets
K_3	Correction for Closely Spaced Shallow Holes	$20 \log_{10} (\coth \frac{A_a}{8.686})$		

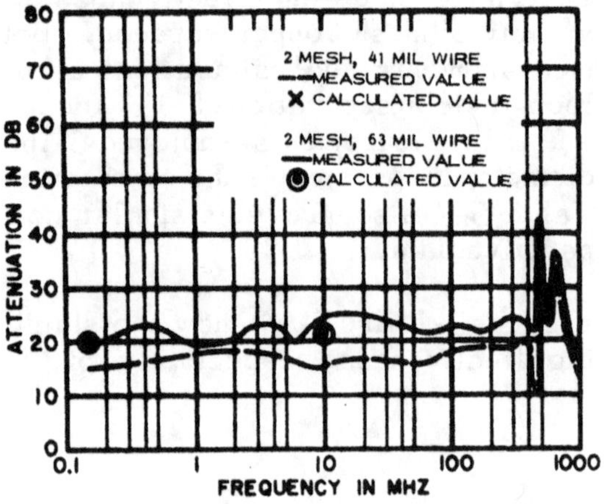

Figure 5-12a

Measured and Calculated Insertion Loss of Copper Screens

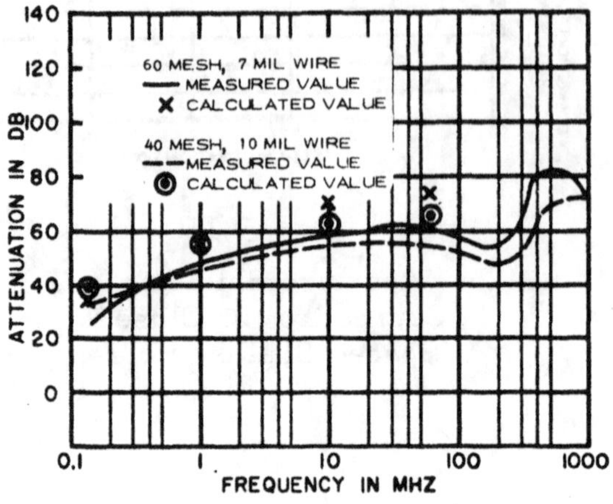

Figure 5-12c

Measured and Calculated Insertion Loss of Copper Screens (cont'd)

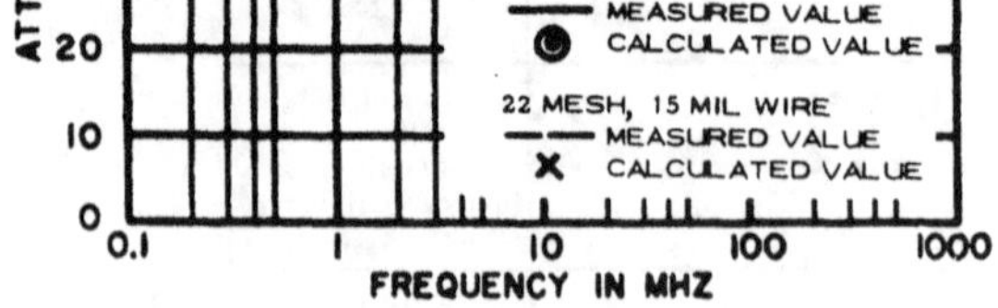

Figure 5-12b

Measured and Calculated Insertion Loss of Copper Screens (cont'd)

Screen Type*	Test Type	Freq (MHz)	Meas Values (DB)	Calc Values (DB)
No. 22 15 mil	Mag field	0.085 1.0 10.0	31 43 43	28 45 49
No. 22 15 mil	Plane Waves	0.2 1.0 5.0 100.0	118 106 100 80	124 110 95 70
No. 22 15 mil	Elec field	0.014 to 60	65	**65
No. 12 20 mil	Elec field	0.014 to 60	50	**53

* All screens made of copper.
** These values assume a wave impedance equal to that of a 30-inch square waveguide.

Table 5-7

Comparison of Measurements and Calculations of Screening Material Shielding Effectiveness

Table 5-8

Effectiveness of Non-Solid Shielding Materials Against Low-Impedance and Plane Waves

Impinging Wave	Form: General	Form: Detail	Material	Thickness (mils)	Nominal Effectiveness (db): 0.1 kHz	1 kHz	10 kHz	85 kHz	1 MHz	10 MHz
Low Impedance	Mesh (Screening)	2 layers 1 inch apart	Cu (oxidized)		2	6	18			
		No. 22	Cu					31	43	43
		No. 16	Bronze					18		
		No. 4	Galvanized Steel					10	17	21
Plane	Perforated Sheet	45 mil dia., 225 sq. inch	Al	20	3040 MHz			9380 MHz		
					60			62		
	Mesh (Screening)	No. 16	Al	dia=13	34			36		
		No. 22	Cu	dia=15	200 kHz		1 MHz	5 MHz	100 MHz	
					118		106	100	80	

Table 5-9

Effectiveness of Non-Solid Shielding Materials Against High-Impedance Waves

FORM: GENERAL	FORM: DETAILED	MATERIAL	THICKNESS (mils)	NOMINAL EFFECTIVENESS (db) 14 kc to 1000 mc	OPEN AREA (%)	AIR-FLOW STATIC PRESSURE (inches of water) 200 cu ft/min	400 cu ft/min
Hexcell	1/4-inch cell, 1 inch thick	Al	3	>90		0.06	.26
TV Shadow Masks (Photo-Etched)	9-mil holes, 28-mil centers	95% Cu 5% Ni	7	>90	12	>2	
					50	0.2	0.4
		100% Ni	3	>90	50	0.2	0.5
Lektromesh	40 count	Cu-Ni	7	>90	36	0.4	1.7
	25 count		5	78	49	0.2	0.5
	40 count	Cu	3	78	57	0.2	0.5
	25 count				56	0.2	0.4

Table 5-9 (Continued)

FORM		MATERIAL	THICKNESS (mils)	NOMINAL EFFECTIVENESS (db)	OPEN AREA (%)	AIR-FLOW STATIC PRESSURE (inches of water)	
GENERAL	DETAILED			14 kc to 1000 mc		200 cu ft/min	400 cu ft/min
Perforated Sheet	1/8-inch dia., 3/16-inch centers	Steel	60	58		0.27	>0.6
	1/4-inch dia., 5/16-inch centers	Al	60	48	46		
	7/16-inch dia., 5/8-inch centers		37	35	45		
Mesh (Screening)	No. 16 16 x 16/sq.in.	Al	20 (dia)	55	36		
	No. 22	Cu		65(14 kc to 60 mc)			
	No. 12		20 (dia)	50	50		
	No. 16	Bronze		45(14 kc to 60 mc)			
	No. 10	Monel	18 (dia)	40			
	No. 4	Galvanized Steel	30 (dia)	35 28 (14 kc to 40 mc)	76		
	No. 2			24	88		

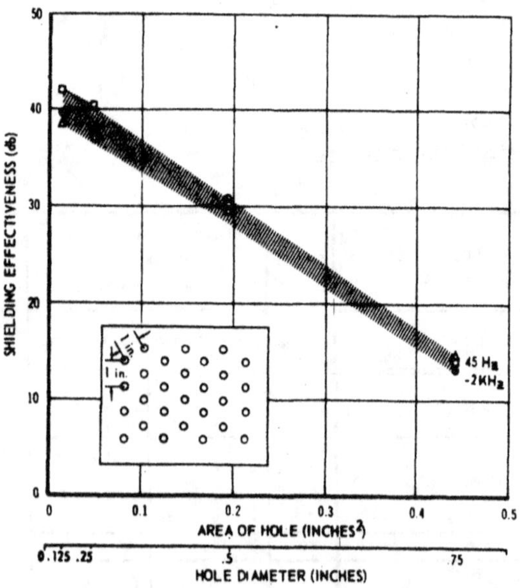

Figure 5-15a

Shielding Effectiveness of Perforated AMPB-65 as a Function of Hole Size

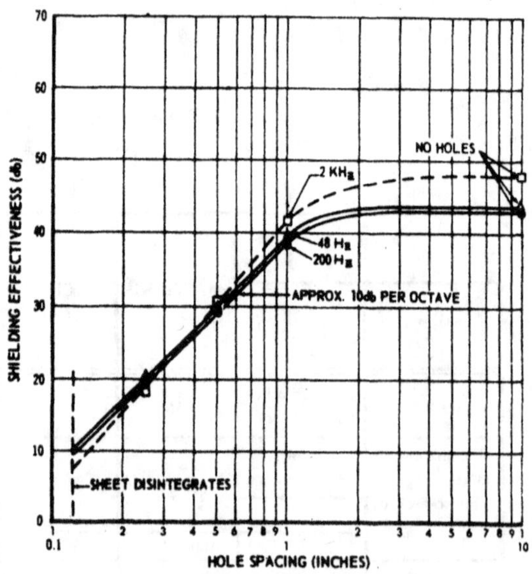

Figure 5-15b

Shielding Effectiveness of Perforated AMPB-65 as a Function of Hole Spacing (hole diameter = 0.125")

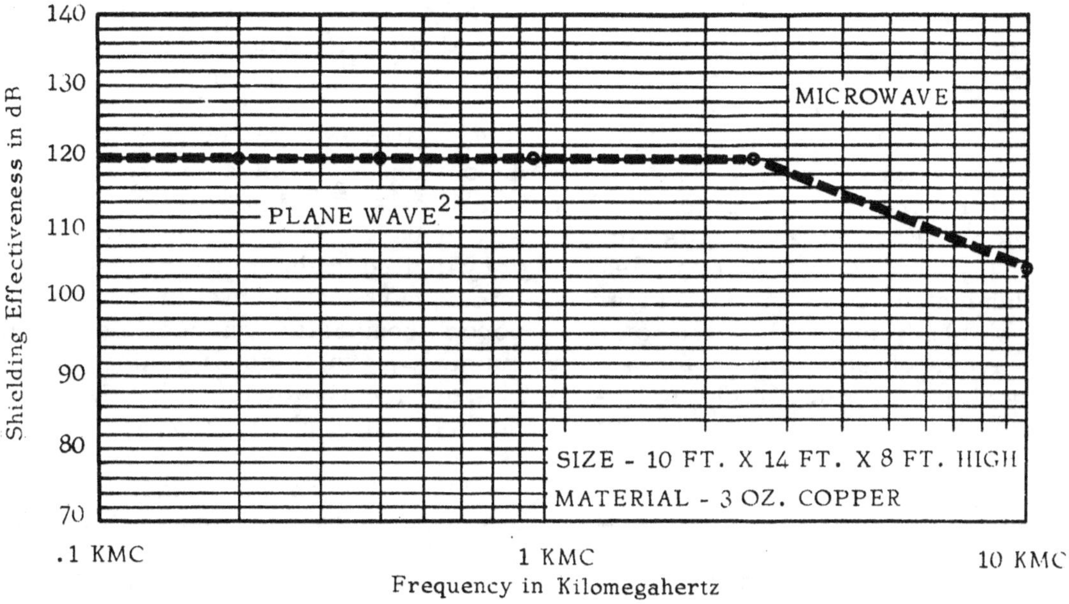

Figure 5-13

Shielded Effectiveness Tests on Double Electrically Shielded Enclosure

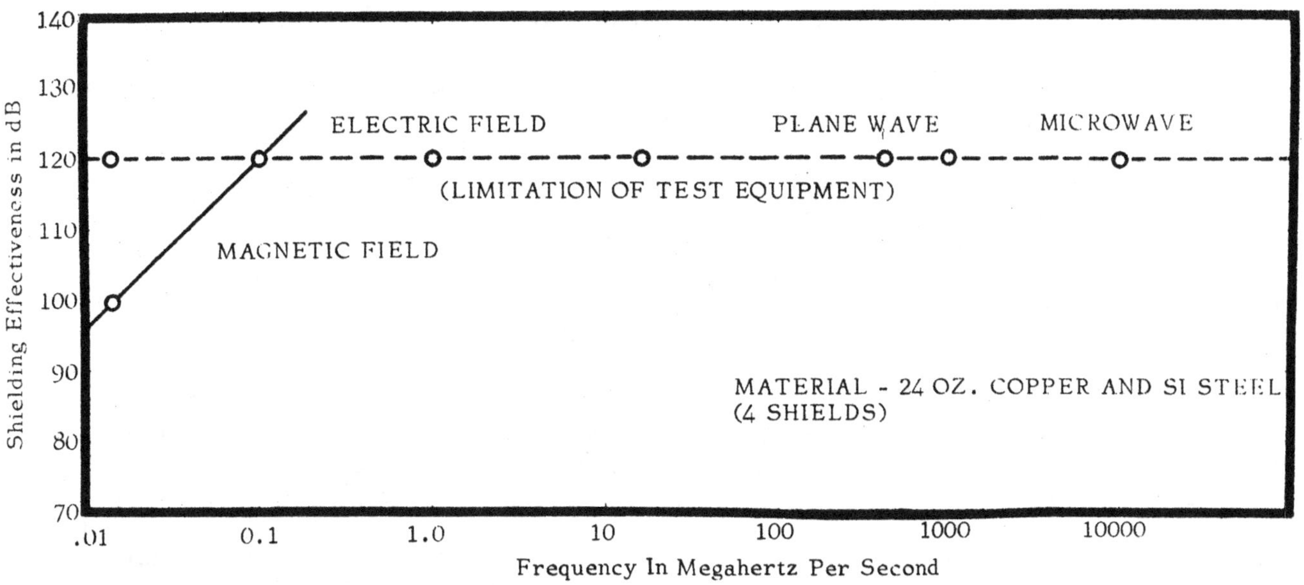

Figure 5-14

Shielding Effectiveness Tests on a Double Electrically Insulated Shielded Cube 34" x 36" x 30" (per square foot) copper screen and 24 oz. (per square foot) copper and silicon steel screen at frequencies up to 10 GHz.[26] The enclosure of Figure 5-13 is made up of two electrically insulated layers of 3 oz. copper sheet backed with saturated tar paper on both interior and exterior sides. The enclosure of Figure 5-14 is made up of two shields. Each shield consists of an outer layer of 24 oz. cold rolled copper laminated to an interior layer 24 gauge M-19 silicon steel, for a total of four layers of shielding material. The superiority of the shielding arrangement of Figure 5-14 is clearly evident.

Representative non-solid sheet shielding effectiveness measurements[2] are shown in Tables 5-8 and 5-9, and in Figure 5-15. The two tables provide data on a variety of material forms, including meshes, perforated sheets, and cellular structures against low-impedance, high-impedance and plane waves. The curves of Figure 5-15 were obtained using source and pick-up coils on opposite sides of a flat sheet of shielding material. Experimental results strongly support the validity of plane-wave shielding theory, even at the extremely low frequencies used. Figure 5-15 illustrates how the low-impedance effectiveness of high permeability perforated sheet material changes with changes in hole size and hole separation.[5]

5.3.3 Composite Materials

In recent years, composite materials have been developed to the point that their strength and weight characteristics make them very attractive as substitutes for metals in various aircraft applications. Composites are a family of polymeric or metal-reinforced materials having organic or inorganic fibers or filaments, woven fabric layering, and epoxy bonding. Boron, graphite, Kevlar 49 and glass fibers are employed, with boron and graphite most frequently used.[1] Applications include C-130 aircraft wing boxes, as well as portions of F-18 and A-7 aircraft.

Boron filaments are produced by vapor deposition on a tungsten wire resulting in a very stiff, high strength, low density fiber. The boron composite is very limited in ability to transport electrical current. This fact makes the boron composite susceptible to high currents, such as lightning, and to charge-accumulated high potentials, such as static electricity.

Graphite filaments are made by pyrolysis of polyacrylonitrile, or decomposition of a material similar to rayon. This produces a strong, stiff low cost filament. The graphite material is more conductive than boron and, therefore somewhat less susceptible to electrical phenomena. In addition, graphite possesses basic electromagnetic characteristics which, although reduced from metals, are improvements over other composites.

The extent to which composites alter the electrical characteristics of the aircraft clearly depends on the amount and location of the material on the aircraft. The areas of major concern as far as composites are concerned are as follows:

Lightning -

The effects of lightning strikes to aircraft are known to be potentially severe even in aircraft of the typical metal variety. Investigations employing both destructive and nondestructive testing techniques have affirmed that lightning-produced currents may adversely affect boron and graphite-reinforced composites that are installed on aircraft. While composites offer significant structural advantages over conventional metals, they are much more easily damaged by the high currents associated with lightning.[2] Methods of improving the lightning protection of composites have included use of wire mesh, metallized paint, flame-sprayed aluminum and dielectric coatings.

Shielding Effectiveness -

The shielding effectiveness of composites is much lower than that of conductive metals. Essentially, the composite provides little protection from flux linkage of lightning energy to internal sub-systems and equipment. In addition, little attenuation is offered to other radiated environmental signals. Typical shielding effectiveness values of boron and graphite composites to electric and magnetic fields are compared against sheet aluminum in Figure 5-15c. The more limited shielding provided by the airframe transfers the shielding burden to internal subsystems of the aircraft.

Electrical System Grounding -

The applications of composites in aircraft will undoubtedly have a significant impact on all aspects of electrical system grounding. Since composites are not capable of supporting significant current flow, they cannot serve as a system ground-plane. Alternative approaches, such as use of wired electrical power returns and ground buses become necessary.

Static Electricity -

Very little is known as to how composites will be affected when exposed to environments which cause static charging. Static electrical charging from various sources, such as rain, ice, dust, air, engine exhaust, etc. can cause equipment malfunction, composite damage, fuel charging, and personal shock. Present techniques for preventing static problems depend almost entirely on low resistance conductive paths for charge dissipation. Currently, consideration is being given to the application of antistatic coatings on composites.

In addition to the above items, composites may have other effects on aircraft operation and performance. These include changes of antenna performance (directivity, gain, VSWR, etc. when a metallic counterpoise is replaced by a composite), changes in aircraft radar cross-section, and others.

A unique problem associated with graphite composites is the contamination effect caused by the conductive filaments when the composite structure is altered. For example, in the event of an aircraft explosion or fire, it is possible for graphite fibers to disperse over a wide area. If some of these fibers come in contact with shipborne or shorebased electronic equipment, it is possible that they can cause circuit shorts, or otherwise degrade performance.

5.4 CABLES AND CONNECTORS

5.4.1 Cable Shielding

There are several methods for shielding cables. These include: (a) braid (b) flexible conduit, (c) rigid conduit, and (d) spirally wound shields of high permeability materials.

Braid, which constitutes woven or perforated material, is used for cable shielding in applications where the shield cannot be made of solid material. Advantages are ease of handling in cable makeup and lightness in weight. However, it must be remembered that for radiated fields, the shielding effectiveness of woven or braided materials decreases with increasing frequency and the shielding effectiveness increases with the density of the weave. The percentage coverage by a braided shield has been a critical parameter in the design of flight-deck operated GSE.

Conduit, either solid or flexible, may also be used to shield weapon system cables and wiring from the RF environment. The shielding effectiveness of solid conduit is the same, for RF purposes, as that of a solid sheet of the same thickness and material. Linked armor or flexible conduit may provide effective shielding at lower frequencies, but at higher frequencies the openings between individual links can take on slot-antenna characteristics, seriously degrading the shielding effectiveness. If linked armor conduit is required, all internal wiring should be individually shielded. Degradation of shielding conduit is usually not because of insufficient shielding properties of the conduit material but rather the result of discontinuities in the cable. These discontinuities usually result from splicing or improper termination of the shield.[9]

Solenoids, or other devices associated with high inrush currents or incorporating switching devices that normally develop high-amplitude transients, might be located in the vicinity of shop-oriented GSE. For protection against this type of energy, shielding materials with high permeability are desirable. These materials cannot be drawn into tubing because they lose their shielding properties when cold worked; therefore, an adequate shield is often developed by wrapping a continuous layer of annealed metal tape

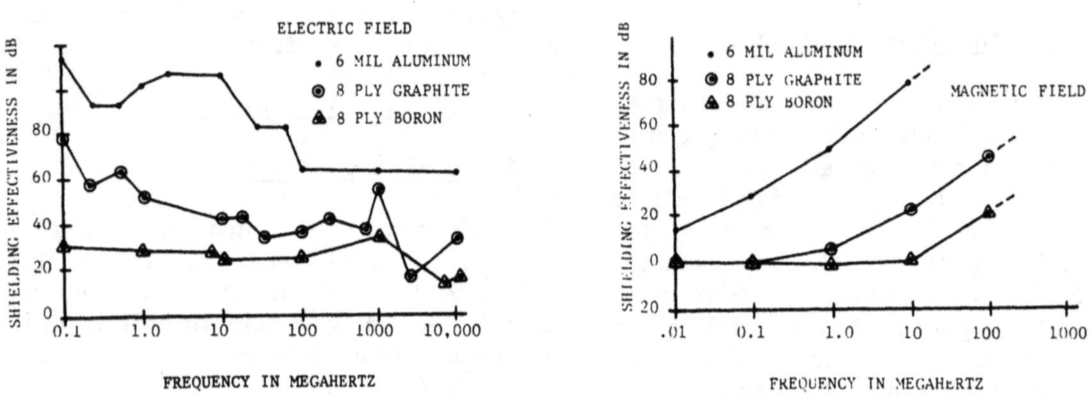

Figure 5-15c

Typical Composite Shielding Effectiveness Characteristics

around the cable. A protective rubber coating is recommended.

The principal types of shielded cables that are available include shielded single wire, shielded multiconductor, shielded twisted pair, and coaxial. Cables are also available in both single and multiple shields in many different forms and with a variety of physical characteristics.

Data on the shielding effectiveness of cables is not readily available. This is primarily because there is no standardized procedure for collecting such information, and because a large number of parameters (some of them external to the cable itself) dictate the particular performance of a cable. The latter includes termination impedances, impinging signal direction and impedance, cable length relative to the interference signal frequency, the particular connectors employed, flexing requirements, and others.

The general characteristics of four classes of shielded cables are identified[9] in Table 5-10. The classes include rigid and flexible conduit, foil-wrapped cable, and braided shielded cable. As indicated previously, shielding effectiveness in most cabling applications is dependent on the percentage coverage of the cable provided by the shield.

For example, MIL-C-7078C is one of several specifications sometimes referenced in GSE procurements. This specification calls for the percent shield coverage (the ratio of metal to total possible shield surface) to be at least 85%. It also imposes a requirement on the angle of the braid wires relative to a plane normal to the cable axis (a parameter that influences percentage shield coverage under cable flexing conditions) of between 10 and 40 degrees.

For GSE flight deck applications, tests made on particular systems indicate it is necessary to impose a shield coverage requirement of 94% on braided cable. The shield braid angle and percent braid coverage are determined in accordance with MIL-C-7078C, as follows:

$$\tan a = 2\pi (D + 2d) P/C \tag{5.17}$$

$$K = 100 (2F - F^2) \tag{5.18}$$

Table 5-10

Comparison of Shielded Cables

	Copper Braid	Foil	Solid Conduit	Flexible Conduit
Shield Effectiveness (audio frequency)	Good	Exc.[1]	Exc.	Good
Shield Effectiveness (radio frequency)	Good	Exc.[1]	Exc.	Poor
Normal % of Coverage	60-95%	100%	100%	90-97%
Fatigue Life	Good	Fair	Poor	Fair
Tensile Strength	Exc.	Poor	Exc.	Fair

[1] may lose its effectiveness when flexed.

where

a = Shield braid angle (angle of braid with axis of cable)

K = Percent coverage

d = Shield strand diameter in inches

C = Number of carriers

D = Diameter of cable under braid, in inches

F = NPd/Sin a

N = Number of strands per carrier

P = Picks per inch of cable length

Note: For 2-conductor cable only

$$D = \frac{(\pi + 2)b}{\pi} \qquad (5.19)$$

where

b = diameter of basic wire

Some of these parameters are defined in Figure 5-16.

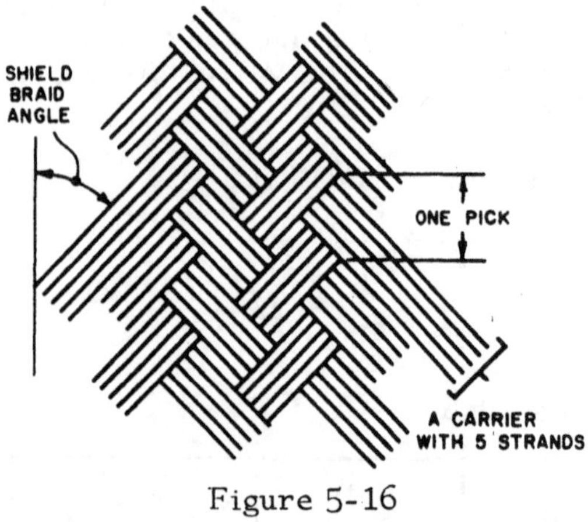

Figure 5-16

Definitions of Cable Shield Parameters

Figures 5-17 and 5-18 illustrate "bad" and "good" cable shielding for flight deck usage. Both pictures were taken with the cable flexed, and show the effects that may be expected from 85% and 94% coverage shields, respectively.

Much of the previous discussion relates to the shielding of cables against plane wave or high impedance fields. For shielding against magnetic fields, the use of annealed high permeability metal strips wrapped around the cable has already been indicated. Multiple layers of counterspiral-wound nickel-iron or silicon iron alloys, or low carbon steel have proven effective under these circumstances. High permeability tape is available with or without adhesive backing. Also, combination high permeability, high conductivity tape is available to provide both electric and magnetic shielding.

Figure 5-17

Braided Shield Cable with 85% Coverage Under Flexing Condition

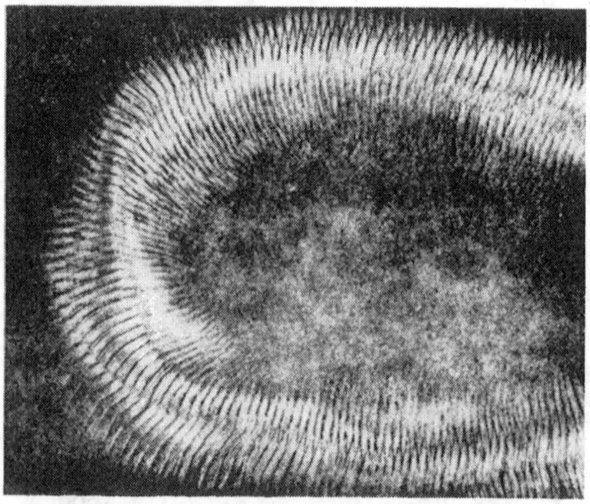

Figure 5-18

Braided Shield Cable with 94% Coverage Under Flexing Condition

5.4.2 Cable Shield Terminations and Connectors

If the effectiveness of a shield is to be maintained, the cable shield must be properly terminated. In an otherwise adequately shielded system, rf currents that are conducted along shields can be coupled to the system wiring from the point of an improper cable termination. This is a particularly important consideration in the case of cables exposed to high power rf fields.

In a properly terminated shield, the entire periphery of the shield is grounded to a low impedance reference, minimizing any rf potentials at the surface of the termination. Solder is undesirable in terminating RF coaxial cables because: (a) too much solder increases the center conductor diameter, thus increasing shunt capacitance and, (b) too little solder increases the current path, thus increasing series inductance. Specification MIL-E-45782B recommends against use of soldering to terminate shields because of the danger of damaging conductor insulation, and suggests a variety of termination methods, all involving crimping operations. A frequently used method of shield termination is illustrated in Figure 5-19. In this arrangement, the cable shield is flared so that it extends over the rear portion of the sleeve, and the crimp ring is slid into place over the sleeve. A crimping tool is then used to crimp the crimp ring onto the sleeve.

An alternative to crimping is shown in Figure 5-20. The shield is placed through the ground ring and flared over and around the ring, and may be secured to the ring with a spot tie (see detail in Figure 5-20). The ground ring is then slid into the rear of the sleeve, which has a tapered base. Tightening the cable clamp onto the end of the sleeve assures positive 360° grounding of the shield, and provides a strain relief for the cable.

The use of silver epoxy or other synthetic conducting material has been found to be unacceptable for shield bonding because of lack of mechanical strength necessary for this application.

For GSE that must operate in high field strength environments, both shields and solid cylindrical members have been terminated using the Magnaforming process. Magnaforming is a metal-forming technique that is used to shrink metal tubes and similar shapes around other forms such as collars, sleeves, rods, etc. The process uses a very intense, short duration magnetic field to induce an opposing current in the tubing or sleeve which is to be shrunk. The magnetic field produced by the induced current in the sleeve opposes the field produced by the current in the magnetic coil, and is of equal magnitude.

The resulting force from the magnetic coil field and the force from the field around the sleeve, combine to produce a uniform compression of the sleeve around the entire sleeve periphery. If the current applied to the magnetic coil is of sufficient amplitude, the combined resultant compressive force on the sleeve will deform the sleeve, reducing its diameter until it conforms to the shape of the underlying material. The basic process is shown in Figure 5-21.

Figure 5-22 shows a cross-section of a fixture in which a workpiece is to be shrunk around a cable sleeve and connector. An insulating "former" is normally used between the magnetic coil assembly and the workpiece, but the insulator does not directly enter into the forming operation.

Figure 5-23 is a cut-away photograph of two connectors terminating opposite ends of a short length of cabling in conduit. The Magnaforming process was used to make the terminations at the points identified in the figure.

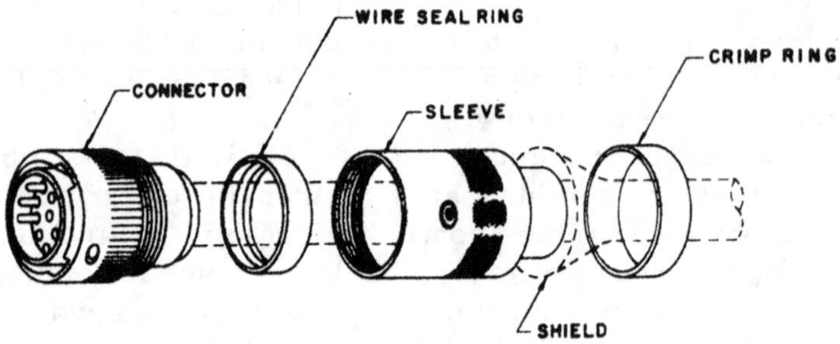

Figure 5-19
Shield Termination Using Crimping

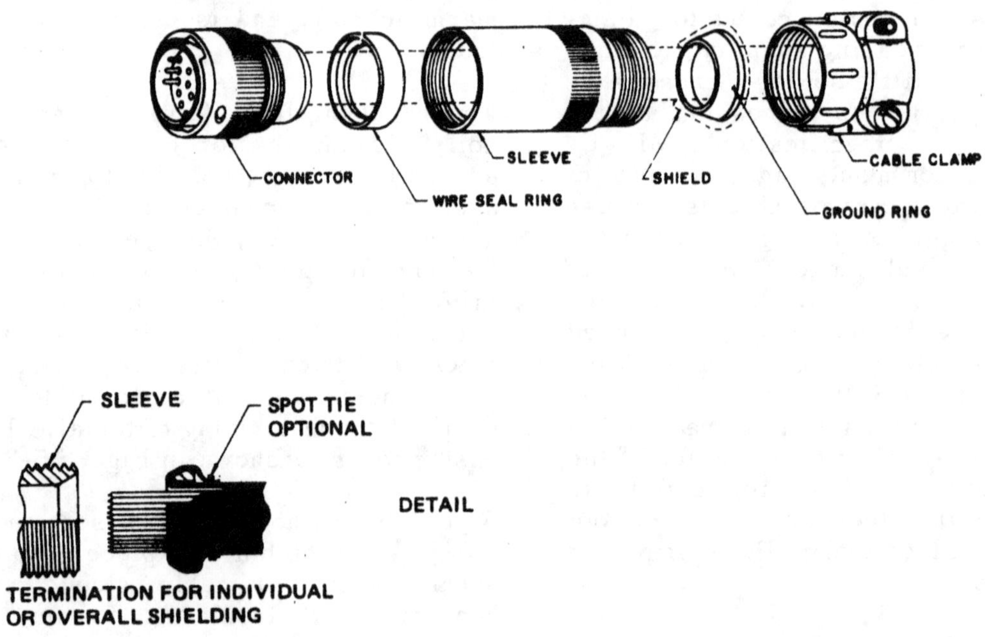

Figure 5-20
Shield Termination Using Threaded Assembly

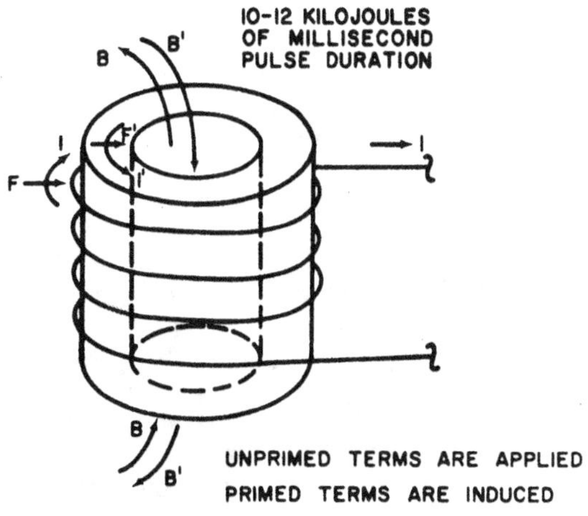

Figure 5-21

Relationship of F, B, and I in Magnaforming Process

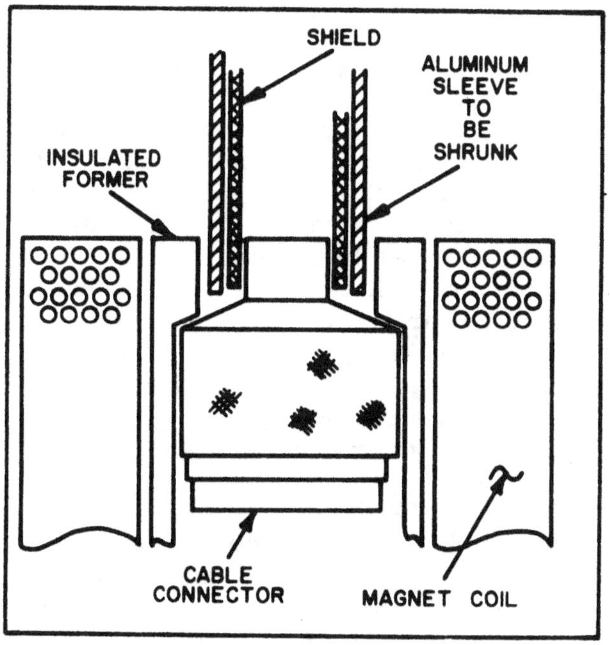

Figure 5-22

Typical Magnaforming Process

When maintaining the shielding integrity of a connector pair (i.e., two interconnecting connectors), a good method to employ (see Figure 5-24) is to place spring contacts inside one portion of one connector so that positive contact is made along the circumference of the mating parts.[9] These contacts are extended so that the shell of the connector mates before the pins make contact on assembly of the connector and breaks after the pins on disassembly. A connector which meets these requirements is available under MIL-C-27599 and is the preferred type to be used in rf-proof designs.

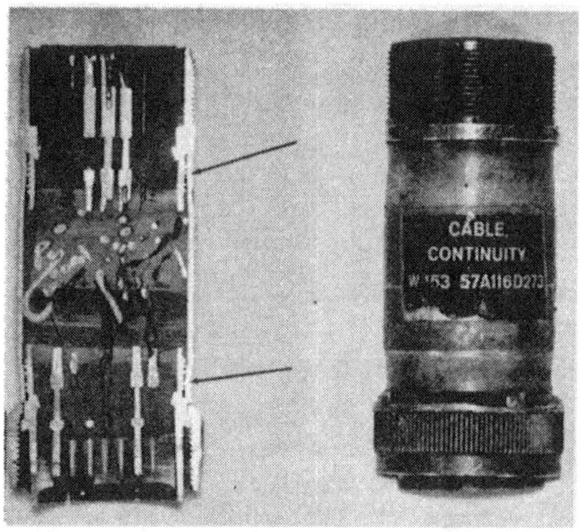

Figure 5-23

Cut-Away View of Magnaformed Cable Transition

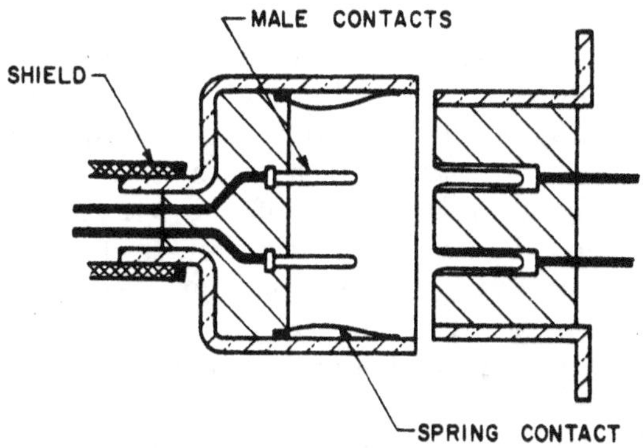

Figure 5-24

RF-Proof Connector

The advantages gained using circumferential spring fingers over bayonet coupling is dramatically illustrated in Figure 5-25. In this case, the spring contacts were of silver-plated beryllium copper.

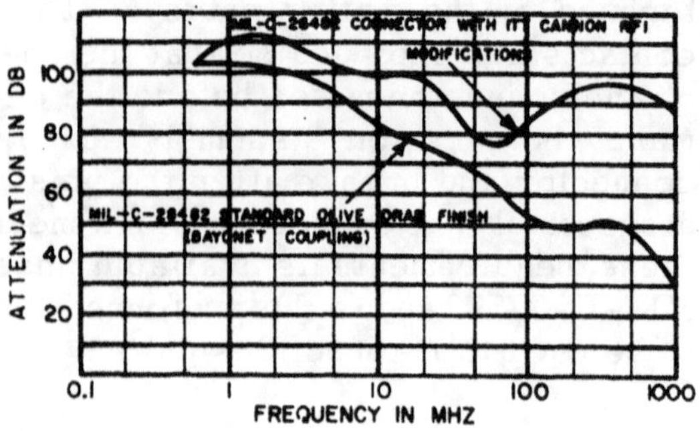

Figure 5-25

Comparison of the RFI shielding at the interface coupling of a MIL-C-26482 connector having standard olive drab finish with the same connector using peripheral spring contact fingers over cadmium plating (RFI type).

Figure 5-26 illustrates the type of connector that should be used when a shielded cable assembly contains individual shielded wires.[9] The practice of pigtailing these shields and connecting them to one of the pins is not recommended. The individual shields should be connected to coaxial pins specifically adapted for this purpose, with the shields of the mating surfaces making contact before the pins.

An alternative method of terminating individually shielded wires (if the shields can be tied together) is shown in Figure 5-27. A pick is used to strip the shield from each of the wires involved. The shields are then laid over a ground ring (item 2 in Figure 5-29), and the entire assembly is secured by screwing items 1 and 4 together.

RF arcing problems can occur in a shield that is adequate at audio frequencies. Induced RF currents can be conducted along cable shields and coupled to the system wiring at points of incorrect cable shield termination. RF potentials at the termination can be minimized by grounding the entire periphery to a low-impedance reference. Grounding may be required every 0.2λ based on the highest frequency at which currents may be present.

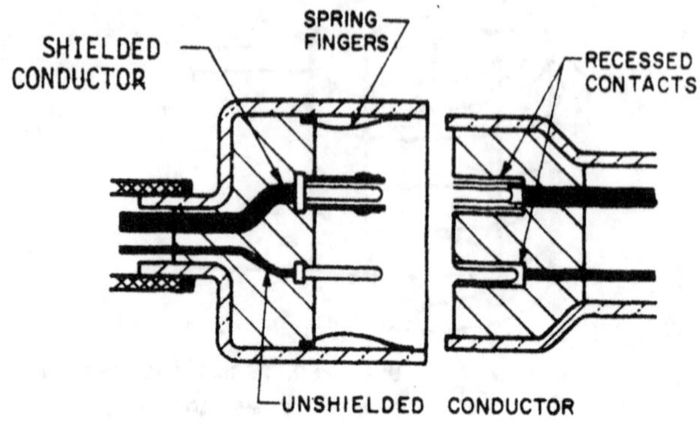

Figure 5-26

Connector for Shield Within a Shield

5.4.3 New Connector Technology

Recent connector developments include two new types with special features that improve shell-to-shell bonding in a high vibration environment. One is the Bendix "Tri-Start" series and the other is the G & H Technology "Breech-Lok" series of connectors. Both are "scoop-proof" connectors that employ metric threads at the rear for attaching accessories. Scoop-proofing prevents shorts and pin damage, and metric threads provide more thread surface and thus a better bond between shells and accessories.

The Tri-Start* connectors are Series III connectors as defined in MIL-C-38999G and DoD-C-38999. They incorporate Series I high density inserts and contacts, and utilize a triple start Acme thread (one full turn to lock) with an anti-decoupling feature for plug-to-receptacle engagement. They also include metallic spring fingers for grounding. They were developed to meet rigid environmental and

*Tri-Start is a trademark of the Bendix Corp.

shock and vibration requirements, but also provide improved EMI protection over the bayonette type Series I and II connectors in a high vibration environment.

The Breech-Lok* connectors are Series IV connectors as defined in MIL-C-38999G and DoD-C-38999. They incorporate Series I inserts and contacts, but utilize a solid-metal locking system (detent and lock in one-half turn) and internal drive threads for plug-to-receptacle engagement. They also include metallic spring fingers for grounding. They too were developed to meet rigid environmental and shock and vibration requirements, but also provide improved EMI protection over Series I and II connectors in a high vibration environment.

Several new accessories that provide improved shield-to-connector bonding have been developed in the past few years. These include the "IRIS" compression ring and "TAG" terminating and grounding ring. These devices provide additional options for terminating overall and individual shields respectively.

*Breech-Lok is a trademark of G & H Technology, Inc.

Figures 5-26a & 5-26b illustrate overall shield bonding by means of "IRIS" compression rings. The compression ring is part of the connector backshell or adapter and encircles the cable at the shield. The "IRIS" is a compressible metal coil spring, conductive elastomer, or other compressible conductive material that provides electrical contact between the overall shield and the backshell/adapter. The advantages of using "IRIS" compression rings are that the installation is simplified since the shield only has to be trimmed, and there is otherwise no disturbance to the shield. Also, the compression ring fits a range of cable sizes and makes a good bond to cables that are out of round.

Figure 5-26c shows a TAG Ring** assembly that provides for terminating groups of individually shielded wires. The individual shields are placed in slots in the main body of the TAG Ring and are held there by a spring device built into the threaded clamp nut. The advantages of the TAG Ring are that installation can be performed without special tools, and that a range of cable sizes and shield thicknesses can be ac-

**TAG Ring is a registered trademark of Glenair, Inc.

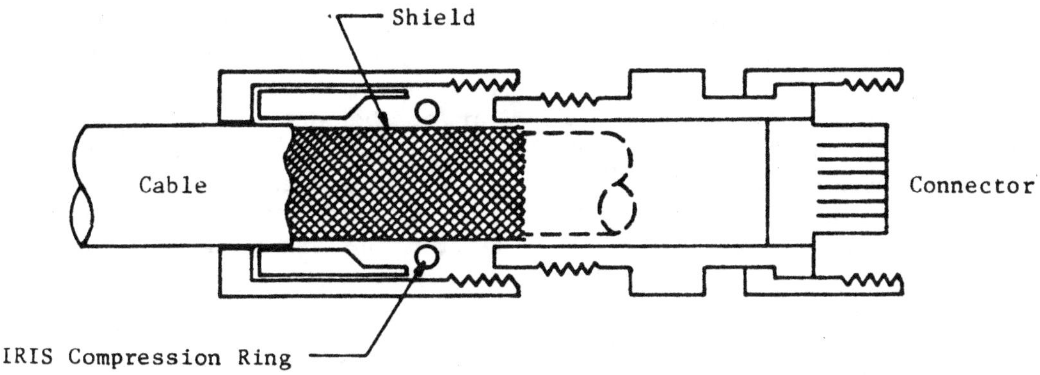

Figure 5-26a

IRIS Coiled Spring Device

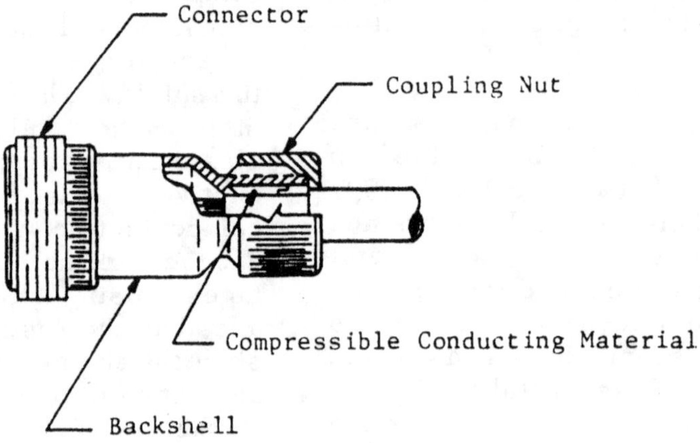

Figure 5-26b

IRIS Conductive Elastomer Device

commodated. Also, a TAG Ring and tapered cone can be installed in the same backshell/adapter to terminate individual shields and an overall shield when both are contained in the cable.

Another example of new connector technology is the "filter connector." Although they were developed in the 1960's filter connectors have only recently been available as production items. Filter connectors are used to supplement shielded cable and shielding connectors when they cannot reduce incoming or outgoing interference to acceptable levels.

A cross section of a filter connector is shown in Figure 5-26d. Individual filter elements are commonly capacitive (bypass), L-section, or π-section. Ferrites and capacitive ceramics provide the electrical properties. Filter element operation is shown in Figure 5-26e. Additional information on connectors having lossy filter elements may be found in Section 8.2.1.7 of this manual.

5.4.4 Fiber Optics

Fiber optics technology is gradually being incorporated into advanced weapons systems and in associated test equipment. It makes possible optical links between subsystems, by replacing hard-wired interfaces with glass fibers; and transmitting voice, video or data signals by light rather than by electricity. Infra-red lasers or light-emitting diodes serve as signal drivers.

The use of fiber optics offers weight and space advantages to the system designer. These benefits accrue largely because of the fibers' broadband transmission capabilities. Figure 5-26f illustrates these differences in a conventional and a fiber-optics cable having the same signal carrying capacity.

From the standpoint of EMC, fiber optics cables are insensitive to electromagnetic field environments (al-

Figure 5-26c

TAG Ring Assembly

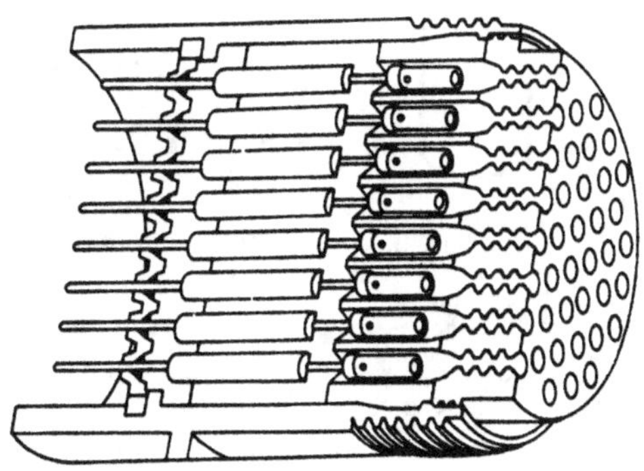

Figure 5-26d

Filter Connector

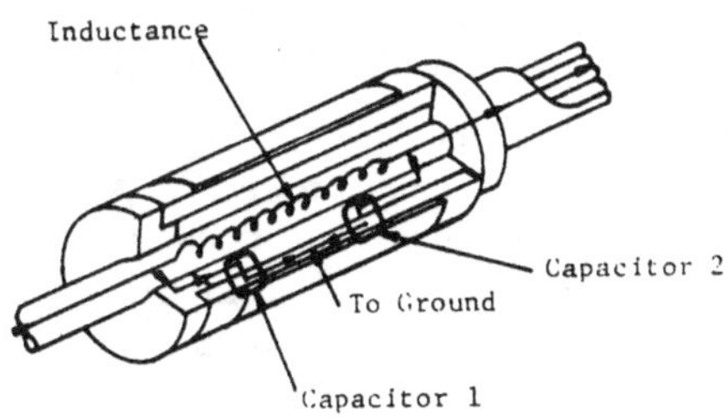

Figure 5-26e

Filter Element

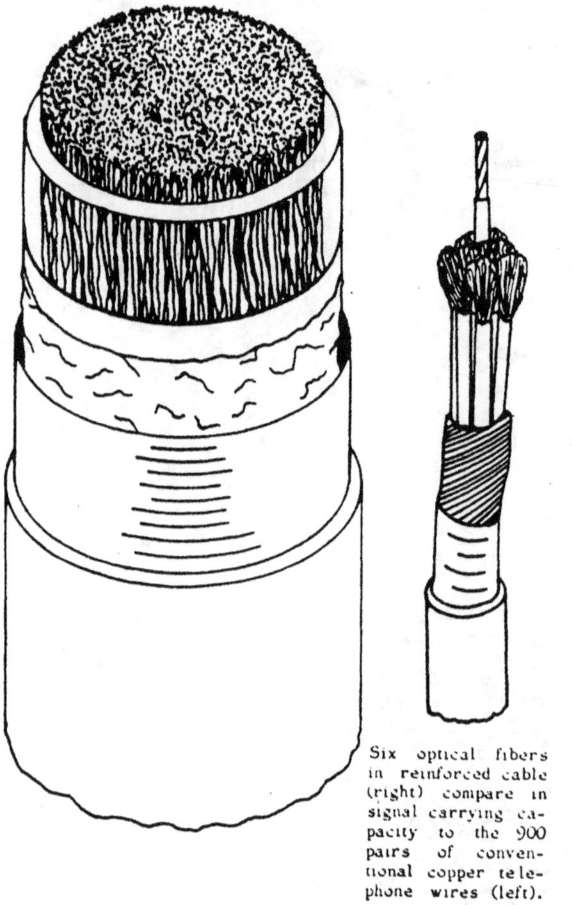

Figure 5-26f

Conventional and Fiber Optic Cables

though that is not necessarily true of the line termination devices) and therefore do not have to be shielded. They provide electrical isolation between source and load and eliminate potential problems of ground loop coupling. Additionally, they are totally safe for use in a fuel vapor or other explosive environment, and offer a secure communication means since they do not radiate detectable energy.

The advantages of broad bandwidth and better aerodynamic and electromagnetic compatibility features must be weighed against other characteristics associated with fiber optics subsystems. These include the more complex and more costly line termination equipment (particularly if a laser is required), and the somewhat higher loss characteristics of the fibers themselves as compared with copper wire.

Typical fiber losses range from 3 to 15 dB/km when measured at 8000 angstroms. Filament bandpass can be 25-400 MHz-km, depending on the type of fiber used, and the fiber diameter. The fibers can be clad and bundled in arrangements very similar to hard-wired cable.

5.5 OTHER DESIGN TECHNIQUES TO MAINTAIN SHIELDING EFFECTIVENESS

5.5.1 Seams Without Gaskets

Seams or openings in enclosure or compartment walls that are properly bonded will provide a low impedance to rf currents flowing across the seam. When good shielding characteristics are to be maintained, permanent mating surfaces of metallic members within an enclosure should be bonded together by welding, brazing, sweating, swaging, or other metal flow process. To insure adequate and properly implemented bonding techniques, the following recommendations should be observed:

- All mating surfaces must be cleaned before bonding.
- All protective coatings having a conductivity less than that of the metals being bonded must be removed from the contact areas of the two mating surfaces before the bond connection is made.
- When protective coatings are necessary, they should be so designed that

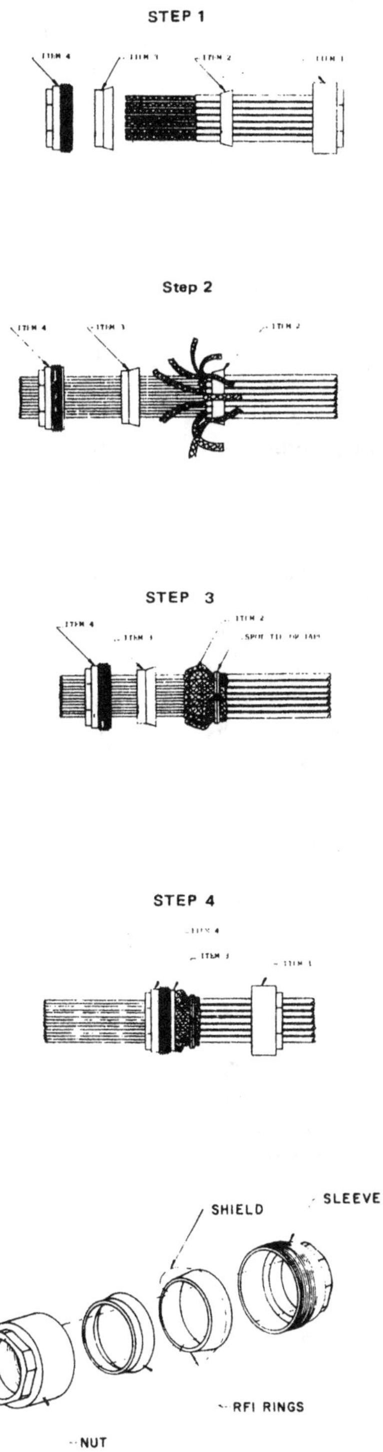

Figure 5-27
Shield Termination of Multiple Shielded Wires

they can be easily removed from mating surfaces prior to bonding. Since the mating of bare metal to bare metal is essential for a satisfactory bond, a conflict may arise between the bonding and finish specifications. From the viewpoint of shielding effectiveness, it is preferable to remove the finish where compromising of the bonding effectiveness would occur.

- Certain protective metal platings such as cadmium, tin, or silver need not in general be removed. Similarly, low-impedance corrosion-resistant finishes suitable for aluminum alloys, such as alodine, iridite, oakite, turco and bonderrite, may be retained. Most other coatings, such as anodizing, are nonconductive and would destroy the concept of a bond offering a low impedance radio frequency path. See Figure 5-28 for shielding effectiveness data on selected surface finishes.[11]

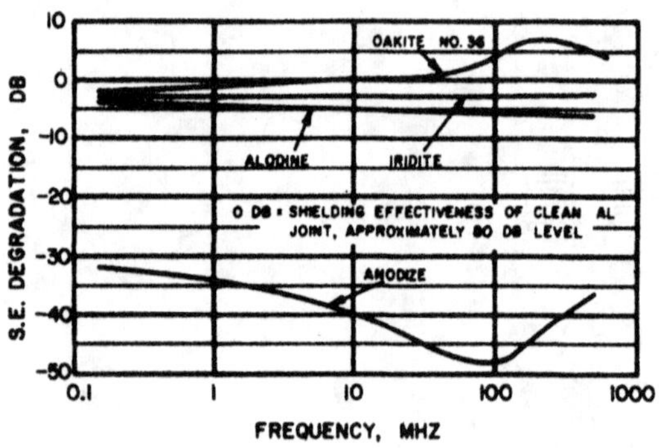

Figure 5-28

Shielding Effectiveness Degradation Caused by Surface Finishes on Aluminum

- Mating surfaces should be bonded immediately after protective coatings are removed to avoid oxidation. Refinishing after bonding is acceptable from the standpoint of shielding effectiveness.

- When two dissimilar metals must be bonded, metals that are close to one another in the electrochemical series should be selected. Elements of the series are shown in Table 5-1 in the Bonding chapter of this design guide.

- Soldering may be used to fill the resulting seam, but should not be employed to provide bond strength.

- The most desirable bond is achieved through a continuous butt or lap weld. Spot welding is less desirable because of the tendency for buckling, and the possibility of corrosion occurring between welds. Rivetting or pinning is even less desirable because of the greater susceptibility of bond degradation with wear.

- An overlap seam, accompanied by soldering or spot welding (see Figure 5-29) provides a relatively effective bond. Other types of crimped seams may be employed, so long as the crimping pressure is maintained.

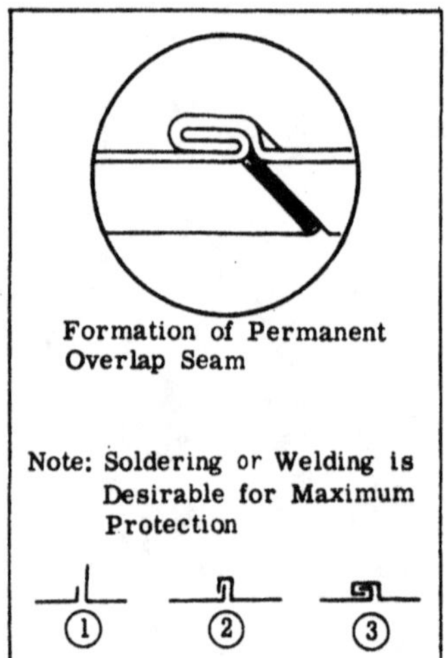

Figure 5-29

Formation of Permanent Overlap Seam

There are often occasions when good temporary bonds must be obtained. Bolts, screws, or various types of clamp and slide fasteners have been used for this purpose.

The same general requirements of clean and intimate contact of mating surfaces, and minimized electrolytic (cathodic) effects apply to temporary bonds as well. Positive locking mechanisms that insure consistent contact pressure over an extended period of time should be used. Lockwashers should be employed that can "bite into" metal surfaces and fasteners and maintain a low bonding resistance. Bolts, nuts, screws, and washers that must be manufactured with material different from the surfaces to be bonded should be higher in the electromotive series than the surfaces themselves, so that any material migration erodes replacable components.

A critical factor in temporary bonds (and in spot-welded permanent bonds as well) is the linear spacing of the fasteners or spot-welds.[12] Figure 5-30 provides an indication of the sensitivity of this parameter for a ½-inch aluminum lap joint. Data is taken at 200 MHz. The shielding effectiveness shown at 1-inch spacing is about 12 dB poorer than the identical configuration incorporating a ½-inch wide monel mesh gasket; the effectiveness at 10-inch spacing is about 30 dB poorer than with the same gasket. Use of conductive gaskets for this and other applications will be discussed shortly.

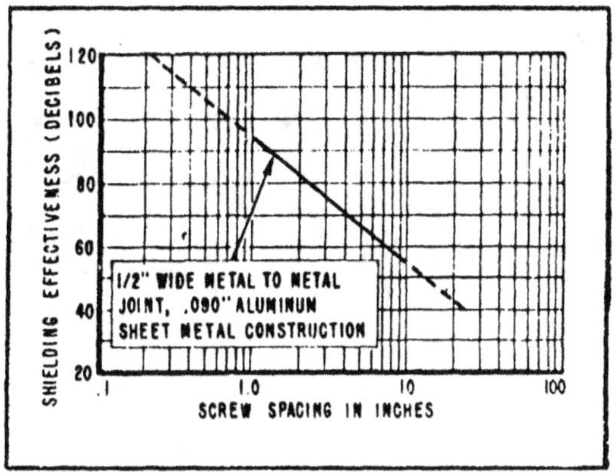

Figure 5-30

Influence of Screw Spacing

Similar techniques to those just described can be employed in connection with seams in magnetic materials.[5] Permanent seams can be butt or lap continuous arc in an argon or helium atmosphere, recognizing that a final material heat treatment will be necessary. Temporary seams are usually screwed or bolted together. Figures 5-31 and 5-32 indicate the change in shielding effectiveness of an AMPB-65 seam at various frequencies as a function of screw spacing and lap joint width, respectively.

Bonding techniques will be discussed in more detail in the next chapter of this design guide.

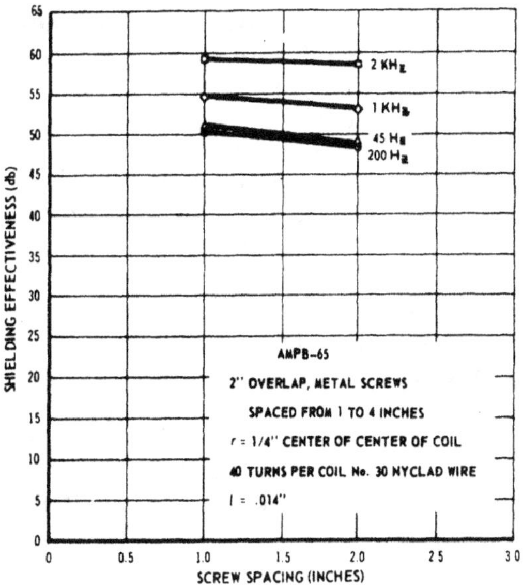

Figure 5-31

Shielding Effectiveness of AMPB-65 Overlap as a Function of Screw Spacing Along Two Rows, 1.5 Inches Apart

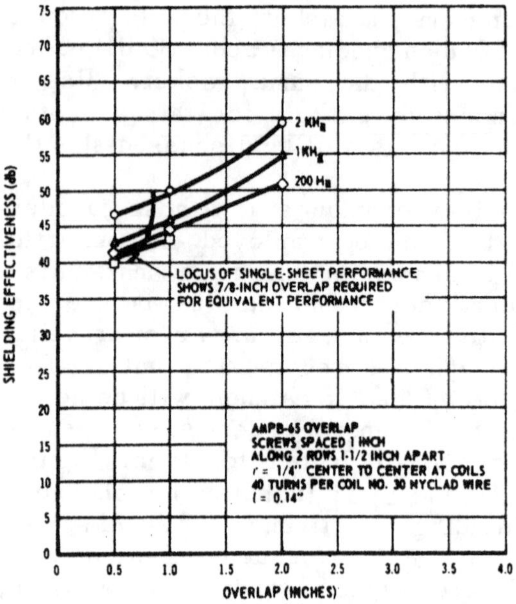

Figure 5-32

Shielding Effectiveness of AMPB-65 Joint as a Function of Overlap

5.5.2 Seams with Gaskets

Considerable shielding improvement over direct metal-to-metal mating of shields used as temporary bonds can be obtained using flexible, resilient electrically conductive gaskets placed between shielding surfaces to be joined. Clean conductive mating surfaces and a good pressure contact are necessary.

The major material requirements for rf gaskets include compatibility with the mating surfaces, corrosion resistance, appropriate electrical properties, resilience (particularly when repeated compression and decompression of the gasket is expected), mechanical wear, and ability to form into the desired shape. On this basis, monel and silver-plated brass are generally the preferred materials, with aluminum used only for gasketing between two aluminum surfaces. Beryllium-copper contact fingers are usually employed, with a variety of platings available, if desired. Mumetal and Permalloy have been used when magnetic shielding effectiveness is of concern.

For applications requiring moisture/pressure sealing as well as RF shielding, combination rubber-metal seals are available. These include metal mesh bonded to neoprene or silicone, aluminum screen impregnated with neoprene, oriented wires in silicone, conductive adhesives and sealants, and conductive rubber. The advantages and limitations of these, as well as non-sealing RF gaskets, are summarized in Table 5-11.

Knitted wire mesh gaskets can be manufactured in a variety of shapes and sizes. Figure 5-33 shows the structure of knitted wire mesh. Figures 5-34 and 5-35 illustrate representative cross sections and shapes, respectively. Gasket thickness is dependent on the unevenness of the joint to be sealed, the compressibility of the gasket, and the force available. Gasket shape depends on the particular application involved, as well as the

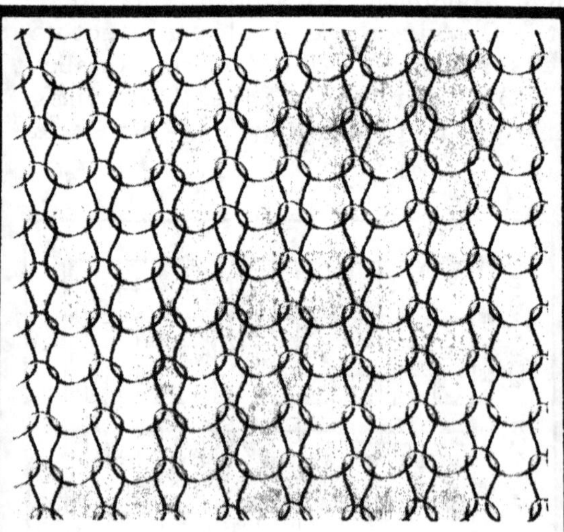

Figure 5-33

An enlarged view of knitted wire mesh. It is made up of many loop-shaped springs interlocked in a hinge-like manner; thus combining resilience, flexibility and cohesion.

Table 5-11
Characteristics of Conductive Gasketing Materials

Material	Chief Advantages	Chief Limitations
Compressed knitted wire	Most resilient all-metal gasket (low flange pressure required). Most points of contact. Available in variety of thicknesses and resiliencies, and in combination with neoprene and silicone.	Not available in sheet (Certain intricate shapes difficult to make). Must be 0.040 in. or thicker. Subject to compression set.
Brass or beryllium copper with punctured nail holes	Best break-thru of corrosion protection films.	Not truly resilient or generally reusable.
Oriented wires in rubber silicone	Combines fluid and rf seal Can be effective against corrosion films if ends of wires are sharp.	Might require wider or thicker size gasket for same effectiveness. Effectiveness reduces with mechanical use.
Aluminum screen impregnated with neoprene	Combines fluid and conductive seal. Thinnest gasket. Can be cut to intricate shapes.	Very low resiliency (high flange pressure required).
Soft metals	Cheapest in small sizes.	Cold flows, low resiliency.
Metal over rubber	Takes advantage of the resiliency of rubber.	Foil cracks or shifts position. Generally low insertion loss yielding poor RF properties.
Conductive Rubber (carbon filled)	Combines fluid and conductive seal.	Provides moderate insertion loss.
Conductive Rubber (silver filled)	Combines fluid and RF seal. Excellent resilience with low compression set. Reuseable. Available in any shape or cross section.	Not as effective as metal in magnetic fields. May require salt spray environmental protection.
Contact fingers	Best suited for sliding contact.	Easily damaged. Few points of contact.

space available, the manner in which the gasket is held in place, and the same parameters that influence gasket thickness.[13]

Silver filled silicone rubber gaskets can be obtained in sheet, die-cut, molded or extruded form. The most popular, and most economical of these types, is the extrusion. Figure 5-36 shows typical extruded shapes and indicates recommended deflection limits for various shapes and sizes. Comments made above concerning thickness, shape and mounting methods for wire mesh gaskets also apply to conductive rubber gaskets.

Figure 5-34
Representative Gasket Cross Sections

Shielding effectiveness of silver filled (or silver plated copper filled) silicones is especially high between 15 kHz and 10 GHz. Plane wave attenuation often improves with higher closure force, especially for die-cut gaskets. Best results are achieved with molded or extruded cross sections held in grooves.

Gaskets may be held in place by sidewall friction, by soldering, by adhesive, or by positioning in a slot or on a shoulder. Soldering must be controlled carefully to prevent its soaking into the gasket and destroying gasket resiliency. Adhesives (particularly non-conductive adhesives) should not be applied to gasket surfaces that mate for rf shielding purposes; auxiliary tabs should be used.

Typical gasket pressures for obtaining effective seals range from 5-100 psi. A usual pressure is 20 psi. Various ways in which wire mesh gaskets may be used are shown in Figure 5-37.

A number of examples of rf gaskets designed for GSE use can be cited. One such design is shown[14] in Figure 5-38. This gasket is used behind the

Figure 5-35
Representative Gasket Shapes

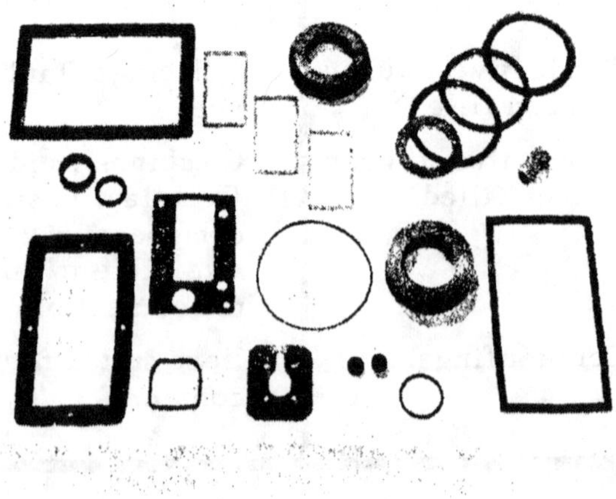

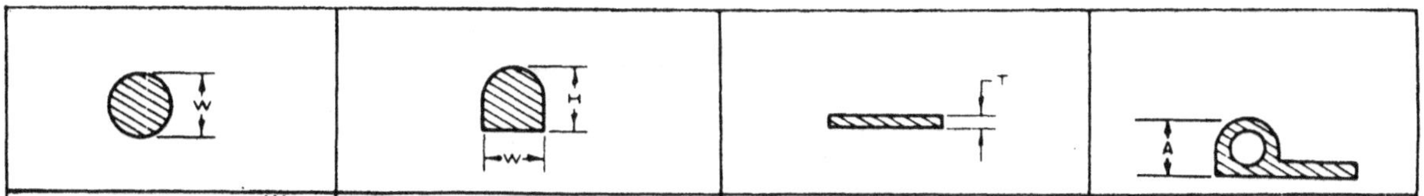

Deflection	W Dia.	Deflection	H	Deflection	T	Deflection	A
.007 – .018	.070	.006 – .012	.068	.001 – .002	.020	.025 – .080	.200
.010 – .026	.103	.008 – .016	.089	.001 – .003	.032	.030 – .125	.250
.013 – .031	.125	.012 – .024	.131	.003 – .006	.062	.075 – .250	.360
.014 – .035	.139	.014 – .029	.156	.003 – .009	.093		
		.016 – .032	.175				

from "EMI/RFI Gasket Design Manual," 1975, Chromerics, Inc.

Figure 5-36
Gasket Deflection Limits (in inches)

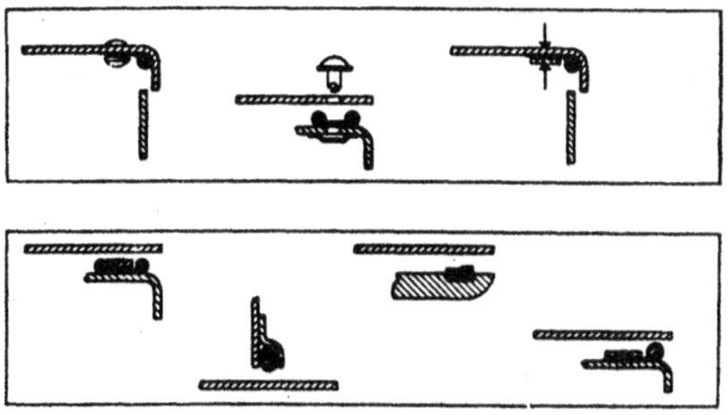

Figure 5-37
Representative Applications
of Wire Mesh Gaskets

front panel of the AN/AWM-54, Aircraft Firing Circuit Test Set. It consists of an outer Neoprene surface to waterproof the unit, plus an inner monel mesh.

Five of the most often encountered rf gasket design problems are illustrated[15] in Figure 5-39, along with the means of overcoming the problems.

A practice that is beginning to receive acceptance is the dating of conductive and pressure gaskets, as well as other items that may suffer wear with continued use. The date is used during maintenance as an indicator of whether the gasket or other tagged item should be replaced.

5.5.3 Use of Waveguide Attenuators

The effects of seams, cracks, openings, holes and other breaks in a shield can often be studied by considering the opening as a waveguide, and determining the guide attenuation. Alternatively, waveguide attenuators can be designed around control shafts, light receptacles and other openings to reduce interference leakage.

If the opening can be treated as a rectangular aperture, its cutoff wavelength, λ_c, is

$$\lambda_c = 2w \qquad (5.20)$$

where

w = the long dimension of the opening.

Slot attenuation, in dB, can be computed from the relationship

$$A = 54.5 \frac{d}{\lambda_c}\left(1 - \frac{\lambda_c}{\lambda}\right)^{\frac{1}{2}} \qquad (5.21)$$

where

A = the attenuation per unit depth

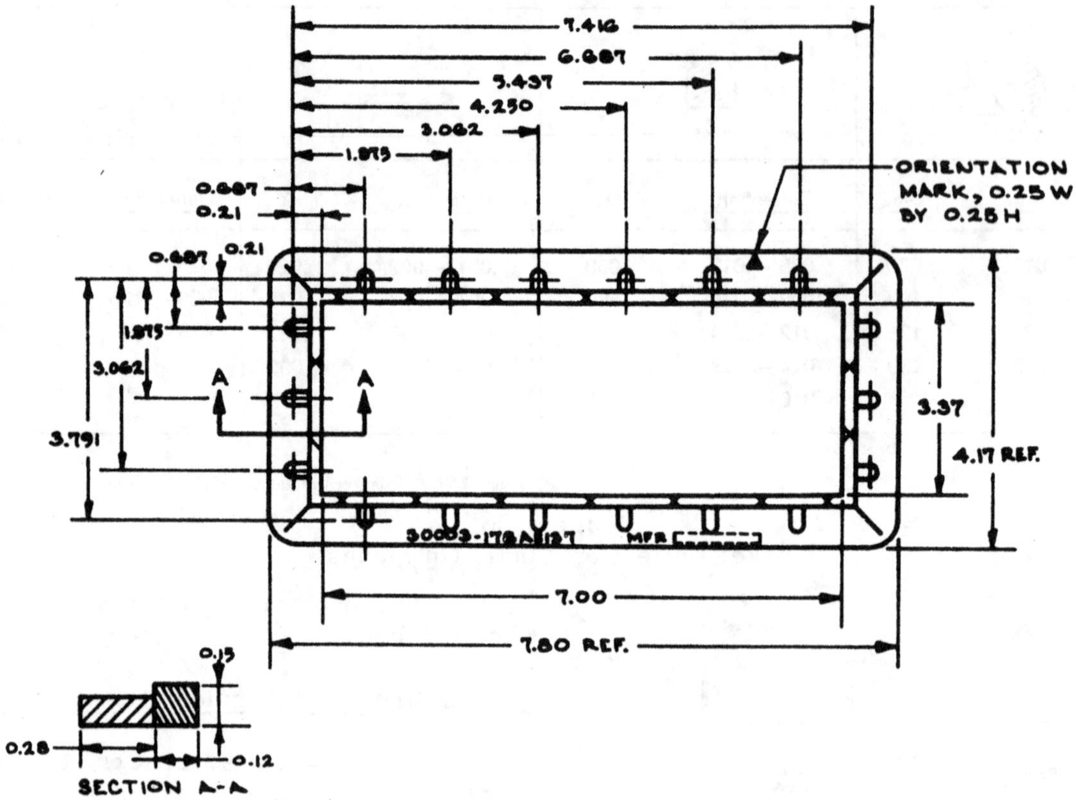

Figure 5-38
Front Panel Gasket for AN/AWM-54

d = slot depth

λ = interference wavelength

Similar relationships can be established for other opening configurations.

A useful rule to follow for circular holes is that, for 100 dB of attenuation, the length of the waveguide must be at least 3 times the diameter of the hole. The maximum permissible diameter (d_{max}) can be obtained by dividing the wavelength for the highest frequency under consideration (λ_{min}) by 3.4, where both diameter and wavelength are expressed in the same units.[2]

$$d_{max} = \frac{\lambda_{min}}{3.4} \qquad (5.22)$$

One example of the use of a waveguide over an opening to obtain attenuation may be found in the AN/AWM-54 Aircraft Firing Circuit Test Set. Figure 5-40 provides a side-view of the test set with its cover removed, and shows two circular guides above the GO/NO-GO indicator lights.

In many cases, shielding screens introduce excessive air resistance and sometimes greater shielding effectiveness may be needed than they can provide. In such cases, openings may be covered with specially designed ventilation panels (such as honeycomb) with openings that operate on the waveguide-below-cutoff principle. Honeycomb-type ventilation panels in place of screening:

detail	why faulty	suggested remedy
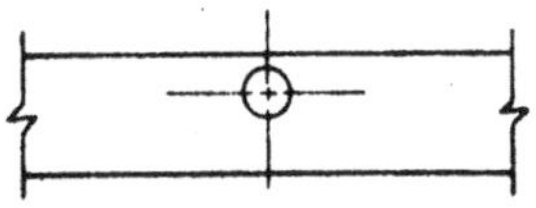 Bolt holes close to edge	Causes breakage in stripping and assembling	Projection or "ear" 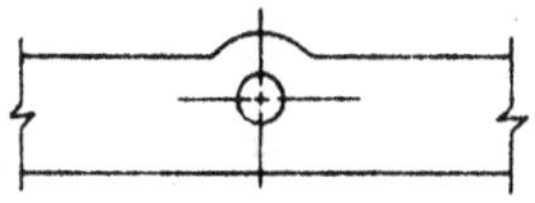 Notch instead of hole 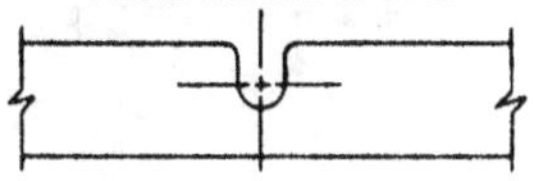
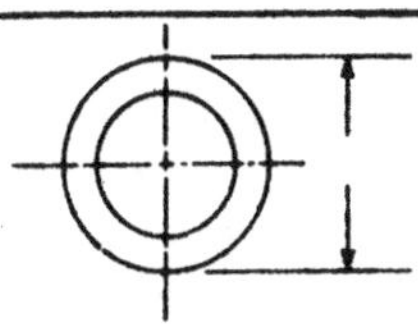 Metalworking tolerances applied to gasket thickness, diameters, length, width, etc.	Results in perfectly useable parts being rejected at incoming inspection. Requires time and correspondence to reach agreement on practical limits. Increases cost of parts and tooling. Delays deliveries	Most gasket materials are compressible. Many are affected by humidity changes. Try standard or commercial tolerances before concluding that special accuracy is required
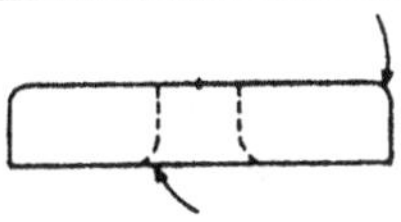 Transference of fillets, radii, etc., from mating metal parts to gasket	Unless part is molded, such features mean extra operations and higher cost	Most gasket stocks will conform to mating parts without preshaping. Be sure radii, chamfers, etc., are functional, not merely copied from metal members
 Thin walls, delicate cross section in relation to over-all size	High scrap loss: stretching or distortion in shipment or use. Restricts choice to high tensile strength materials	Have the gasket in mind during early design stages
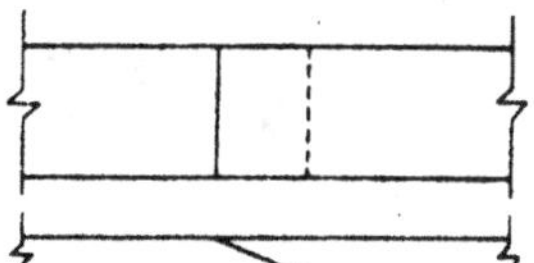 Large gaskets made in sections with beveled joints	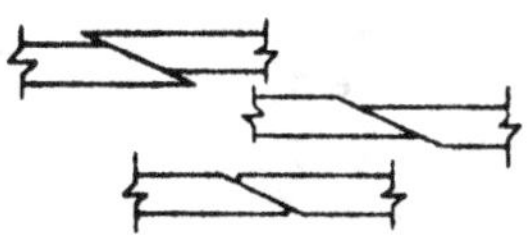 Extra operations to skive. Extra operations to glue. Difficult to obtain smooth, even joints without steps or transverse grooves	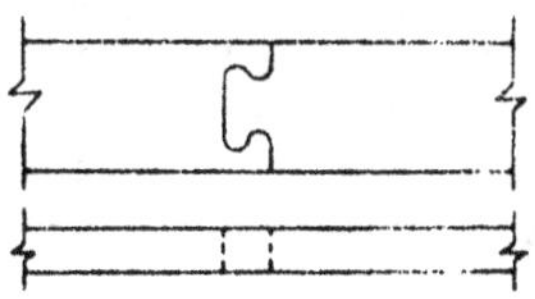 Die-cut dovetail joint

Figure 5-39

Common Errors in Gasket Design

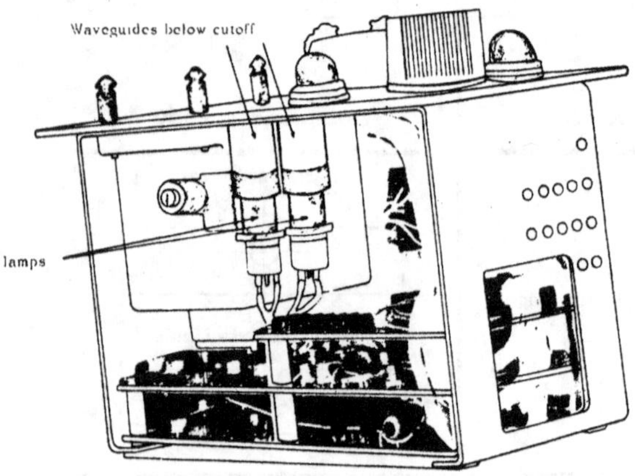

Figure 5-40
Application of Waveguide Below Cutoff Principle to AN/AWM-54 Aircraft Firing Circuit Test Set

a. Allow higher attenuation than can be obtained with mesh screening over a specified frequency range.

b. Allow more air to flow without pressure drop for the same diameter opening.

c. Cannot be damaged as easily as the mesh screen, and are therefore more reliable.

d. Are less subject to deterioration by oxidation and exposure.

All nonsolid shielding materials, such as perforated metal, fine mesh copper screening, and metal honeycomb, present an impedance to air flow. Metal honeycomb is the best of these materials because it enables very high electric field attenuations to be obtained through the microwave band with negligible drops in air pressure. However, honeycomb has the disadvantages of occupying greater volume and costing more than screening or perforated metal. Also, it is often difficult to apply honeycomb paneling because flush mounting is required. Thus, screening and perforated sheet stock sometimes find application for purely physical design reasons, although honeycomb panels can achieve attenuations to 136 dB above 10 MHz.[3]

5.5.4 Panel Openings

Panel openings that must accommodate control shafts may be shielded in one of several ways. A waveguide attenuator may be used around the panel opening as discussed under Section 5.5.3, so long as the shaft within the guide is non-conducting. Alternatively, the portion of the control that is behind the panel may be shielded to separate the control from the remainder of the equipment, and the control leads filtered.

A military requirement exists regarding RFI-shielded rotary-shaft seals (MIL-B-5423/9A). The requirement specifies a brass, nickel-plated nut with a knitted monel wire mesh insert to ground the control shaft. The nut is shown in Figure 5-41.

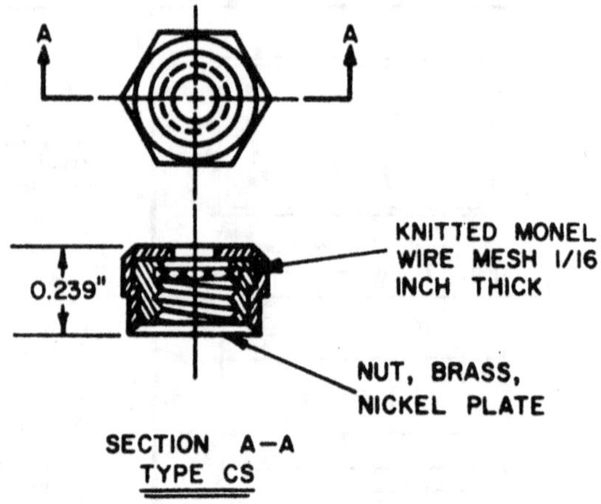

Figure 5-41

RFI-Shielded Rotary Shaft Seal

Fuseholders, phonejacks, panel connectors not in use, and other receptacles can be fitted with a metallic cap that provides an electrically continuous

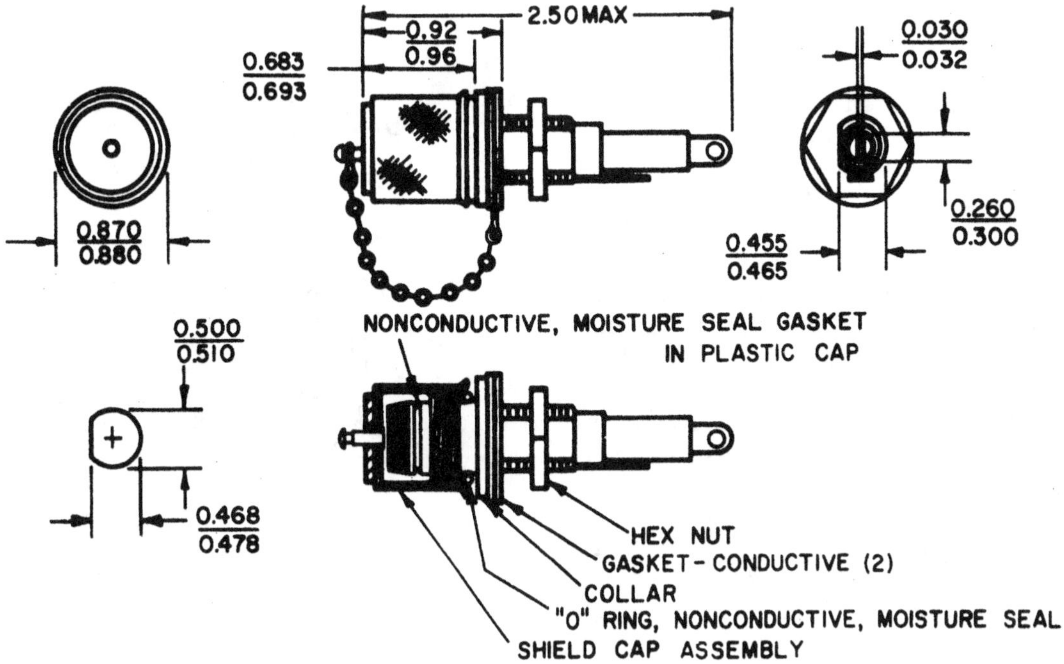

Figure 5-42

RFI-Shielded Extractor Post Type Fuseholder

cover and maintains case integrity. An example of this type of cap for fuseholder applications is shown in Figure 5-42, as specified by MIL-I-49207/36A. The two conducting gaskets are knitted monel impregnated with silicone rubber.

Pushbutton switches are recommended to be shielded using boots having a mesh gasket insert (MIL-B-5423/7A). This approach (see Figure 5-43) appears to be more effective than the alternative of a conductive rubber boot.

5.5.5 Required Visual Openings

Often it is necessary to provide rf shielding over pilot-light bulbs, digital display faces, meter faces, strip chart recorder outputs, oscilloscope faces, or similar devices that must be observed by the equipment user. The alternatives available include:

- use of a waveguide attenuator
- use of screening material
- providing a shield behind the assembly of concern, and filtering all leads to the assembly
- use of conducting glass

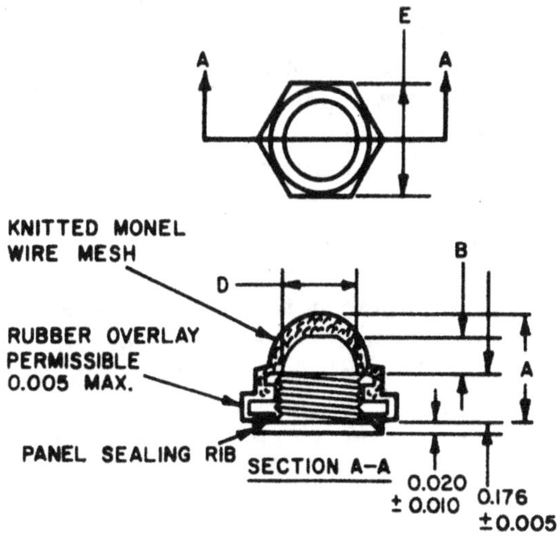

Figure 5-43

RFI-Shielded Pushbutton Switch

A waveguide attenuator is a practical approach for rf shielding of lamps, and an example has already been provided in Figure 5-40 This technique has the advantage of not introducing light transmission loss. However, it is not particularly suitable for most meter openings or larger apertures, because of the space requirements involved.

Use of screens over meter faces and other large apertures has often been employed for shielding purpose. A typical screen introduces a minimum of 15-20% optical loss, and can create difficulties in reading meters. If the device being shielded has a scale (such as an oscilloscope reticule), bothersome zoning patterns can result. However, these potential deficiencies are counterbalanced by good shielding efficiencies at a fairly low cost.

Figure 5-44 illustrates one method of mounting such screens, when they are not incorporated directly into the device to be shielded,[1] and Figure 5-45 indicates the shielding effectiveness provided by commercially available screen assemblies.[16] In the latter case, data on either .0045 or .002 inch diameter monel, with either 12x12 or 6x4 openings per inch are presented. The screen may be imbedded in the center of a single pane of acrylic, or incorporated into a glass sandwich. It should be tinned or otherwise bonded around its periphery to achieve good mating to the metalic plate. A variety of screen materials are available. Slight variations in the dimensions of the mesh openings are intentionally made to reduce the meter-reading and zoning problems.

A screened meter face on a prototype GSE test set is shown in Figure 5-46.

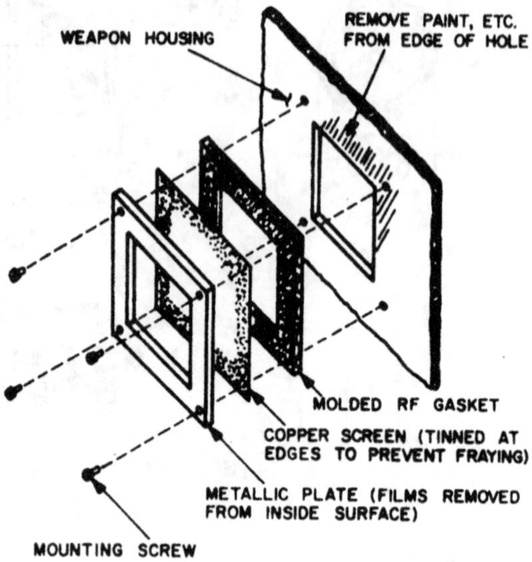

Figure 5-44

Method of Mounting Wire Screen Over a Large Aperture

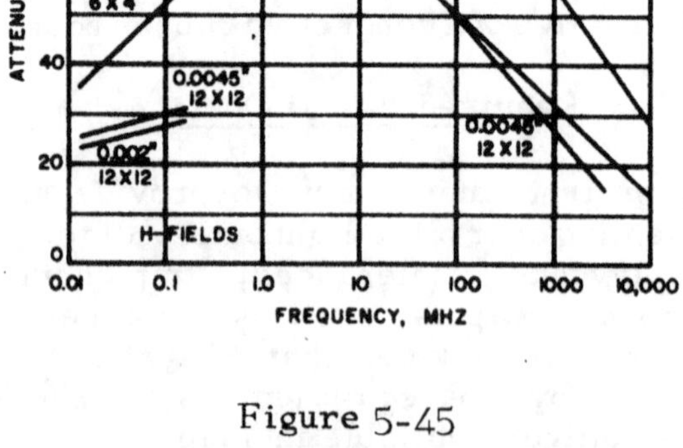

Figure 5-45

Shielding Effectiveness of Meter Screening Assembly

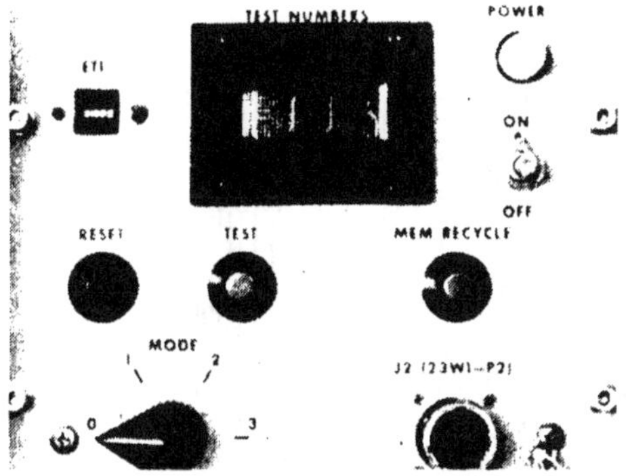

Figure 5-46

GSE With Shielded Digital Meter Face

Two approaches can be employed when shielding behind an assembly and then filtering all leads passing through the shield. These approaches are shown in Figure 5-47 a and b, in the case of a meter. One method, when the meter involved is essentially an off-the-shelf item, is to build a supplementary enclosure and to pass the meter leads through feed-through capacitors or other appropriate filters to eliminate interference that may have been picked up through the meter face. The other method is to procure a meter whose back can be used in place of a supplementary shield, and one which incorporates the necessary lead filtering. In either case, it is assumed that external fields will not cause adverse effects to the operation of the meter itself.

Glass coated with conducting material such as silver can provide shielding across viewing surfaces with some loss in light transmission. Conductive glass is commercially available from a number of glass manufacturers.

Figure 5-48 provides shielding effectiveness data on 50 and 200 ohms per square* silver-impregnated glass against electric arc discharges.[17] Figure 5-49 indicates similar characteristics against plane waves in the frequency range of 0.25 to 350 MHz.[17] The light transmission characteristics of this type of glass as a function of surface resistance[18] is presented in

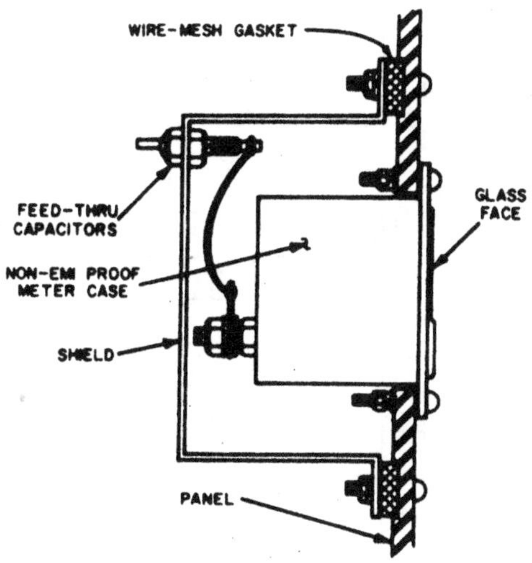

Figure 5-47 (a)

Meter Shielding Techniques

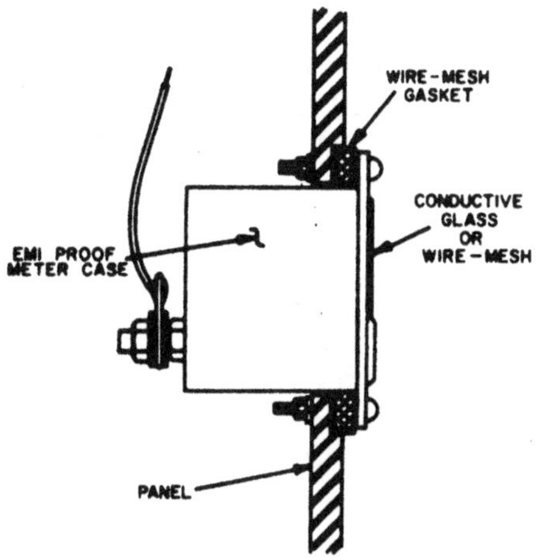

Figure 5-47 (b)

Meter Shielding Techniques

*The resistance across a square section of uniformly conductive material is independent of the size of the section - hence the unusual dimension of "ohms per square."

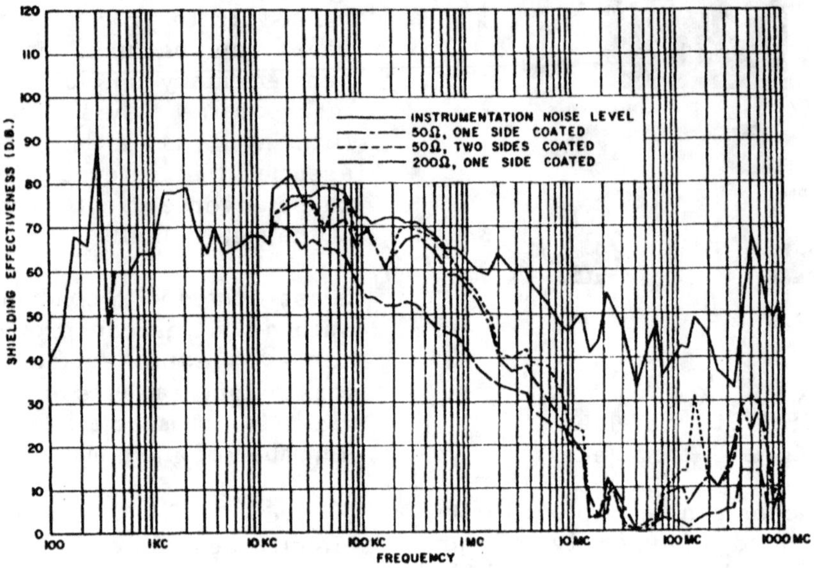

Figure 5-48

Shielding Effectiveness of Conductive Glass to High Impedance Waves

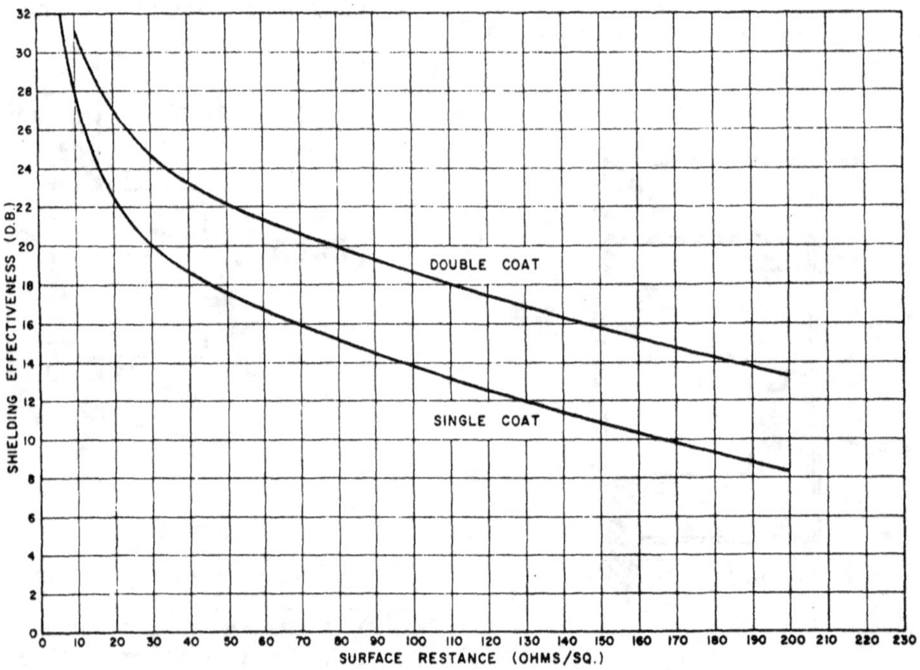

Figure 5-49

Shielding Effectiveness of Conductive Glass to Plane Wave Propagation

Figure 5-50. For effective shielding, good contact to the conducting surface of the glass must be maintained around its periphery.

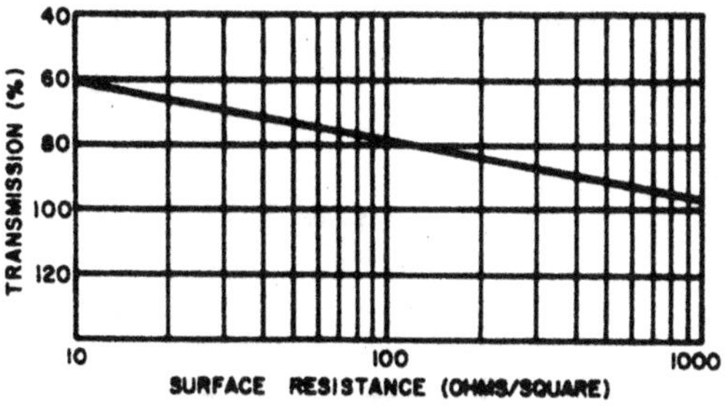

Figure 5-50

Light Transmission vs Surface Resistance for Transparent Conductive Glass

5.6 SHIELDING TESTS

5.6.1 General

The variety of test methods available for evaluating shielding effectiveness are due, at least in part, to the many different factors that can affect material shielding capabilities. These factors include the configuration of the shield (is it a sheet of material, or is it a box?), the frequency range of concern, whether or not the impinging wave is planar, the wave impedance, and others. This section will discuss the most frequently employed and generally applicable shielding effectiveness tests. While they are not all appropriate to evaluating a particular GSE design, they provide a set of established procedures from which designers can select applicable techniques. These tests include:

- Low Impedance Magnetic Field Testing Using Small Loops
- Low Impedance Magnetic Field Testing Using a Helmholtz Coil
- High Impedance Electric Field Testing Using Rod Antennas
- High Impedance Electric Field Testing Using a Parallel Line Radiator
- Plane Wave Testing Using Antennas
- Plane Wave Testing Using a Parallel Plate Transmission Line
- MIL-STD-1377 Testing

A number of the above tests are very similar to tests designed to measure equipment and system EMC in accordance with MIL-STD-462.[19] They also are similar to tests performed to evaluate EM effectiveness of shielded enclosures used for testing purpose in accordance with MIL-STD-285.[20] The GSE design engineer who is concerned with the measurement of shielding properties should familiarize himself with both of these standards.

The MIL-STD-1377[21] tests to be discussed represent recommended procedures for evaluating the shielding (and filtering) effectiveness of Navy weapons systems. The specification contains a unique approach to shielding measurements, and its cable effectiveness evaluation methods are good illustrations of how cable and connector performance tests can be performed.

It should be pointed out that a high degree of measurement accuracy cannot generally be expected from shielding tests. Typically, wave impedances are not established when the tests are performed, antenna correction factors used for calibration purposes are based on plane-wave assumptions even though the test condition may not warrant this assumption, the degree of radiated field distortion by proximal structures is not known, and other factors limit the accuracy of the measurement. However, the tests can be expected to provide guidance to the

GSE designer on the shielding design approaches he should take and the general effectiveness to be expected of those approaches.

Tabulated procedures are provided for each shielding test. These procedures are included for descriptive purposes only, and are not meant to be rigid rules for conducting a particular test. The reader is referred to the cited references for further details.

5.6.2 Low Impedance Magnetic Field Testing Using Small Loops [20, 22]

5.6.2.1 Objective

This test is designed to indicate the GSE enclosure or compartment effectiveness in reducing the intensity of predominantly magnetic field radiation impinging upon it. It employs two small loop antennas, and evaluates loop coupling with and without an intervening shield.

MIL-STD-285 incorporates a similar magnetic field small loop measurement procedure to evaluate the shielding effectiveness of shielded enclosures used for electronic testing purposes. The specific parameters of this procedure will be noted where applicable.

5.6.2.2 Test Setup

In this test, a pair of identical small loop antennas are used, one on one side of the shield and one on the other, spaced equidistant from the shield. If an enclosure is being tested, the usual practice is to have the test signal source within the enclosure and the receiving loop and detector outside the enclosure.

Figures 5-51 and 5-52 show the two basic loop orientations. In Figure 5-51 the loops are coaxial, that is, both loops are normal to a common loop axis. In Figure 5-52 the loops are coplanar, that is, the loop surfaces lie on the same plane. Tests using at least these two orientations should be employed, but other orientations that may result in a lower effectiveness figure should not be ignored.

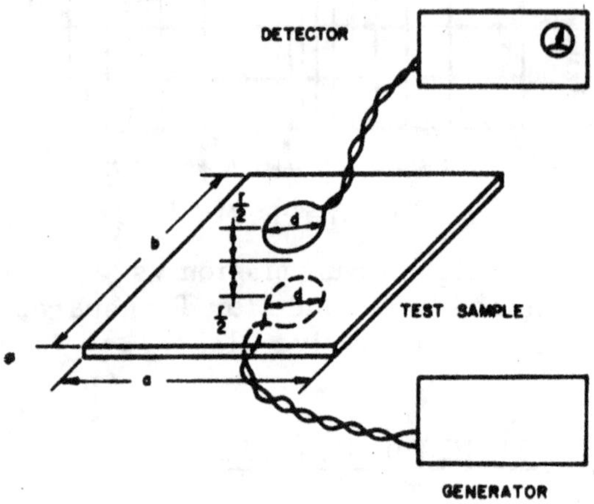

Figure 5-51

Small Loop to Loop Test Setup (Coaxial Arrangement) $r << a, b$

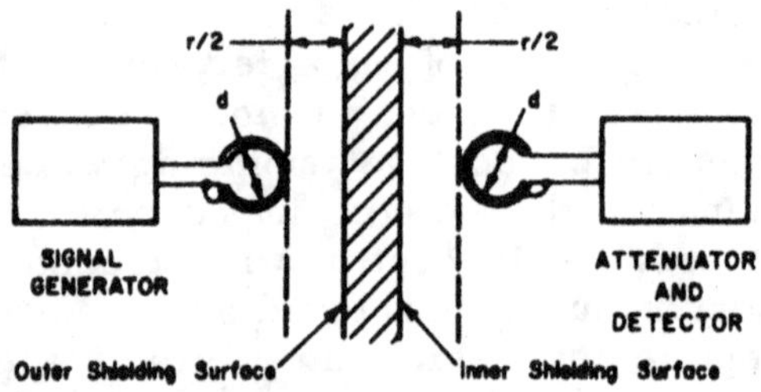

Figure 5-52

Small Loop to Loop Test Setup (Coplanar Arrangement)

Both the loop diameters and the loop separations should be significantly less than the shortest dimension of the box, container, or enclosure being tested. Since this will result in only a small section of the shield being illuminated at one time, it will be necessary to move the loop over the entire surface of the shield to establish the effectiveness of the shield.

The frequency range over which this test can be performed is a function of the level of shielding effectiveness that must be measured (measurement system dynamic range), the sensitivity of the test equipment, the available power to drive the test transmitting loop, and the loop-to-shield separations. The limiting factors are usually the areas of the loops and the number of turns in the loops, since these establish the self-resonance frequency of the loop.

If the detection system is sensitive (0.1 uV input sensitivity) and the available test signal power is large (+10 dBm or greater), then small diameter loops with one, or at most a few, turns will be necessary. To compensate for low test signal level and detector insensitivity, or for small test sample size, smaller loops with more turns may have to be used. Loop-to-shield separation should not be closer than the loop diameter. Test personnel will have to establish loop parameters on a case by case basis, and wind test loops accordingly.

The signal source may be a signal generator or power oscillator of sufficient CW, MCW, or Pulsed CW output. When the radiating loop is connected to the output of a signal generator or power oscillator, a tuning capacitor may be connected in series with the loop to obtain resonance at the test frequency used.

The detector can be any receiver or field strength meter having a low impedance input, or a high input impedance detector with an appropriate matching device. Tests should be performed in a shielded enclosure (screen room) to obtain maximum dynamic test range, and not be limited by the ambient environmental level at the test site.

If the signal generator has an accurate output attenuator, it may be used instead of the attenuator indicated in Figure 5-52.

The small loop-to-loop test setup specified in MIL-STD-285 is as shown in Figure 5-52, with the following parameter values employed:

Loop diameters (d) = 12 inches

Loop-to-shield separations (r/2) = 12 inches

Loops - One turn of No. 6 AWG Copper Wire

The test setup in this specification is intended to provide a minimum of 70 dB measurement range.

5.6.2.3 Test Procedure

To perform this test, set up the loops as described previously, and proceed as follows:

- At each test frequency, increase the generator output level until the detector exhibits at least a 3 dB increase (and preferably a 10 dB increase) in output over noise.
- Adjust the loop orientations to maximize the detector output.
- Record the input signal amplitude (E_2).
- Either remove the shield from between the loops, or move the loop antennas and set them up nearby in the same configuration that resulted in the previous maximum output condition.
- Decrease the generator output level

until the previous detector level is obtained. It may be necessary to add additional attenuation at the receiver input to sufficiently reduce the level.

- Record the input signal amplitude (E_1) and the amount of added attenuation.
- Compute the shielding effectiveness from the relationship

$$SE(dB) = 20 \log \left(\frac{(E_1)(10^{-\frac{A}{10}})}{E_2} \right) \tag{5.23}$$

where A = added attenuation, in dB.

- Position the loops over various portions of the enclosure, particularly those incorporating seams, panels, and other discontinuities, and repeat the test.

5.6.2.4 Advantages/Disadvantages

- The test equipment is easily assembled and fabricated, but the test itself may be very time-consuming for large enclosures.
- The antenna sizes are convenient to use at low frequencies.
- A strong localized field is available.
- By controlling the distance between the loops and the shield, the effectiveness of shielding structures can be evaluated at different impedance levels of low-impedance fields.
- The method lends itself to measuring enclosures of widely varying size.
- Voltages induced in the small loops due to reflections from the enclosure are minimized.
- The method is applicable whether the signal source or the detector is inside the enclosure, thus providing considerable flexibility in the test setup.
- Since the radiating loop illuminates only a small section of the shield, a probing procedure must be employed.
- Dynamic range may be poor or, alternatively, the frequency range of the test may be limited.
- The test is fairly sensitive to small changes in loop spacing, and the test results are difficult to repeat.
- Loop diameter must be kept less than 0.01λ. Thus, a 12 inch receiving loop will provide accurate measurements at frequencies less than 10 MHz.

5.6.3 Low Impedance Magnetic Field Testing Using a Helmholtz Coil[22, 23]

5.6.3.1 Objective

This test is also designed to measure enclosure magnetic shielding effectiveness. It uses a large radiating coil so that it bathes the entire enclosure in a magnetic field.

5.6.3.2 Test Setup

In this test, the test sample is placed inside a Helmholtz coil (large loop), with a small detection loop inside the test enclosure. The small loop is usually located near the center of the enclosure, but may also be used as a probe for leakage energy. The use of a Helmholtz coil enables a large portion of the enclosure to be illuminated at one time. Various orientations of the sample relative to the loop should be tried.

The frequency range is limited by the

test sample size, which affects the size of the Helmholtz coil. Increasing coil size increases its inductance, reduces its self-resonant frequency, and decreases the frequency range over which its magnetic field intensity remains constant. The coil diameter should be at least two and preferably three times the longest test sample dimension. The upper frequency limit is typically 100-500 kHz.

The test arrangement is shown in Figure 5-53. On occasion, two coaxial coils instead of one have been used, with the test sample placed between the coils. The two coils are for the purpose of providing a more uniform field over the volume of the sample[23].

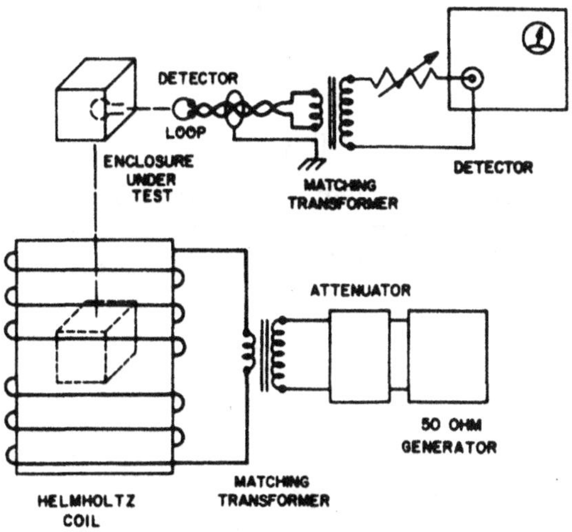

Figure 5-53

Helmholtz Coil Test Arrangement

The leads from the detector loop should be twisted, shielded cable. When possible, the shield should be grounded circumferentially where it leaves the enclosure under test, and should be single-point grounded at the detector.

5.6.3.3 Test Procedure

The test procedure using the Helmholtz coil is basically the same as the one employed for the two small loops method. However, instead of the two-loop probing operation, different orientations of the enclosure in the field of the Helmholtz coil are tested. Also, as indicated previously, some probing with the detector probe may be employed.

5.6.3.4 Advantages/Disadvantages

- The field generated by the Helmholtz coil can be made very uniform over a wide area or within a specified volume.
- The intensity of the generated field is reasonably strong; hence dynamic range is good.
- It provides a convenient way to detect joint and seam leakage, however it does not locate the particular leakage source.
- Measurement repeatibility is high.
- The test setup is quite complex and bulky.
- The maximum frequency at which this procedure can be used is considerably less than that for the two-loop method.
- Uniformity of the field as frequency increases can only be maintained by a reduction of the Helmholtz coil dimensions, thus limiting the sample size.
- Leaky joints of the enclosure that are parallel to the direction of the current flow in the large loop may not be detectable. Thus, to detect the effect of all possible seams and discontinuities, the enclosure must be measured for different orientations in the field.

5.6.4 High Impedance Electric Field Testing Using Rod Antennas[20, 22]

5.6.4.1 Objective

This test is intended to evaluate whether or not a GSE enclosure, or an internal compartment, affords adequate protection for the enclosed equipment against high impedance electric field energy. This is accomplished using rod antennas located on both sides of the shield. The frequency range covered by this test is typically 15 kHz to 30 MHz.

5.6.4.2 Test Setup

Figure 5-54 indicates a representative test setup for measuring high impedance wave affects. Rod antennas with passive impedance matching trans-fromers are employed. The antennas must be matched to the respective terminating devices. Antennas such as the VR-105 and VA-105 rod antennas supplied with the NF-105 Field Intensity Meter can be used directly, since these antennas incorporate matching step-up autotransformers.

Typically, the radiating rod antenna is 41 inches long. The detector antenna length is variable, depending on the particular test application. MIL-STD-285 specifies a detector antenna length of 41 inches, while other tests have used a 10 inch antenna. Antenna-to-shield spacing is also dependent on the test involved; the MIL-STD-285 specification on spacing is 12 inches, which limits the tests to frequencies below about 15 MHz.

The signal source should be capable of delivering at least +10 dBm into a 50 ohm load. Attenuators may be incorporated into the test setup ahead of the detector input as required.

The detector system is located external to the shielded enclosure containing the unit being tested, to prevent the radiated field from interfering with that system. Connection is made to the detector with coaxial cable that passes through a waveguide-below-cutoff filter in the shielded enclosure wall.

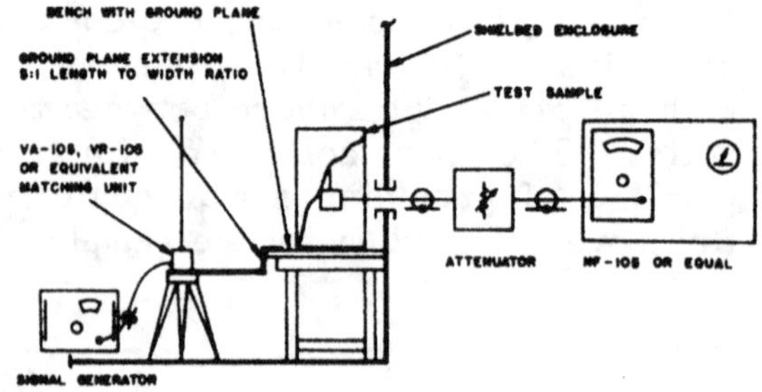

Figure 5-54

Test Equipment Setup for Performing High Impedance Electric Field Shielding Effectiveness Tests

5.6.4.3 Test Procedure

- Adjust the test signal generator frequency to the desired value, set the generator output level to its maximum value, and tune the detector for a received signal indication.
- Adjust the detection receiver in gain and input attenuation to give mid-range output.
- Orient the shield around the detection antenna until the highest output indication is obtained from the detection system. Also, position the source antenna anywhere around the

enclosure under test, in any orientation so long as it remains parallel to the enclosure wall, for highest output indication. Continue to adjust the attenuator and gain control settings to maintain a mid-range output reading. Record the final attenuator and gain control settings.

- Remove the shield under test from around the detection antenna, and place the test shield sample outside the test area.

- Decrease the signal generator output level until the detection system has the same output indication as with the shield in place around the detection antenna. It may also be necessary to increase the attenuator setting. The shielding effectiveness is then computed in accordance with Equation (5-23).

5.6.4.4 Advantages/Disadvantages

- The wave impedance generated by this procedure is high impedance so long as the general separation conditions of Figure 2-6 are maintained.

- The equipment used is readily available and the test setup is straightforward.

- The method is flexible, and allows measurement of different sample sizes.

- The test field can be generated inside or outside the enclosure under test, depending on sample size and screenroom availability.

- The field generated is nonuniform and thus has the same uncertain effects as a nonuniform field generated by a small loop.

- The reduction of measured data is usually based on plane wave theory. High-impedance field measurements may not be directly verifiable through application of plane wave considerations.

- Monopole antennas are susceptible to reflections from surrounding objects, observers, etc., to the extent that the accuracy of the results becomes largely a function of the experience and capability of the testing personnel.

5.6.5 High Impedance Electric Field Testing Using a Parallel Line Radiator[24]

5.6.5.1 Objective

This test is also for the purpose of testing enclosure shielding effectiveness against high-impedance fields. It is identical to the procedure of Section 5.6.4 using rod antennas, except that the source antenna configuration is now a parallel line radiating device.

5.6.5.2 The Parallel Line Radiator

The Parallel Line Radiator is merely a parallel-wire line, with the high-impedance wave being the fringing field from this line. Such a radiator is shown end-on in Figure 5-55.

There are many ways in which the line itself can be constructed; one such configuration is also shown in Figure 5-55. The length of the line (if an end-fed line is used), should be $\lambda/8$ at the highest test frequency. The line separation is not critical, but a separation of 0.5 to 1 meter is suggested to get a fairly planar field of high impedance at a 1 meter distance to the test sample. Line impedance is computed from the relationship:

$$Z_1(\text{ohms}) = 276 \log_{10}\left(\frac{2S}{d}\right), \quad \frac{S}{d} \gg 1 \qquad (5.24)$$

where

S = line separation

d = wire diameter, in same units as S

Using this equation, lines can be designed to work into other impedances. Lines of low impedance may not be able to provide adequate electric field levels, while lines of high impedance require a matching transformer having a higher turns ratio. The latter may reduce the maximum frequency capability of the line.

Courtesy of DON WHITE CONSULTANTS, INC.

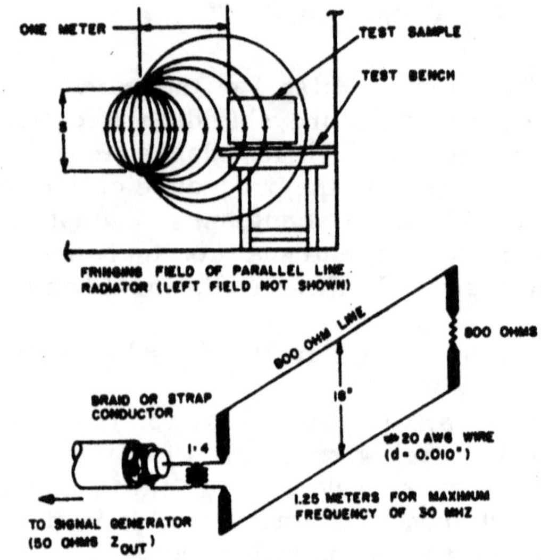

Figure 5-55

Parallel Line Radiator for Generating High Impedance Fields

5.6.5.3 Test Setup and Procedure

The test setup and procedure is the same for the parallel line radiator as for the rod antennas (Section 5.6.4).

5.6.5.4 Advantages/Disadvantages

- The field impedance using this method will be less than that achieved using the radiating rod antenna.
- The parallel line does not require a ground plane reference.
- The line is easily set up, but requires somewhat more space than the radiating rod antenna.

5.6.6 <u>Plane Wave Testing Using Antennas</u>[20, 22]

5.6.6.1 Objective

This test is designed to subject the enclosure under test to plane waves over the frequency range from 30 MHz to 35 GHz. The test is indicated as a "plane wave" test because, for the majority of the frequency range covered, the test distances are such that the wave characteristic approaches an impedance of 120 π ohms. The test uses a variety of radiating antennas, depending on frequency range. The 30-200 MHz range is generally covered with a broadband bi-conical antenna; while the range from 200 MHz to 12 GHz is covered with circular log-periodic conical antennas, (200-1000 MHz and 1-12 GHz). Horn antennas may also be used.

Because of test sample size limitations, it probably will not be practical to employ a detector antenna identical to that of the radiating antenna. A short dipole or monopole antenna is often used as the receiving antenna, located inside the test sample enclosure. In some cases, a horn antenna might also be used for receiving purposes.

5.6.6.2 Test Setup

The following equipment is required for this test:

- Signal generators covering the frequency range from 30 MHz to 35 GHz.
- A tuned voltmeter or detector/re-

ceiver covering 30 MHz to 35 GHz.

- Radiating antennas covering the frequency range of interest. The VSWR of these antennas should be less than 3:1 at all test frequencies.

- A shielded enclosure, preferably anechoic (non-reflective), in which to perform the test may be desirable.

- A short dipole or monopole having elements extending to a total distance less than the shortest inside dimension of the enclosure under test.

Test equipment should be arranged as shown in Figure 5-54. The source antenna should be located sufficiently far from the test sample so that far-field conditions apply. A similar test defined by MIL-STD-285 designates that the minimum separation shall be 6 feet.

5.6.6.3 Test Procedure

The procedure to be followed for this test is essentially the same as for the high-impedance rod antenna test (Section 5.6.4).

5.6.6.4 Advantages/Disadvantages

- The indicated antenna types are easily obtained or simple to construct, and generally have broadband characteristics.

- Conical spiral antennas, being circular polarized devices, are insensitive to GSE enclosure seam geometry.

- Log-periodic and horn antennas have relatively high gain, and therefore can provide a very high level of electric field and a subsequent wide dynamic range measurement capability.

5.6.7 Plane Wave Testing Using a Parallel Plate Transmission Line[24]

5.6.7.1 Objective

This test is designed to subject a GSE enclosure to a plane wave over a fairly wide frequency range, for the purpose of determining enclosure shielding effectiveness. The test uses a parallel plate transmission line for forming the field and controlling wave impedance. The field generated is planar. Parallel plate lines are normally used to 30 MHz[19], but can be designed to measure to .1 GHz if small test samples are involved.

5.6.7.2 Test Setup

The Parallel Plate Transmission Line is essentially a strip line consisting of two rectangular sheets of conductive material, of equal length but one often wider than the other, and separated from each other. The sheets are driven and terminated at their respective narrow ends. A TEM mode field is created between the two sheets, as illustrated in the end-view of Figure 5-56. The test sample is placed between the sheets, where the electric field lines are essentially normal to the two surfaces.

Courtesy of DON WHITE CONSULTANTS, INC.

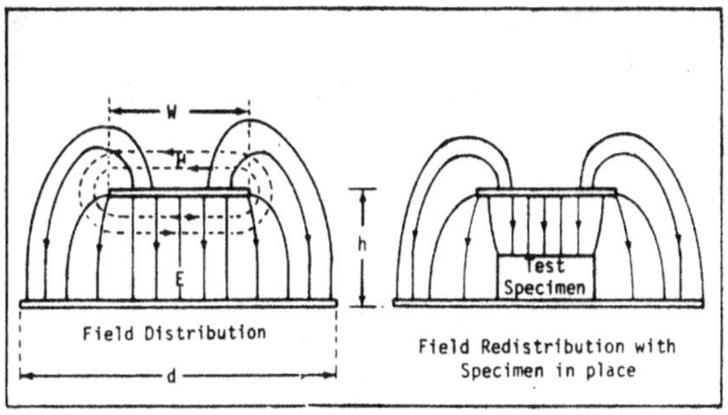

Figure 5-56

TEM Field Inside of Strip Line

The major advantage of the parallel plate line is that it is capable of establishing a reasonably high planar electric field strength. Field strength (in volts/meter) can be computed from the relationship

$$FS = 39.4 \frac{V_t}{h} (10^{-\frac{A}{10}}) \quad (5.25)$$

where

V_t = input voltage to the line, in volts

A = matching network loss, in dB

h = separation between transmission line plates, in inches

The design parameters for this type of line are shown in Figure 5-57. The figure points out that the lower plate should be at least three times wider than the upper plate, and that the characteristic impedance of the line is a function of upper plate width and plate separation [24].

Courtesy of DON WHITE CONSULTANTS, INC.

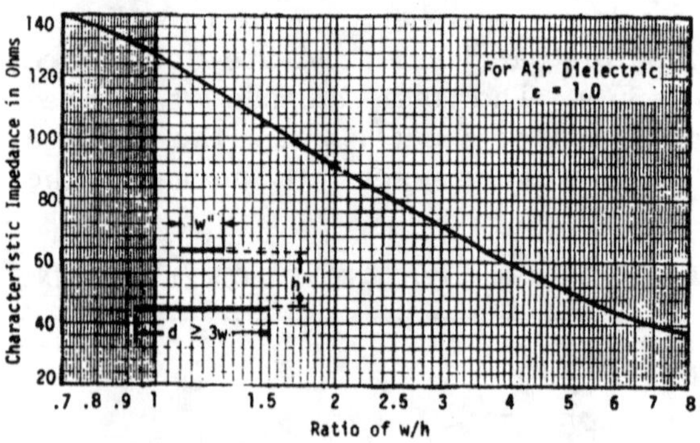

Figure 5-57

Characteristic Impedance of Strip Line vs. w/h Ratio

Impedance matching is relatively unimportant so long as the line length is less than 0.1λ. However, at frequencies when this relationship does not apply, the line acts like a transmission line, and standing waves will exist unless careful matching is accomplished.

When a test sample is inserted into the line, it will perturb the field, and two effects will result. First, the VSWR of the line will change. This change may be approximated from the following equations:

$$VSWR = \frac{R_L}{Z_D} \quad (5.26)$$

$$Z_D = \frac{-jX_c R_L}{R_L - jX_c} \quad (5.27)$$

$$X_c = \frac{1}{2\pi f C_d} \quad (5.28)$$

$$C_d = 0.22A\left(\frac{1}{h_D} - \frac{1}{h}\right) \quad (5.29)$$

where

R_L = characteristic impedance of the line, in ohms

A = area in square inches of the test specimen face parallel to the plates

h_D = separation between transmission line plates, in inches

Second, the electric field strength in the region above the test specimen will increase because of the effective increase in capacitance in that portion of the structure. As an approximation, the increase in field strength would be inversely proportional to the separation between the upper plate and the top face of the test sample. Thus if the separation is reduced by one-third (to 2/3 of its normal spacing), the field strength will increase to 1.5 times its normal value.

An example of a 90-ohm parallel plate

transmission line capable of plane wave measurements to 200 MHz is shown in Figure 5-58. The line can test samples of cigar-box size while maintaining a VSWR of less than 1.2:1. The upper plate tapers to an input connector and a 50-to-90 ohm resistive matching network. The termination resistors in this arrangement are three 270 ohm distributed resistance cards connected in parallel.

Courtesy of DON WHITE CONSULTANTS, INC.

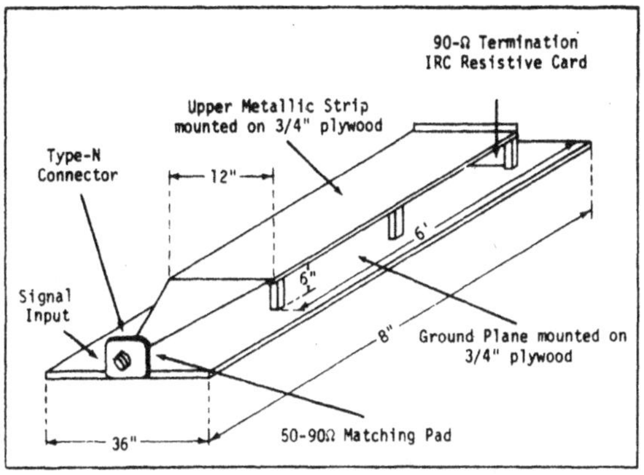

Figure 5-58

90 Ohm Strip Transmission Line

The lower plate of the parallel plate transmission line should be bonded to the shielded enclosure wall. The equipment under test should be placed as near to the center of the parallel plate line as possible. No other grounding scheme should be employed.

A small loop antenna or short stub monopole antenna should be placed inside the enclosure under test. The output of the antenna inside the enclosure should be routed through a double shielded cable to a detector.

5.6.7.3 Test Procedure

The following steps should be used in performing this test:

- Connect a signal generator to the input of the parallel plate line. Be sure that the line is properly terminated for the generator and that the generator output level is sufficient to produce the desired field intensities.
- Place the small antenna inside the enclosure under test and connect it to the detector.
- Adjust the signal generator frequency and detector frequency appropriately.
- Increase the signal generator output level until an indication of received signal is given by the detector. Continue to increase the generator output level until the detector output is at least 3 dB (and preferably 10 dB) above the system noise level. Note the output indication of the detector and record the generator input level (E_2).
- Remove the test enclosure from the test fixture, keeping the receiving antenna in the same position in the transmission line as when the test enclosure was in place.
- Reduce the signal generator output level until the detector indication is the same as that recorded with the test enclosure in place.
- Record the output signal generator level (E_1).
- Calculate the shielding effectiveness measured from the relationship:

$$SE(dB) = 20 \log \left(\frac{E_1}{E_2}\right) \qquad (5.30)$$

5.6.7.4 Advantages/Disadvantages

- The parallel plate line is easily obtained or simple to construct, and has broadband characteristics.

- The field generated is uniform and of known impedance level.
- High field intensities may be established, with nearly all energy concentrated within the line.
- Design parameters limit the capabilities of this device to the testing of medium to small-size enclosures.
- The strip line concept is adaptable to generating planar fields having a wide range of wave impedance, by controlling line termination impedance.

5.6.8 MIL-STD-1377 Testing[21, 25]

This standard is employed by weapon system developers to evaluate shielding and filtering effectiveness. It defines applicable shielding effectiveness measurement techniques covering the frequency ranges of 100 kHz to 30 MHz, and 1000 MHz to 10 GHz; and filtering effectiveness measurement techniques covering the frequency range of 100 kHz to 10 GHz. Although not originally intended for this purpose, it is being brought to the attention of the GSE designer, since he may want to consider an approach taken in MIL-STD-1377 to evaluate the equipment or system he is developing.

At frequencies below 30 MHz, shielding in MIL-STD-1377 is evaluated on the basis of Surface Transfer Impedance (STI). STI is the ratio of the longitudinal voltage drop on the outer surface of the shield to the current creating that voltage drop. If a cable or cable/connector is under test, it is terminated in a short, and driven by a signal source. The STI is measured with an rf voltmeter, as shown in Figure 5-59. Shield enclosure discontinuities are measured in a similar way. An STI of about 10 milliohms has been correlated to a level of 15% of Maximum No Fire Current for typical electroexplosive devices.[25]

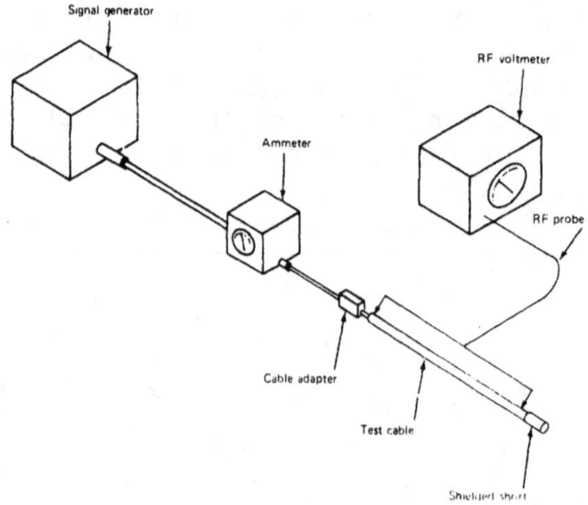

Figure 5-59

STI Measurement Configuration for Cables

At frequencies above 1 GHz, a special test enclosure is employed. The enclosure is a conductive container into which the test sample is placed. If a cable/connector shielding effectiveness test is being performed, the cable assembly is mounted in the enclosure. One end is terminated in a short, and the other end is brought out of the test enclosure to a power meter. A wire strung diagonally within the enclosure is used to generate a test field. Another wire strung diagonally within the enclosure is used for calibration purposes. A paddle wheel tuner is employed as a mode converter, so that a maximum level environment can be established.

Figures 5-60 and 5-61 illustrate the test setup and enclosure for this test. A similar procedure is employed for determining the shielding effectiveness of equipment and system enclosures.

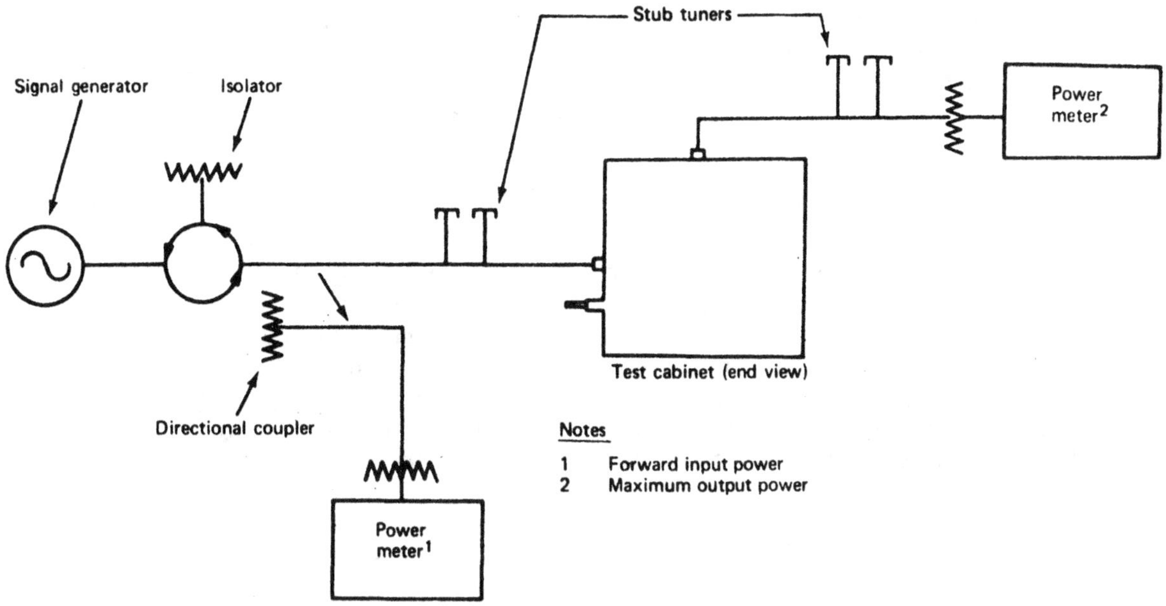

Figure 5-60
Arrangement of Test Apparatus for MIL-STD-1377 1-10 GHz Tests

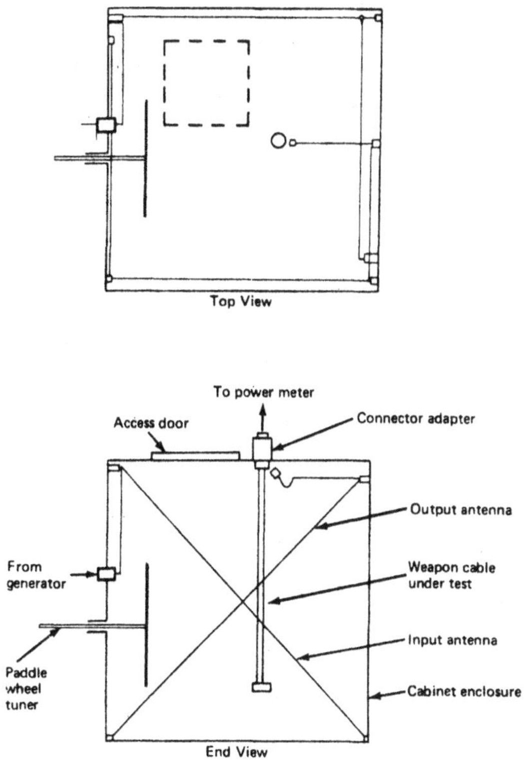

Figure 5-61
Test Cabinet (With Cable Under Test)

5.7 SUMMARY OF GOOD SHIELDING PRACTICES

The following represent what might be considered the more salient points on GSE shielding design considerations covered or implied in the previous discussion:

a. Good conductors such as copper, aluminum, and magnesium should be used for high-frequency electric-field shields to obtain the highest reflection loss.

b. Magnetic materials such as iron and Mumetal should be used for low-frequency, magnetic-field shields to obtain the highest penetration loss.

c. Any shielding material strong enough to support itself will usually be thick enough for shielding electric fields at any frequency.

d. In the case of thin-film shields, the

effectiveness of the shield is fairly constant for material thicknesses below $\lambda/4$, and increases markedly above that thickness.

e. Multiple shields (for both enclosures and cables) can provide both higher shield effectiveness and extended shielding frequency range. Cost considerations will probably be the deciding factor between use of multiple shields and using other means of achieving EM compatibility, although factors such as reduced cable flexibility with double braids may also come into play.

f. All openings or discontinuities should be treated in the design process, to assure minimum reduction in shield effectiveness. Particular attention should be paid to selection of materials that are not only suitable from the shielding standpoint, but from the electrochemical corrosion viewpoint as well.

g. When other aspects of system design will permit, continuous butt or lap weld seams are most desirable. The important consideration is to get intimate contact between mating surfaces over as much of the seam surfaces as possible.

h. Surfaces to be mated must be clean and free from nonconducting finishes unless the bonding process positively and effectively cuts through the finish. When EMC and finishing specifications conflict, it is important that the finishing requirement be modified.

i. The critical factors in flight-deck cable shielding are shield coverage under operational cable flexing conditions, and cable shield termination at the connector. A minimum of 94% shield coverage is recommended for these applications. Shields should be peripherally bonded to connector back shells to maintain shielding effectiveness at mating surfaces.

j. Conductive gaskets and spring fingers, waveguide attenuators, screens and louvres, and conducting glass are the major devices and mechanisms available for maintaining enclosure shield effectiveness. Many factors in addition to shielding capabilities per se, ranging from space availability to cost, and from air circulation requirements to visibility factors, will affect particular methods employed in particular situations.

k. Shielding represents only one method of reducing equipment EM interactions, and should not be considered without also considering tradeoffs of filtering, grounding, and bonding techniques that simplify or eliminate requirements for shields.

SHIELDING BIBLIOGRAPHY

1. NAVWEPS OD 30393, "Design Principles and Practices for Controlling Hazards of Electromagnetic Radiation to Ordnance (HERO Design Guide)," June 6, 1965.

2. Department of the Air Force, "Mutual Electromagnetic Interference," AFM 100-35, September 1, 1960.

3. Naval Ships Engineering Center, "Electronics Installation and Maintenance Book, Electromagnetic Interference Reduction," NAVSHIPS 0967-000-0150, June 1970.

4. S.L. O'Young, et al., "Survey of Techniques for Measuring RF Shielding Enclosures," IEEE Transactions on Electromagnetic Compatibility, Vol. EMC-10, No. 1, March 1968, p. 73.

5. R.B. Schulz, "ELF and VLF Shielding Effectiveness of High Permeability Materials," IEEE Transactions on Electromagnetic Compatibility, Vol. EMC-10, No 1, March 1968, pp. 95-100.

6. R.A. Week, "Thin-Film Shielding for Microcircuit Applications and a Useful Laboratory Tool for Plane-Wave Shielding Evaluations," IEEE Transactions on Electromagnetic Compatibility, Vol. EMC-10, No. 1, March 1968, pp. 105-112.

7. W. Jarva, "Shielding Efficiency Calculation Methods for Screening Waveguide Ventilation Panels and Other Perforated EM Shields," Proceedings of Seventh Conference on Radio Interference Reduction and EMC, November 1961, pp. 478-498.

8. R.B. Schulz, et al., "Shielding Theory and Practice," Ninth Tri-Service Conference on Electromagnetic Compatibility, October 1968, pp. 597-636.

9. Headquarters, U.S. Army Material Command, "Engineering Design Handbook, Hardening Weapon Systems Against RF Energy," AMC Pamphlet AMCP 706-235, February 1972.

10. F.W. Schor, "Measurement of RF Leakage in Multipin Electrical Connectors," IEEE Transactions on Electromagnetic Compatibility, Vol. EMC-10, No. 1, March 1968, pp. 135-141.

11. Headquarters, Air Force Systems Command, "Handbook of Instructions for Aerospace Systems Design (HIASD)," AFSCM 80-9, Vol. IV, April 20, 1967.

12. C.B. Pearlston, "Case and Cable Shielding, Bonding and Grounding Considerations in Electromagnetic Interference," IRE Transactions on RFI, Vol. RFI-4, No. 3, October 1962, pp. 1-16.

13. O.P. Schreiber, "RF Tightness Using Resilient Metallic Gaskets," Proceedings of the Second Conference on Radio Interference Reduction, March 1956, pp. 343-359.

14. Naval Avionics Facility, Indianapolis, Drawing No. 178AS137.

15. O.P. Schreiber and S. Nellis, "Design of Combination Fluid or Pressure and RF Gaskets," Proceedings of the Third Conference on Radio Interference Reduction, February 1957, pp.340-355.

16. Metex Corporation, "EMI Shielding Materials, Gaskets & Components, Catalog and Handbook."

17. H.M. Sachs, et al., "Evaluation of Conductive Glass in Fluorescent Light Shielding Applications," Unclassified Proceedings of the Sixth Conference on Radio Interference Reduction and Electronic Compatibility, October 1960, pp. 281-294.

18. H.A. Lassitter, "Low Frequency Shielding Effectiveness of Conductive Glass," IEEE Transactions on Electromagnetic Compatibility, Vol. EMC-6, No. 2, July 1964, pp. 17-30.

19. Department of Defense, Washington, D.C., MIL-STD-462, "Electromagnetic Interference Characteristics, Measurement of," 31 July 1967.

20. Department of Defense, Washington, D.C., MIL-STD-285, "Attenuation Measurements for Enclosures, Electromagnetic Shielding, for Electronic Test Purposes, Method of," 25 June 1956.

21. Department of the Navy, MIL-STD-1377, "Effectiveness of Cable, Connector, and Weapon Enclosure Shielding and Filters in Precluding Hazards of Electromagnetic Radiation to Ordnance; Measurement of," 20 August 1971.

22. S.L. O'Young, et. al., "Survey of Techniques for Measuring RF Shielding Enclosures," IEEE Transactions on Electromagnetic Compatibility, Vol. EMC-10, No. 1, March 1968, pp. 72-81.

23. W.W. Cooley, "Low-Frequency Shielding Effectiveness of Nonuniform Enclosures," IEEE Transactions on Electromagnetic Compatibility, Vol. EMC-10, No. 1, March 1968, pp. 34-43.

24. D.R.J. White, "A Handbook Series on Electromagnetic Interference and Compatibility, Volume 4, EMI Test Instrumentation and Systems," 1971.

25. J.R. Pollard, "A Method for Evaluating the RF Shielding of Weapon Systems Below 30 MHz," 1967 IEEE Electromagnetic Compatibility Symposium Record, July 18-20, 1967, pp. 433-438.

26. E.A. Lindgren, "Contemporary RF Enclosures," 1967.

27. C. L. Blake, "Composites - Their Electrical and Electromagnetic Impact," IEEE 1976 International Symposium on Electromagnetic Compatibility, July 1976, pp. 170-173.

28. F. A. Fisher, and W. M. Fassell, "Technical Report AFAL-TR-72-5," Lighting Effects on and Electromagnetic Shielding Properties of Boron and Graphite Reinforced Composite Materials.

29. C. D. Skouby, "McDonnell Final Report," Electromagnetic Effects of Advanced Composites, January 1975.

30. D. F. Strawe,and L. D. Piszker, "Technical Report AFML-TR-75-141," Interaction of Advanced Composites with Electromagnetic Pulse (EMP) Environment.

6.0 BONDING

6.1 GENERAL

Bonding is the establishment of a low impedance path between two metal surfaces. This path may be between two points on a system ground plane, or between ground reference and a component, a circuit or a structural element. The purpose of the bond is to make the structure homogenous with respect to the flow of RF currents, thus avoiding the development of electric potentials between metallic parts which can produce interference.

Generally there are two types of bonding: direct bonding, where there is metal-to-metal contact between the members to be bonded; and indirect bonding through the use of conductive jumpers.

Permanent joints of metallic parts made by welding, brazing, sweating, swaging and soldering are the best direct bonds. Semi-permanent joints of machined metallic surfaces rigidly held together provide excellent direct bonds, as long as the contact areas are clean and all non-conductive coatings are removed prior to assembly. Joints that are press-fitted or jointed by self-tapping or sheet metal screws cannot be relied upon to provide a low-impedance bond at high frequencies. Riveted joints on 3/4" centers are acceptable if the rivet holes are bare. Direct bonds must always be made through continuous contact to bare or conductively finished metals.

An indirect bond (or bonding jumper or strap) is an intermediate electrical conductor used to connect two isolated items. Because jumpers often have significant impedance at frequencies above 10 MHz, their use is less preferable than direct bonds. However, there are obviously circumstances when direct bonds cannot be used (such as when equipment must be removable, or when mechanical shock-mounting is necessary), and it is these situations that warrant indirect bonds.

Good bonding between equipment and a ground reference plane is essential to minimizing interference.

- Good bonding enables the design objectives of other methods of EMI suppression, such as shields and filters, to be more nearly achieved.
- Good bonding minimizes the buildup of rf voltage differences and ground current loops.
- An adequate bond deters the buildup of static charges in equipment operation.

In addition to reducing interference, good bonding minimizes damage which might be caused by lightning strikes, and protects personnel from the shock hazard that could result if primary power were inadvertently shorted to an enclosure.

An example of the effects of a poor bond is shown in Figure 6-1. If the impedance of the bond is significant, the interference signal, which is supposed to be bypassed by the filter capacitors, will be coupled into the susceptible equipment.

6.2 SURFACE TREATMENT

Both direct and indirect bonding connections require metal-to-metal contact of bare surfaces. It is frequently necessary to remove protective coatings from metals to provide

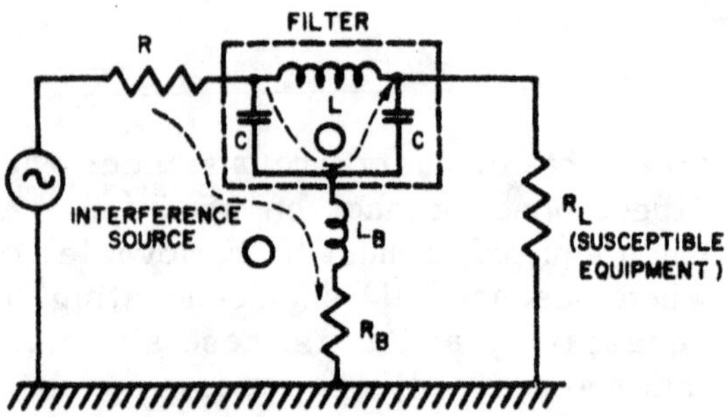

L_B = INDUCTANCE OF A POOR BOND

R_B = CONTACT RESISTANCE OF A POOR BOND

NOTE - WHEN $(1/j\omega C + R_L) < (R_B + j\omega L_B)$ INTERFERENCE CURRENTS WILL FOLLOW PATH (2) TO SUSCEPTIBLE EQUIPMENT.

Figure 6-1

Circuit Representing Poor Bonding Between a Filter and Ground

a satisfactory bond. The area cleaned for bonding should be slightly larger than the area to be bonded. Ridges of paint around the periphery of the bonding area can prevent good metal-to-metal contact. Washers or fittings must fit inside the cleaned area. Immediately prior to bonding, all chips, paint, grease, or other foreign matter must be removed with a proper cleaning solution.*

After bonding, the exposed areas should be refinished as soon as possible with the original finish. However, if the paint used is too thin, refinishing paint may seep under the edges of bonded components and impair the quality of the bond.

A suitable conductive coating may be used when removable components must be provided with a protective finish. Where aluminum or its alloys are used, corrosion resistant finishes that offer low electrical resistance are available. Refer to Figure 5-30 of Chapter 5 for data showing the degradation of shielding effectiveness of metal-to-metal joints caused by anodizing aluminum, and the characteristics of various conductive aluminum finishes.

6.3 CORROSION**

Corrosion occurs between two dissimilar metals in solution, since they form an electrochemical cell. The extent of corrosion depends on the metals comprising the electrochemical cell and the conditions under which the dissimilar metals come into contact with each other. By properly modifying these two factors, the extent of corrosion in the vicinity of a bond can be reduced.

No appreciable corrosive action occurs between two metals in the same group in the electrochemical series shown in Table 6-1. If the two metals are in different groups, the metal coming first in the list of Table 6-1 will form the anode and be relatively heavily corroded, whereas the metal coming later will form the cathode and be relatively free from corrosion and will be protected.

The greatest degree of corrosion occurs when dissimilar metals are openly exposed to salt water, salt spray, rain, gasoline, jet fuel, or other liquids that may act as an electrolyte. Minimum corrosion occurs

* See also Paragraph 5.5.1 for additional information in connection with the bonding of enclosure seams.

** See also Paragraph 5.5.1 for additional information in connection with the bonding of enclosure seams.

Table 6-1

Electrochemical Series Ordered by Decreasing Sensitivity to Corrosion

Anodic End (most easily corroded)	
Group I	Magnesium
Group II	Aluminum, aluminum alloys, zinc, cadmium
Group III	Carbon steel, iron, lead, tin, tin-lead solder
Group IV	Nickel, chromium, stainless steel
Group V	Copper, silver, gold, platnum, titanium
Cathodic End (least corroded)	

when metals are kept dry and completely free from exposure to moisture. The following three exposure conditions are defined:

Exposed:

Metal has an open, unprotected exposure to weather.

Sheltered:

Milder exposure than above, and metal surfaces receive limited protection from direct action of weather.

Housed:

Metal surfaces of equipment are housed in weatherproof buildings.

For a given pair of dissimilar metals in contact under the three exposure conditions enumerated above, the extent of corrosion can be minimized. Table 6-2 may be used for this purpose. The table indicates the protection to be applied to a bond for the expected environmental conditions in which the particular type of bond is to be used. The table must be used in conjunction with Table 6-1 to identify the group associated with each bond metal.

Condition of Exposure	Anode I	Anode II	Anode III	Anode IV	Cathode
Exposed	A	A			II
Sheltered	A	A			II
Housed	A	A			II
Exposed	C	A	B		III
Sheltered	A	B	B		III
Housed	A	B	B		III
Exposed	C	A	B	B	IV
Sheltered	A	A	B	B	IV
Housed	A	B	B	B	IV
Exposed	C	C	C	A	V
Sheltered	A	A	A	B	V
Housed	A	A	B	B	V

Table 6-2

Groups of Materials Recommended for Providing Protective Bond Between Two Dissimilar Metals Used as Anode and Cathode

Notes: Bond Protection Code

A. The couple must have a protective finish after metal-to-metal contact has been established so that no liquid film can bridge the two elements of the couple.

B. The two metals may be joined with bare metal exposed at junction surfaces. The remainder must be given an appropriate protective finish.

C. This combination cannot be used except where short life expectancy can be tolerated or when the equipment is normally stored and exposed for only short intervals. Protective coatings are mandatory.

Consideration should be given to the electrochemical series to assure that, where possible, corrosion occurs only to replaceable elements of a bond. For example, if an aluminum (Group II) equipment case is to be bonded to a stainless steel (Group IV) frame, it is good practice to interpose a tin- (Group III) or cadmium - (Group II) plated washer between the two metal surfaces. If the protective coating is then chipped, the washer instead of the aluminum case will be attacked by corrosion.

6.4 BONDING EFFECTIVENESS CHARACTERISTICS

6.4.1 Bond Jumper Equivalent Circuit

The use of bonding jumpers in indirect bonding is equivalent to the problem of maintaining low impedance paths. At low frequencies, bonding jumpers do not present any special problems except for resistance, and any reasonable length jumper can be used. At higher frequencies, however, the rf impedance of the bond becomes a critical design consideration.

The equivalent circuit of a bond strap at frequencies where the length of the strap is short compared to a wavelength is a simple parallel-tuned circuit, as illustrated in Figure 6-2(a)[2]. As such, the bonding jumper has the usual electrical parameters of resistance, inductance, and capacitance. Of these parameters, its resistance is an inherent property of jumper resistivity, depending on the material selected; its capacitance is dependent upon the physical configuration and separation from the bonded units; and its inductance is dependent upon the physical dimensions of the bonding jumper.

The dc resistance of a bonding strap per unit length of the bond can be obtained from the relationship:

$$R_{dc}(\text{ohm-cm}) = \frac{\rho}{A} \qquad (6.1)$$

where

ρ = the specific resistivity of the material, and

A = the cross-sectional area of the bond.

If ρ is in units of ohm-centimeters (for copper, $\rho = 1.724 \times 10^{-6}$ ohm-cm) and A is in square-centimeters, then R_{dc} is in ohms-per-centimeter of bond length.

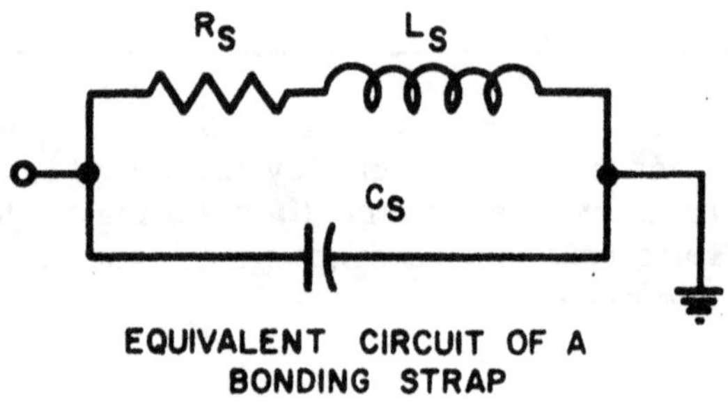

Figure 6-2a
Bonding Equivalent Circuits

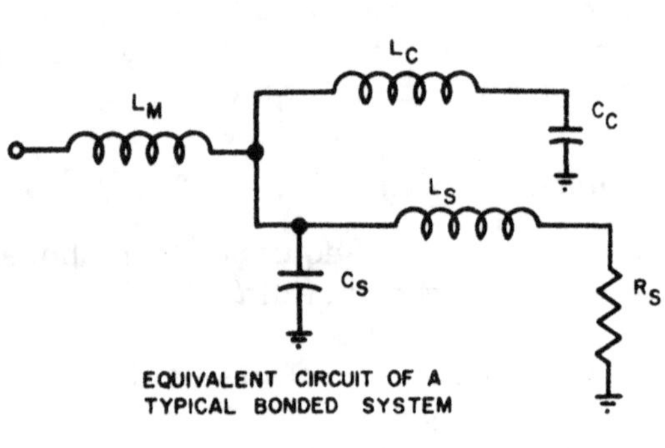

Figure 6-2b
Bonding Equivalent Circuits

At frequencies where skin effect becomes significant, the ac resistance of the bond can differ significantly from its dc value. Skin effect is defined as the depth at which the current density is 1/e (about 37%) of its surface current density. This depth, in centimeters, is defined as[3]:

$$\delta = 5033\sqrt{\frac{\rho}{uf}} \qquad (6.2)$$

where

f = the frequency, in Hertz

u = material permeability, relative to copper

The assumption that all of the current in a conductor is within the skin depth defined by equation (6.2) results, in the case of circular conductors, in the following equation for ac resistance:

$$R_{ac}(\text{ohms/cm}) = \frac{\rho}{2\pi r \delta} \qquad (6.3)$$

where

r = the conductor radius, in centimeters.

Figure 6-3 indicates the ratio of ac-to-dc resistance for several sizes of copper wire over the frequency range of 1-1000 MHz.

For a straight bonding strap of non-magnetic metal, bonding inductance, L, is given by[3]:

$$L = 0.00508\,a\left[2.303\,\mathrm{LOG}_{10}(2a/b+c) + 0.5 + 0.2205(b+c/a)\right] \qquad (6.4)$$

If the bond is made with a straight piece of wire of circular cross section, L is given by:

$$L = 0.00508\,a\left[2.303\,\mathrm{LOG}_{10}(4a/d) - 0.75\right] \qquad (6.5)$$

In the above two equations,

a = the length of the strap in inches

b = the width of the strap in inches

c = the thickness of the strap in inches, and

d = the wire diameter in inches

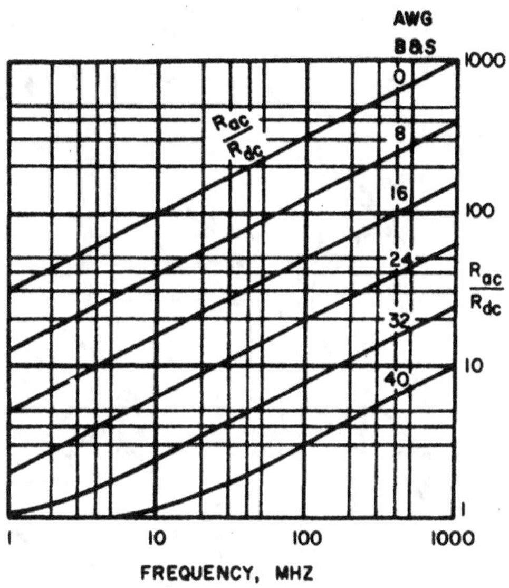

Figure 6-3

Ratio of AC to DC Resistance for Various Wire Sizes

Figure 6-4 is a plot of the calculated inductive reactance of three specific bonds as a function of frequency. Two of the bonds are copper wire, while the third is a copper strap.

At frequencies where the length of the bond approaches or exceeds the wavelength of concern, the bond can act as a transmission line. The result is the development of standing-waves on the bond, and a deterioration in bond effectiveness.

Typical bond straps may be manufactured as flat, solid straps; as braided straps; or a solid wire. The material used is generally copper or aluminum, but in the case of flat straps can also be phosphor bronze.

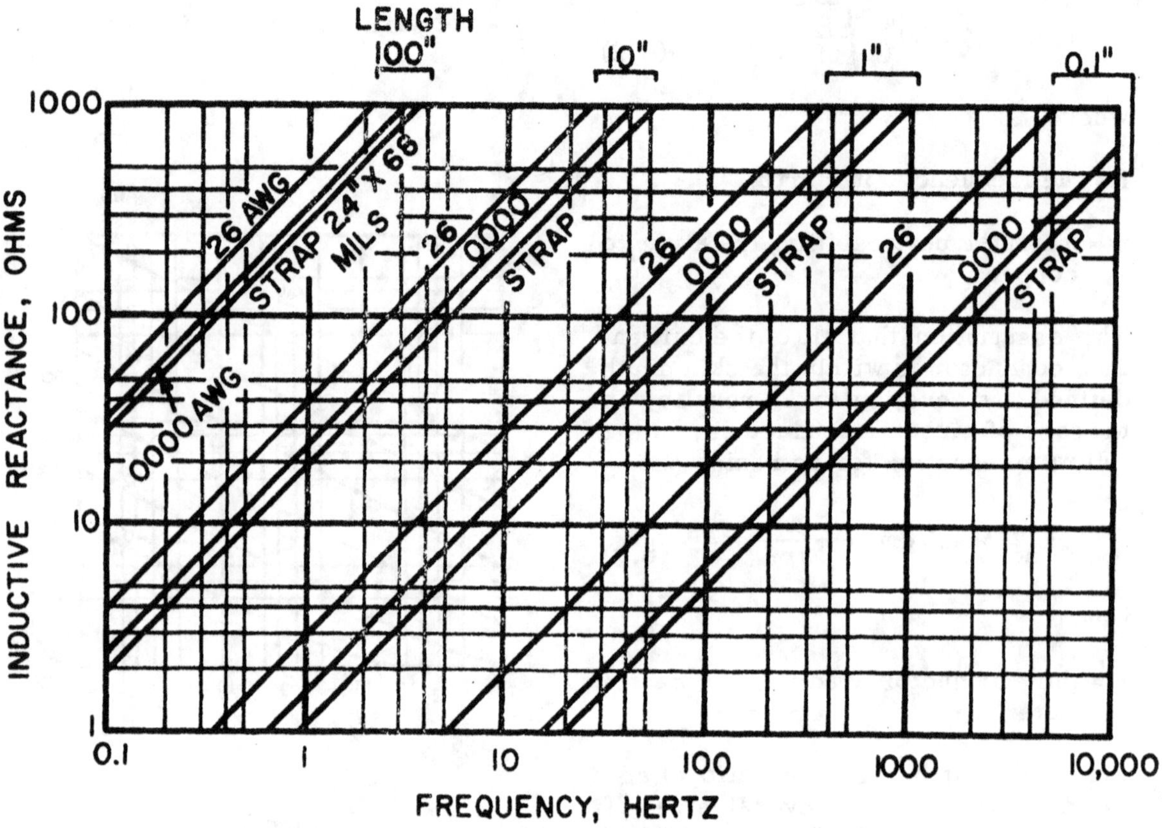

CALCULATED INDUCTIVE REACTANCE, IN UH, FROM,

$$L = 0.00508\, a\left[2.303 \log_{10}(4a/d) - 0.75\right], \quad \text{OR}$$

$$L = 0.00508\, a\left[2.303 \log_{10}(2a/b+c) + 0.5 + 0.2205(b+c/a)\right]$$

WHERE a = Length, (Inches)
b = Width, (Inches)
c = Thickness, (Inches)
d = Diameter, (Inches)

Figure 6-4

Inductive Reactance of Wire and Strap Bond Jumpers

6.4.2 Equipment Effects on Indirect Bonds

The representation of the equivalent circuit of a bond as a parallel-tuned circuit ignores the effects of the equipment enclosure or other item being bonded, and the latter can significantly affect bond strap performance. Figure 6-2(b) shows an equivalent circuit when taking the system being bonded into account. In Figure 6-2(b), the inherent inductance of an equipment case is represented by L_c, its capacitance to other equipments and the reference plane is indicated by C_c, and the impedance contributed by the insertion loss measuring circuit is represented by L_m. Measurements confirm this representation, but additionally denote that, under certain circumstances, the dip in shielding effectiveness caused by resonant effects can go below 0 dB (-10 to -20 dB have been noted), indicating that the induced voltage on the unit being bonded can be increased by attachment of a bond strap.

Figures 6-5 and 6-6 contain plots of measured bonding effectiveness of two bond straps of different lengths that are connected between an equipment case and a ground plane[2]. The proximity of the bond strap to the ground plane is also a parameter in the plots. The bonding effectiveness is the difference (expressed in dB) between the induced voltages on the equipment case with and without the bond strap. A negative value of bonding effectiveness in the plots indicates that the bond strap increases the amount of voltage developed on an equipment case at the indicated frequency.

It should be noted from Figures 6-5 and 6-6 that above about 100 MHz for the 9 1/2" straps and above 400 MHz for the 2 3/8" straps (but below the frequencies where the length of the bond is a significant part of a wavelength), a bond will generally not affect the voltage developed on the equipment case. Also, the bonding effectiveness of the 2 3/8" straps are typically 20 dB better at 10 MHz than with the longer straps.

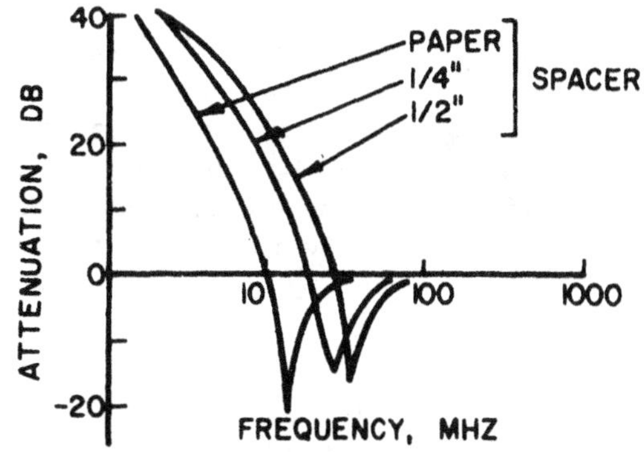

Figure 6-5

Bonding Effectiveness of a 9 1/2-inch Bonding Strap, Measured Using Shunt-T Insertion Loss Technique

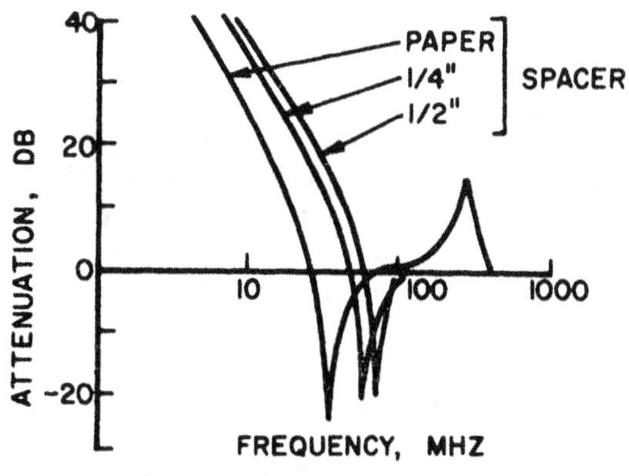

Figure 6-6

Bonding Effectiveness of a 2 3/8-inch Bonding Strap, Measured Using Shunt-T Insertion Loss Technique

The data in Figures 6-5 and 6-6 show a significant frequency range for each bond where the bond creates the negative bonding effectiveness condition cited earlier. This range can be increased in frequency (and reduced somewhat in frequency range) by employing shorter bonds located away from the ground plane.

In selecting a bond strap, the resonant frequency of the strap must be well above the highest interfering frequency which is expected to be encountered by the GSE. This resonant frequency corresponds closely to the frequency at which the worst bonding effectiveness value is obtained.

To reduce the RF impedance of the bond, and thus increase its bonding effectiveness, a high C to L ratio should be achieved, by minimizing the case to ground spacing and the length to width ratio of the bond strap. The inductance of the bonding strap can be minimized by selecting a strap whose length is less than 5 times its width[4]. A good bond should have an inductance of less than 0.025 microhenry[5].

6.4.3 Bonding Resistance

Although measurement of the dc resistance of a bond is obviously not an indication of the bond's ac characteristics, it is often used as a guide to the anticipated performance of the bond. Depending on the purpose of the bond, military specifications designate the maximum dc resistance allowable for a good bond.

For example, bonds that are installed to prevent shock hazards are required by both MIL-B-5087B and MIL-STD-1310B (SHIPS) to have a resistance of less than 0.1 ohms. Radio-Frequency bond requirements in MIL-B-5087B specify a resistance of less than 2.5 milliohm. Additionally, in areas prone to explosion or fire hazards, maximum values of bond resistances are designated; these values are functions of anticipated maximum fault current in the event a power line to ground short occurs. A guideline as far as a good rf bond is concerned, is a dc resistance value of between 0.25 and 2.5 milliohm[4].

Measurements of various types of bond straps indicate the flat, solid strap provides less inherent ac resistance than other types[6]. However, this advantage is often a minor one, and may not offset the advantages of flexibility of braid and lower cost of solid wire.

6.5 BONDING TESTS

6.5.1 Introduction

The bonding tests described below are intended to identify practical test techniques for measuring the parameters of an electrical bond. The specific tests to be discussed include:

- DC Resistance Measurement
- Swept Frequency/Shunt-T Insertion Loss Measurement

Tabulated procedures are provided for each of these tests. They are included for descriptive purposes only, and may have to be altered depending on specific test conditions.

6.5.2 DC Resistance Measurement

6.5.2.1 Objectives

This test is intended to give design personnel a general indication of bond adequacy after installation, based on the dc resistance of the bond. The test is often performed to establish whether or not the bond resistance is within preestablished design limits. However, as pointed out previously, measurement of bond resistance is not a complete indicator of the electrical characteristics of the bond.

6.5.2.2 Instrumentation and Test Set-Up

Equipment required for this test is as follows:

- A dc resistance bridge capable of measuring to about 0.001 ohm. The bridge should be portable and easily moved, and should not be position-sensitive. Connection of the test sample to the bridge terminals should be easily performed without cumbersome adaptors or special tools.

- A pair of heavy-duty spring clip leads for connection between the bridge and the bonded junction. Clip leads may be connected to braided straps and plugs to make connection to bridge. Total resistance of external connectors and leads should not be greater than 0.001 ohms.

Equipment set-up is as follows:

- Using the heavy-duty spring clips and braid, connect the leads to the bridge.

- Place the bridge in operation according to the manufacturer's operation manual.

- Zero the bridge, including leads, and connect the clip-leads across the bonded junction.

6.5.2.3 Test Procedure

- Adjust the bridge balance until a null is obtained.

- Compare the indicated resistance to the tolerance or design limit.

- If the measured resistance is above the required value, check to see that the bond has been properly installed. Tighten to design limit any bolts, clamps, etc. and repeat the test.

- If bond resistance is still not within limits, redesign of the bond may be necessary.

6.5.2.4 Special Precautions

- Be sure that the clip-leads and connections are not contributing significantly to the total resistance measured.

- Be sure that the bridge is zero balanced to take into account the effects of the external connectors and straps.

6.5.2.5 Advantages/Disadvantages

- A dc resistance measurement of a bond will give an indication of bond performance under dc or low-frequency high current flow, but does not provide rf impedance levels.

- Variations in bond rf impedance can severely affect system performance and safety. Since the dc resistance test does not give such data, its usefulness is restricted to an indicator of good mechanical union between metallic surfaces. Physical form factors of the bond, external to the bond junction, affect bond rf performance. Additional tests may be necessary to insure adequate rf performance.

6.5.3 <u>Swept Frequency/Shunt-T Insertion Loss Measurement</u>

6.5.3.1 Objective

The shunt-T insertion loss test, using a swept frequency source and wideband detector, is designed as a rapid method of characterizing bonding impedance over the frequency range where bonding performance degradation might ordinarily occur. The test, with careful test fixture design, can measure bond impedance up to about 400 MHz[2, 7].

6.5.3.2 Discussion

The insertion loss-shunt-T test is based on the fact that if the shunt arm impedance of a T-attenuator is low in comparison to the series arm impedances, the current through the shunt arm will be essentially constant for varying values of shunt arm impedance. For a fixed voltage input to the T-attenuator, variations in the T-attenuator output voltage will be proportional to changes in the shunt arm impedance.

Figure 6-7 is an equivalent circuit of an insertion loss impedance measurement device. Two isolation resistors, R_s, are placed in series with the connections to the unknown impedance to form a T-network. The isolation resistors are chosen to be large with respect to the unknown impedance. The symbols r_1, r_2, L_1 and L_2 denote the connection impedances.

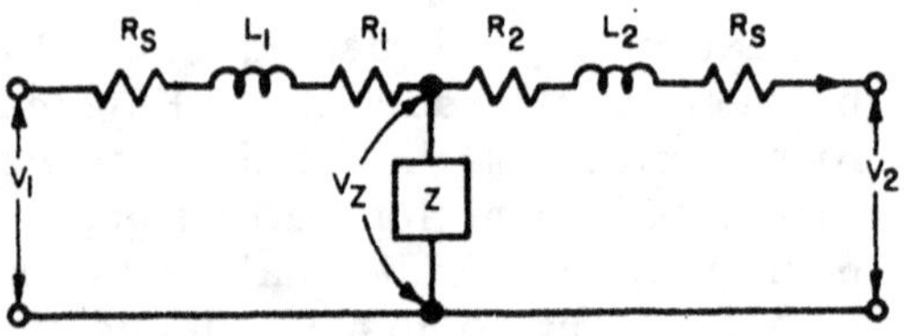

Figure 6-7

Equivalent Circuit of the Impedance Measuring Device

The relationship between the unknown impedance and the insertion loss through the test fixture is given as:

$$Z = \frac{R_s (R_s + 50)}{50} \left| \frac{V_{out}}{V_{in}} \right|, \text{ ohms} \quad (6.6)$$

This equation assumes a 50 ohm measurement system. It also applies so long as the absolute value of the bond impedance is less than 0.1 R_s. In addition, with the measurement terminals open circuited, a minimum attenuation value will exist; the equation applies only for values that are well above this minimum.

The overall construction of a device for use in performing shunt-T insertion loss measurements is shown in Figure 6-8. The unit is intended to hold all of the test components and connectors, and maintain short lead lengths in the process. The bolt on the test fixture is connected to the bond strap under test. The case of the test fixture must be connected to the ground plane using threaded holes on the bottom of the fixture.

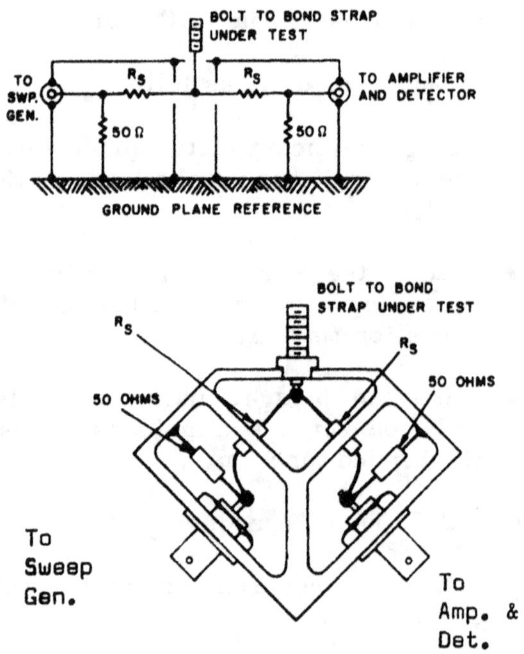

Figure 6-8

Schematic Diagram and Mechanical Details of Bonding Impedance Measurement Test Fixture

The two 50 ohm resistors in the test fixture are not shown in Figure 6-7. They are used for properly terminating the test equipment, and are incorporated in the fixture only when the value chosen for R_s is large compared to 50 ohms. In that case, the bond impedance becomes 4 times the value computed using equation (6.6).

6.5.3.3 Instrumentation and Test Set-Up

The test equipment is arranged as shown in the block diagram of Figure 6-9. The following equipment is required for the test:

- A swept-frequency source covering the desired frequency range. The output signal level of the source should be constant versus frequency, and the X-axis sweep should be a known function of frequency, preferably either linear or logarithmic. Its output signal level should be between 0 dBm and +10 dBm for achieving a reasonable dynamic range.

- A wideband amplifier, with gain according to the following scale:

Anticipated Minimum Bond Impedance, Ohms	0.01	0.1	1.0	10	100
Amplifier Gain Required, dB	80	60	40	20	Not required

The amplifier should be impedance-matched to the sweep generator and to the detection system. The test fixture design has assumed both the sweep generator output and amplifier input impedances are 50 ohms.

- A wideband detector, which may be a diode/low-pass filter combination, or a wideband logarithmic amplifier with dc output that is a function of $\log(E_{in})$. The latter type is

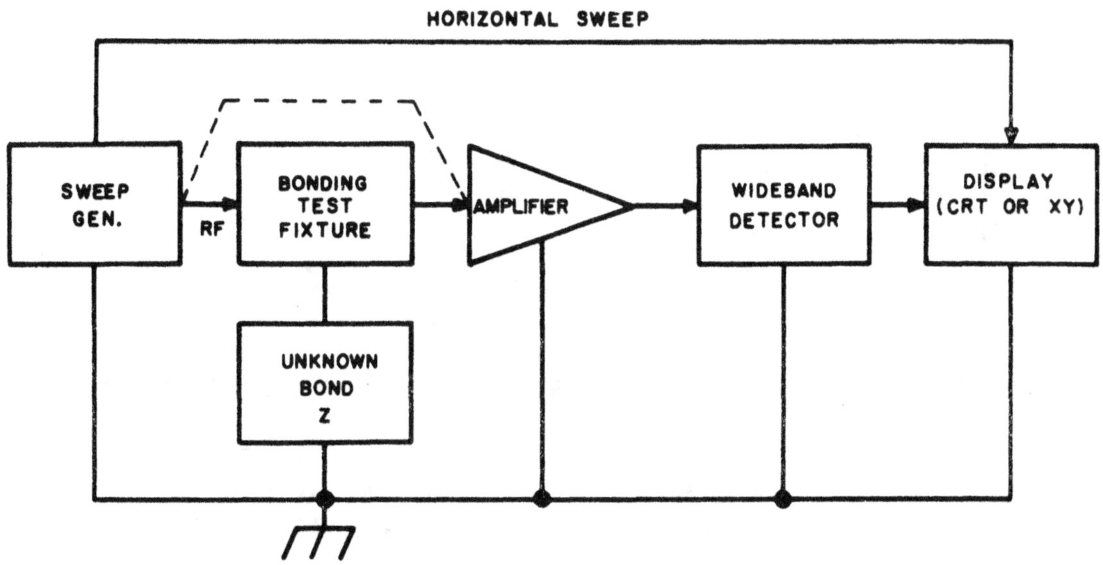

Figure 6-9
Swept Frequency System for Measuring Bonding Impedance

preferred.

- Display and readout may be either an oscilloscope or X-Y plotter. The input to the display device should be matched to the output of the wideband detector.

- A bonding test fixture, as shown in Figure 6-8. The test can be performed using commercially available 50 ohm resistive splitters without the shunt-matching 50 ohms, or a test fixture can be fabricated from commercially available components.

6.5.3.4 Test Procedure

- Attach the bonding test fixture to the bond to be tested, using bolts or clamps to insure good connection to the ground plane.

- Connect the test equipment as shown in Figure 6-9. Initially by-pass the test fixture, and connect the sweep source directly to the amplifier.

- Adjust the sweep generator for maximum desired deflection on the display/readout device and for the desired frequency range to be swept. Adjustment of wideband amplifier gain and display/readout deflection (X and Y) may be necessary to optimize the display.

- Break the connection between the sweep generator and the wideband amplifier, and connect the sweep generator output to the input of the test fixture. Connect the output of the test fixture to the input of the wideband amplifier.

- A display similar to those shown in Figure 6-10 should be obtained. The attenuator on the sweep generator output, or the input attenuator on the display input can be used to calibrate the vertical display in terms of insertion loss.

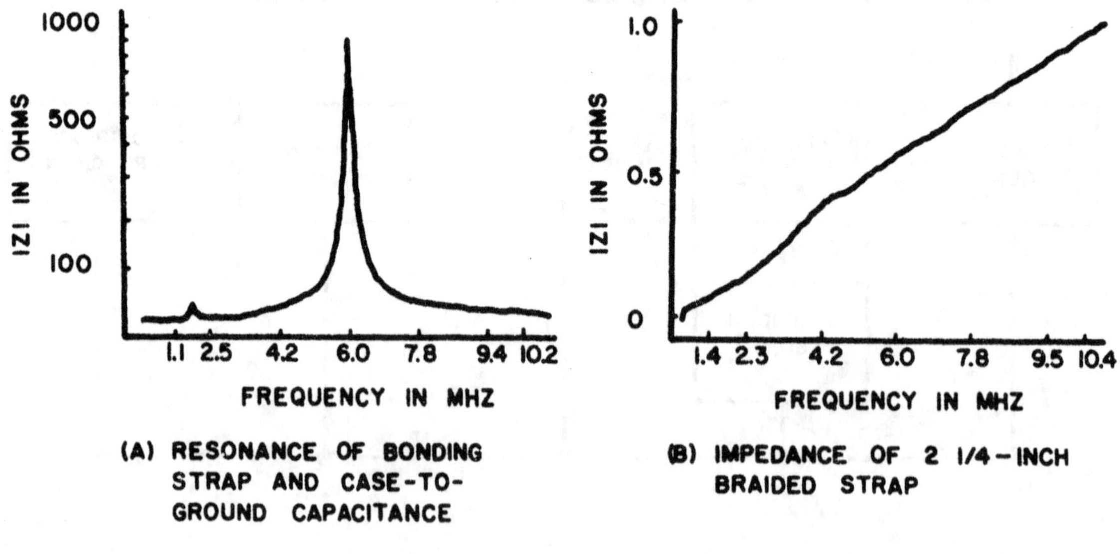

Figure 6-10

Typical Bond Impedance Vs. Frequency Measurements

- From equation (6.6), determine the bond impedance at a frequency of interest. Don't forget the 12 dB (4X) correction factor to be used if the test fixture contains 50 ohm shunt resistors. The impedance corresponding to a particular value of insertion loss should be compared with both the value of R_s in the test fixture and with the minimum value of impedance obtained with the measurement circuit open. The measured value should be less than 0.1 R_s and well above the minimum impedance value.

- If the above conditions are not met, a test fixture with a different value of R_s must be used.

6.5.3.5 Advantages/Disadvantages

The shunt-T insertion loss test technique is convenient and easy to use, once the test fixture is available. With a suitable, leveled output sweep generator and a wideband detector, and a carefully constructed test fixture, the method can be accurate to 400 MHz, and down to absolute values of bond impedance of 0.01 ohm, although the low limit of impedance is frequency-dependent. The frequency range of the test is limited by the series reactance of the test fixture and by the amount of insertion loss introduced as a result.

6.6 SINGLE VS. MULTI-POINT BONDING

At frequencies sufficiently low that the entire structure is less than $\lambda/20$ in size, a single point bond or ground generally is most desirable since it avoids the problems of common-mode impedance coupling. In fact, efforts are made to assure that contact between two ground planes can take place at only one point to avoid noise that may result when irregular multiple point contacts are permitted, as when two large metallic surfaces are in surface contact under only moderate pressure.

Problems arise in implementing single point ground contacts when parasitic capacitance is significant between equipment housings or between different subsystem grounds. Problems also arise when interconnecting cables are used, especially those having cable shields longer than $\lambda/20$ between the source and the load. When cable shields interconnect subsystems, multiple grounding paths from a given subsystem to the system ground (reference) point exist. Furthermore, at frequencies high enough that the lengths in the system exceed $\lambda/20$, parasitic capacitive reactance can constitute a low impedance path while bond inductance can provide a comparable, but inductive, reactance. Anti-resonance can even occur, thus effectively isolating a circuit from its intended ground point at the anti-resonant frequency of the bond and the parasitic capacitance.

Thus while efforts should be made to preserve a single-point bond for audio frequency applications, a multi-point bond is desirable at high frequencies.[8] The purpose of the multi-point bond is to obtain a low enough bonding inductance (reactance) that no significant impedance can appear between the two points that are connected by the bond.

6.7 BONDING DESIGN GUIDELINES

The effectiveness of a bond depends on its application, frequency range, magnitude of current, and environmental conditions such as vibration, temperature, humidity, fungus, and salt content in the ambient. Figures 6-11 through 6-15 provide examples of direct and indirect bonds, to illustrate the variety of techniques that have been employed to obtain low impedance connections. Many other examples are available in the literature, and particularily in MIL-STD-1310B.

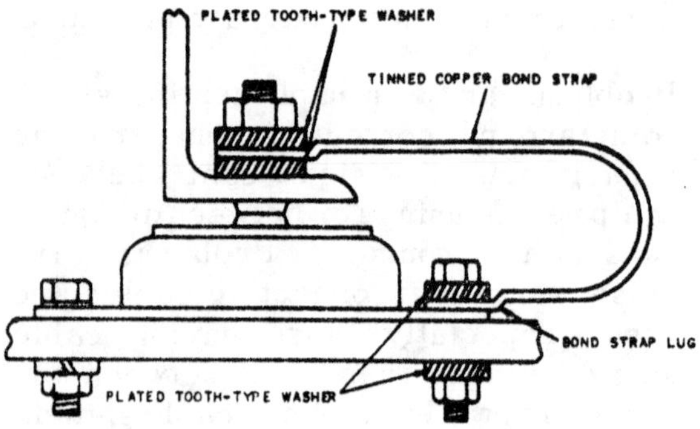

Figure 6-11
Typical Shock Mount Bond

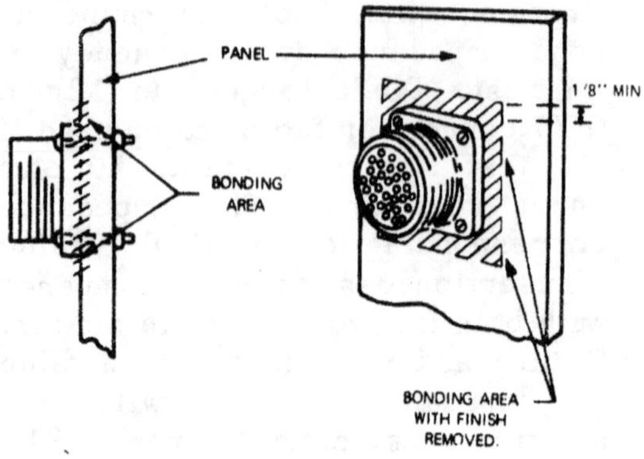

Figure 6-13
Bonding of a Connector

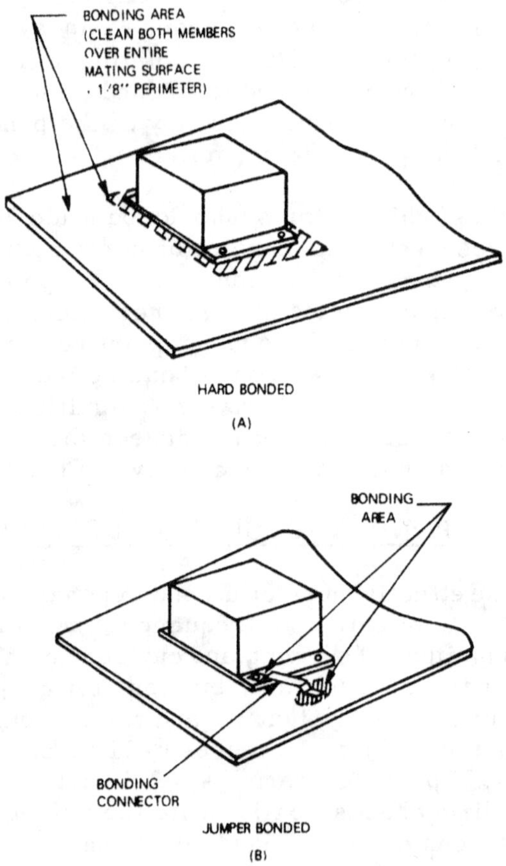

Figure 6-12
Base Mount Components

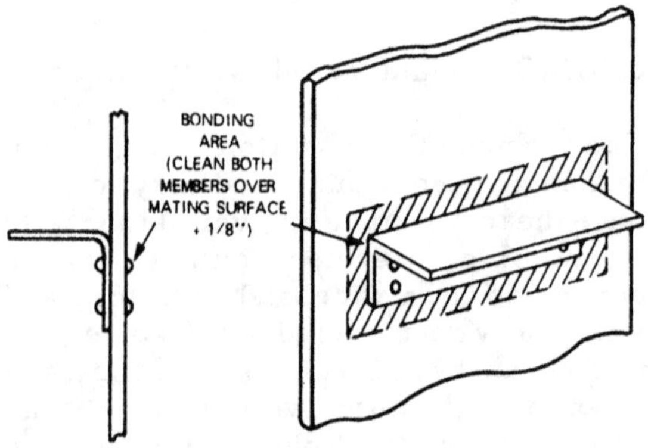

Figure 6-14
Bolted Members

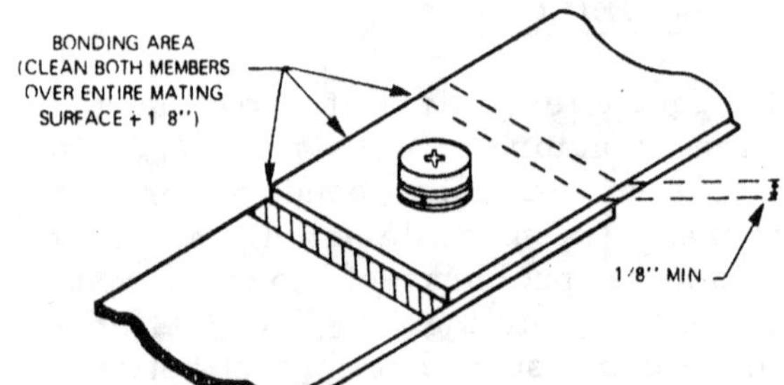

Figure 6-15
Bracket Installation (Rivet or Weld)

Some general guidelines to obtaining good bonds can be provided.

The secret to good bonding is intimate contact between metal surfaces. Surfaces must be smooth and clean, and not coated with a nonconductive finish. The fastening method must exert sufficient pressure, to hold the surfaces in contact in the presence of deforming stresses, shock, and vibrations associated with the equipment and its environments.

Bonds are always best made by joining similar metals. If this is not possible, special attention must be paid to the possibility of bond corrosion, through the choice of the materials to be bonded, the selection of supplementary components (such as washers) that will assure any corrosion will affect replaceable elements only, and the use of protective finishes.

Solder should not be used to provide mechanical strength to a bond.

Protection of the bond from moisture and other corrosion effects must be provided where necessary.

Bonding jumpers are only a substitute for direct bonds. The jumpers should be kept as short as possible, have a low resistance and low L/C, and not be lower in the electrochemical series than the bonded members. A good rule to use is that the jumper should have a length-to-width ratio of less than 5.

Jumpers should be bonded directly to the basic structure, rather than through an adjacent part. They should not be connected with self-tapping screws, or by any other means where screw threads are the primary means of bonding.

It is always important in the broadest types of bonding application that the bonding jumper or direct bond is sufficient to be able to carry the currents that may be required to flow through it.

Use single point bonds at low frequencies (circuit dimensions $< \lambda/20$) but use multi-point bonds at high frequencies.

BONDING BIBLIOGRAPHY

1. NASA SP-3067-1971, "Radio Frequency Interference Handbook."

2. H.W. Denny, "Some RF Characteristics of Bonding Systems," IEEE Transactions on Electromagnetic Compatibility, February 1969.

3. Naval Air Systems Command, Electromagnetic Compatibility Manual, NAVAIR 5335.

4. U.S. Army Electronics Laboratories, Ft. Monmouth, N.J., "Interference Reduction Guide for Design Engineers, Volume I," August 1964.

5. NASA, Electromagnetic Compatibility Principles and Practices, Apollo Design Reliability Series, May 1965.

6. R.M. Soldanels, "Flexible RF Bonding Configuration-Analysis, Measurement and Practical Applications," IEEE Transactions on Electromagnetic Compatibility, December 1967.

7. H.W. Denny and K.G. Byers, Jr., "A Sweep Frequency Technique for the Measurement of Bonding Impedance Over an Extended Frequency Range," 1967 IEEE Electromagnetic Compatibility Symposium Record, July 1967, pp. 195-202.

8. D.R.J. White, "A Handbook Series on Electromagnetic Interference and and Compatibility, Volume 3, EMI Control Methods and Techniques," 1973.

7.0 GROUNDING

7.1 GENERAL

Grounding involves the establishment of an electrically conductive path between two points, with one point generally being an electrical/electronic element of a system and the other being a reference point. When the system element of concern is an equipment enclosure (such as one rack of the VAST AN/USM-247(V) System), then the reference point may be a ship structural member. When the system element of concern is a circuit within a GSE, then the reference point can be the equipment case or a ground plane that may or may not be isolated from the case.

A good, basic ground plane or reference is the foundation for obtaining reliable, interference-free equipment operation. An ideal ground plane should be a zero-potential, zero-impedance body that can be used as a reference for all signals in the associated circuitry, and to which any undesirable signal can be transferred for its elimination. An ideal ground plane would provide equipment with a common potential reference point anywhere in the system, so that no voltage would exist between any two points. However, because of the physical properties and characteristics of grounding materials, no ground plane is ideal, and some potential always exists between ground points in a system.

The extent to which potentials in the ground system can be minimized and ground currents can be reduced will determine the effectiveness of the ground system. A poor system, by enabling these spurious voltages and currents to couple into a circuit, subassembly, or equipment, can degrade the shielding effectiveness of well-shielded units, can essentially bypass the advantages of good filters, and can result in EMI problems that may be rather difficult to resolve after-the-fact.

Various ground systems must meet requirements for personnel safety as related to the electrical power system, for lightning protection of personnel and property, and for providing a signal ground bus as a common electronic circuit return. Emphasis in this chapter will be placed on the establishment of good signal grounds, since this aspect of grounding will be of most concern to GSE design engineers.

It is important to note that the GSE designer, even if he is developing a small, portable tester, must consider grounding from a system point of view. The system involved includes not only the test device, but the equipment or system that the device will be testing. The ground systems must be compatible under the interfaced condition.

7.2 GROUNDING TECHNIQUES

There are three fundamental grounding concepts that can be employed, as illustrated in Figure 7-1. The approaches can be used separately or in combination in any given equipment or system. They are:

a. Floating Ground System

b. Single-Point Grounding System

c. Multiple-Point Grounding System

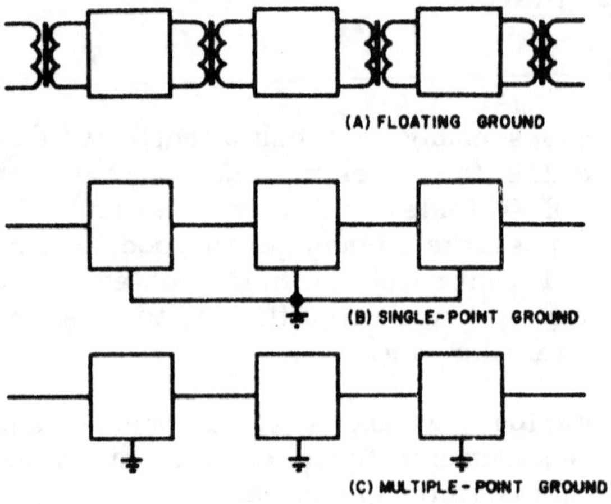

Figure 7-1

Grounding Methods

The Floating Ground System is a method to electrically isolate circuits or equipments from a common ground plane, or from any common wiring that might introduce circulating currents. Floating grounds depend, for their effectiveness, on truly "floating".

For many situations, this complete isolation may be very difficult to achieve. This is particularly true of a piece of GSE that will be hard-wired to an aircraft or weapon in order to perform its test function. Also, certain hazards exist in the use of floating systems, in that static charges or lightning potentials may accumulate between the floating grounds and other accessible grounds such as the flight deck or other portions of the ship structure, power line neutrals, the skin of the weapon system under test, etc.

Examples of isolation techniques are shown in Figures 7-2 and 7-3. Figure 7-2 employs transformer isolation, while Figure 7-3 obtains isolation by optical means.

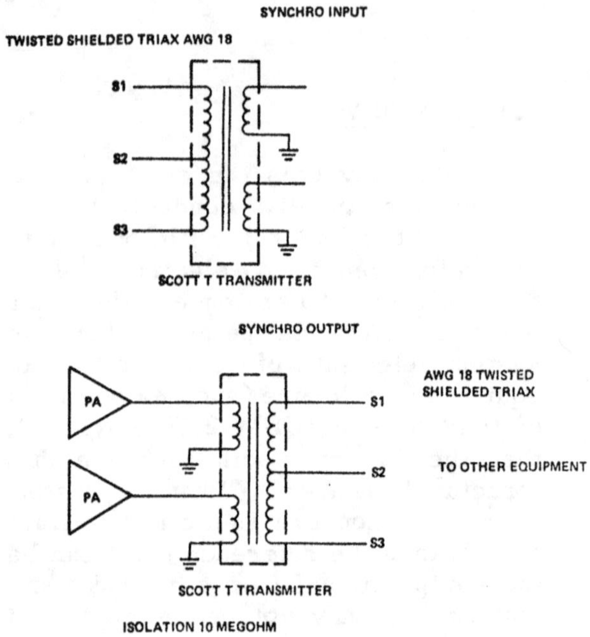

Figure 7-2

Typical Synchro Input, and Synchro Output Isolation Circuitry

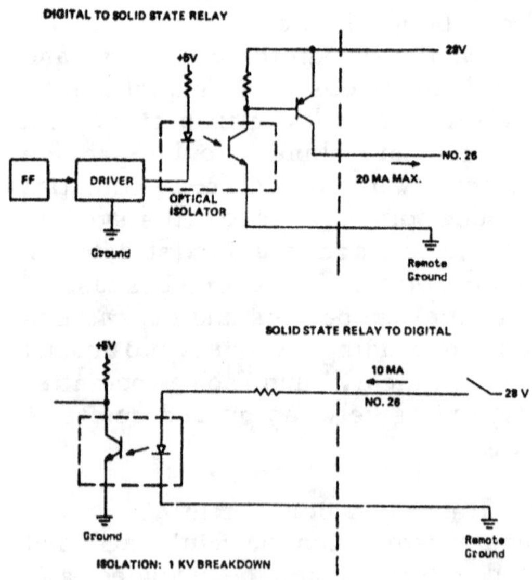

Figure 7-3

Digital to Solid-State Relay, and Solid-State Relay to Digital Isolation Circuitry

In a Single Point Grounding System, a single physical point in the circuitry is defined as a ground reference point. All ground connections are tied to this point.

For multiple cabinet configurations, the cabinet and electronic circuit grounds are often kept separate, with the single point grounding concept being used independently for each ground system. The interconnection between ground systems is then made only at the reference point. This isolates the rack or cabinets, and prevents circulating currents in one ground system from affecting the other[1].

Again, because of the appendage nature of most GSE, it would be extremely difficult to maintain a complete single-point grounding system. However, under certain conditions, it may be advantageous to use single-point grounding for the GSE and combination grounding for GSE/weapon system.

At high frequencies (frequencies whose wavelengths approach equipment ground plane dimensions or cable lengths), single-point ground systems are no longer practical. Therefore, it is important for the equipment/system designer to have an understanding of the frequency-susceptibility of his device, as well as the internal and environmental frequencies that are around to create EMI, before he finalizes on his grounding system design.

The Multipoint Grounding System is one in which a ground plane is used instead of individual return wires for each circuit. The ground plane might be an equipment chassis, or a ground wire that is carried throughout a system, or (in the case of a very large system) building structural members.

The advantages of a multipoint grounding approach are that circuit construction is easier, and it is the only way to avoid standing-wave effects in the ground system at high frequencies. Also, a large conductive mass can be selected for the ground plane. However, since multiple-point grounding creates many ground loops, the quality of the ground system becomes very important. Also, the ground plane must be carefully maintained, particularly with reference to corrosion, vibration, and mechanical damage, which can introduce high impedances into the ground system.

7.3 CIRCUIT GROUNDING CONSIDERATIONS[1, 2, 3]

If the ground plane is fairly large, significant potential differences may exist between two points on the ground plane. These voltages must be considered when defining the permissible ambient interference level in the system, and when determining the expected signal-to-interference ratio of the signal transmission circuits.

The simplest and most direct approach to keeping potential differences introduced by the circuit ground plane to a minimum is to arrange circuit components physically so that ground return paths are short and direct, and have the fewest possible crossings of these paths. In this way, the inter-circuit coupling of the ground currents will be low and isolated.

The effect of ground potential can be cancelled by electrically isolating the circuits using a floating ground system. This method is especially effective at audio and low radio frequencies. Above these frequencies, its effectiveness progressively diminishes because, as the frequency of equipment operation increases, coupling paths appear that bypass the isolation transformers.

Band pass filters are another method of obtaining circuit isolation. These filters are only effective when the frequency of the interference energy is outside of the frequency band of the desired signal. Then the band pass filter will reject the interference signal across the load, causing the voltage drop to occur elsewhere in the circuit and not at the load. When using band pass filters, care must be taken to insure that "ringing" is not initiated by the undesired signal when it is processed by the filter.

Differential or balanced circuitry can help reduce the effects of ground circuit interference. Since a differential circuit responds only to the potential difference between its input leads, the noise voltage at the source may be above ground potential by a considerable amount without degrading circuit performance. Figure 7-4 illustrates such a differential circuit. The input voltage, V_g, is the voltage to which the device responds. The interference voltage, V_n, is simultaneously impressed on both input leads but is balanced out in the input to the device because each input lead has the same impedance to ground. Thus, the device does not respond to the ground circuit signal. In theory, the ambient noise voltage is cancelled out, assuming the impedance of V_g is zero ohms. In practice, there is always some unbalance in the differential device or associated circuitry, and some part of V_n will appear as a difference voltage across an equivalent resistance R. The noise voltage differential results in a reduced output signal-to-noise ratio. Figure 7-5 shows how the unbalance causes a portion of V_n, ΔV_n, to appear across the input terminals of the device.

Figure 7-6 summarizes the situation just discussed. It shows the ground circuit voltage, V_n, introducing an incremental voltage, ΔV_n, at the input to the differential circuit.

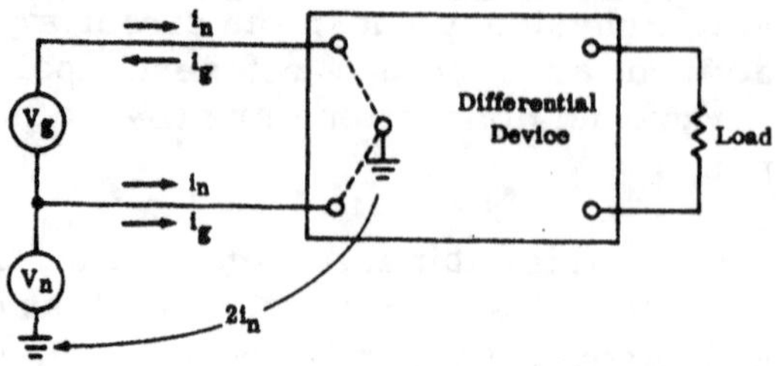

Figure 7-4

Schematic Diagram of a Differential or Balanced Circuit

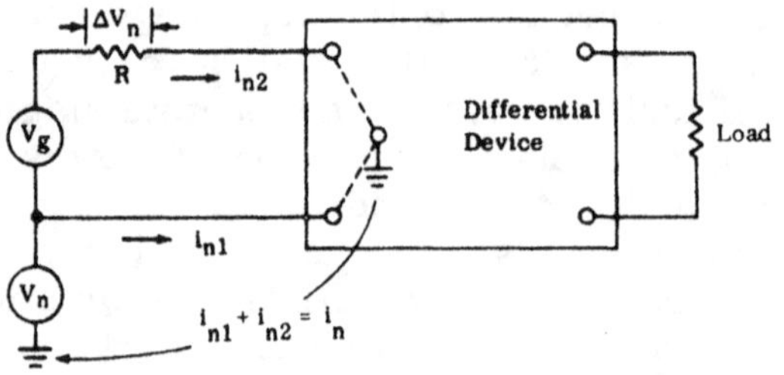

Figure 7-5

Effect of Unbalance in a Differential Circuit

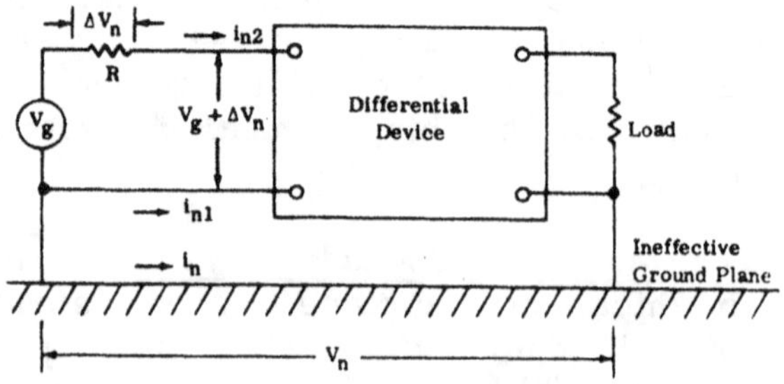

Figure 7-6

Common-mode Voltage Generated by Current Flowing in a Finitely Conducting Ground Plane

In transistorized digital circuits, the input and output impedances of the circuits are generally relatively low. This makes the circuits more susceptible to the effects of low impedance (magnetic) fields than to the effects of high impedance (electric) fields. One of the important parameters controlling interference due to a low impedance field is the loop area of the circuits causing and picking up the interference. By reducing the loop area of these circuits to a minimum, much of the interference problem can be eliminated.

Where modular type construction is used, one method of minimizing the loop area of grounded circuits is to mount modules on a sheet of good conducting material, such as copper or aluminum, that is connected to the circuit ground as directly as possible. All intra-module wiring should be run as close to this sheet as possible. This technique can reduce the loop area of these circuits to an extremely small value, but the capacitive coupling of the circuits is increased. The effect of this increase in capacitive coupling on circuit operation should be considered if this technique is used.

The use of high impedance input or output circuit impedances when signals must be transmitted over even a few inches should, in general, be scrupulously avoided. Where the use of such signals cannot be avoided, the interconnecting lead must be shielded and the shield grounded at each end.[4]

7.4 POWER SUPPLY CONSIDERATIONS[1]

The ideal method of utilizing ac power supplies in GSE would be to have an individual power supply for each electronic circuit. This would give complete decoupling of circuits viewed from the power supply, and since this common conducting path would be eliminated, interference from this source would also be eliminated. However, this is obviously a very impractical solution in most cases, both from the standpoints of cost and space requirements. Other alternatives must be considered.

It is generally good practice to isolate power and signal grounds from each other. This will go a long way to minimizing the possible coupling of signals between these two major types of paths.

Groups of electronic circuits can be connected to one power supply based on a pre-arranged logical pattern. Critical circuits must be kept separated (i.e., critical with respect to interference susceptibility), and each given their own power supply when possible. If it is decided that these critical circuits cannot be kept separate, a voltage regulator can be introduced to act as a decoupling component; Zener diodes can usually be used quite effectively in such applications.

Use of a large capacitor or an RC network at the load end of power supply leads can offer some degree of decoupling. This is particularly helpful in digital circuits to reduce the inductive effects of long power supply leads, but can introduce "ringing" at frequencies where the line and capacitor resonate. This condition can be circumvented by shunting the decoupling capacitor with a small rf bypass capacitor.

When deciding on which circuits to connect to which power supplies, the physical location of the electronic circuits must be considered. If no other factors are involved, all circuits fed by a given power supply should be grouped in the same area. However, in cases where various circuits will have to operate simultaneously (as in a synchronous digital system) some spatial separation may be advantageous to prevent coupling of radiated energy.

Alternating currents of high amperage are potential sources of interference coupling to adjacent lines. Transposed or twisted lines should be used for all ac power circuits. Routing of these lines should be away from susceptible lines. AC power circuits in which switching transients are expected may beneficially use shielding to contain the higher frequency components of the transient. Such shielding should be over the entire power wiring bundle and should be grounded at both ends.

The power system that drives electromechanical relays should not generally be used as a supply voltage for any electronic circuits because of the large amount of noise in this system due to relay operations. In addition to providing power to relays, this power source may also furnish power for indicator lights, small motors, electromagnetic clutches and brakes, and these devices may also be sources of noise. The power supply itself generally consists of a transformer-rectifier combination or a motor generator set. In either case, the source is usually regulated to insure a constant voltage over a wide range of loads.

The primary cause of relay interference is due to the de-energizing of relays. When a relay circuit is opened, the collapsing magnetic field of the relay coil induces a large voltage of the opposite polarity across the coil. This induced voltage initiates an arc across the contacts that are breaking the circuit. The arc contains energy, in a wide band of frequencies, that may be transmitted as interference both by conduction and radiation. There are many standard arc suppression techniques, but even when these are used, a certain amount of noise will be conducted or radiated from the source.

Another cause of conducted interference can be the changes of current in the relay power supply busses due to the continually changing demands when the equipment is in normal operation. The conducted interference can be reduced by isolation of the relay circuits from the rest of the equipment.

To maintain isolation of the relay circuits and to prevent the system from floating and possibly reaching a high potential above ground, the relay system should preferably be grounded at a single point. Because of the unusually wide distribution of relay circuits, and because it is the point of maximum current, the ground connection should be made at the source.

In designing the power distribution system for relay systems where large currents are expected, it is important to consider the IR drops of the conductors in addition to their current-carrying capacities, although there is relatively low voltage involved. Also, the generation of varying magnetic fields due to the variation of the relay power supply currents must be considered. The effects of these variations can be minimized by running the supply and return leads as close together as possible throughout the system, thereby reducing the loop areas of the various relay circuits.

To minimize the possibility of dc supply lines coupling interference to other circuits and to protect them from receiving interference from other lines, the best procedure is to use a twisted-pair for the supply line and its return. Shielding offers little advantage on lines of this low impedance. The routing should be away from ac power and control lines. When possible, run separate supply lines from the various circuits using the supply to the supply output. Do not share supply return lines with other circuit returns.

7.5 PRIME POWER CONSIDERATIONS[5]

There are two primary concepts regarding power returns. These are ground or structure common return and wired return. Although the GSE designer will probably have no control over the prime power configuration into which his equipment must work, he should nevertheless be familiar with these two concepts.

In the ground or structure common concept, one side of the power system is grounded at the power source, and all loads use the vehicle frame or structure as the return conductor. In three-phase connected AC systems the neutral is grounded and all single phase loads use the structure for the return circuit. The principal advantage of this system is the reduced weight resulting from the elimination of a great many heavy power return wires.

The disadvantage of the common return system is that it does not distribute the power efficiently. The flow of currents through structure produces voltage drops in the structure. These voltages are normally small compared to the operating level of the power system, but are large compared to the operating level of electronic systems. This can create potential interference problems in any electronic system using the structure as a power return. Even systems using structure as a ground plane for shield grounds are subject to induced voltages in susceptible circuits.

In the wired return concept, all systems are grounded at one point only and have wires for all return circuits. The wired return system eliminates the vehicle structures as an impedance common to all systems and therefore eliminates the complex ground loops which exist with the common return system.

7.6 CABLING CONSIDERATIONS[5]

The problem of electromagnetic compatibility in a complex electrical or electronic system is in many cases dependent on the treatment of the shielding and the grounding of the shields of interconnecting leads. Injudicious application of a grounded shield to a wire may cause coupling problems that otherwise would not exist.

Grounding of the shields may be accomplished as single-point or multipoint grounding. Factors that influence the selection of single-point or multipoint grounding include the interference signal frequencies involved, the length of the transmission line, and the relative sensitivity of the circuit to high- or low-impedance fields.

a. Single-point shield grounding. For multilead systems, each shield may be grounded at a different physical point as long as individual shields are insulated from each other. Single-point grounding is more effective than multipoint shield grounding only for short shield lengths. Single-point grounding is ineffective in reducing magnetic or electrostatic coupling when conductor-length-to-wavelength (L/λ) ratios are greater than 0.15, where the wavelength is that of the highest frequency to be used (or the highest frequency interference to be expected) on the wire or in the system[4].

b. Multipoint shield grounding. For L/λ ratios greater than 0.15, multipoint grounding at intervals of 0.15λ is recommended, since the shield can act as an antenna that is relatively efficient at $L/4\lambda$ when one end is grounded. When such grounding of the shield at intervals of 0.15λ is impracticable, shields should at least be grounded at each end. Multipoint shield grounding is effective in reducing all types of electrostatic coupling, so long as large ground currents do not exist. In general, multipoint shield grounding solves most problems, but in audio circuits single-point shield grounding may be more effective because of the ground current problem.

An interesting comparison of the magnetic interference susceptibility of cable-connected circuitry is shown in Figure 7-7. The evaluation was performed at 100 kHz, with the measurement parameters having to do with the type of cable used, whether or not the load is grounded, and how the cable shield is grounded. Among other results, the comparison shows (for the cases indicated) the disadvantage of returning the load to the ground plane, and the advantages gained using tightly twisted leads.

The impact of cable shield and termination grounding on crosstalk has also been explored. Figure 7-8 shows a coaxial cable test setup for investigating single- versus multiple-point grounding, with single-point grounding always applied at the input ends of two 17-foot long cables[6]. The induced voltage in the receiving cable was measured for four conditions:

a. Sending and receiving cable resistor termination floating at the receiving end.

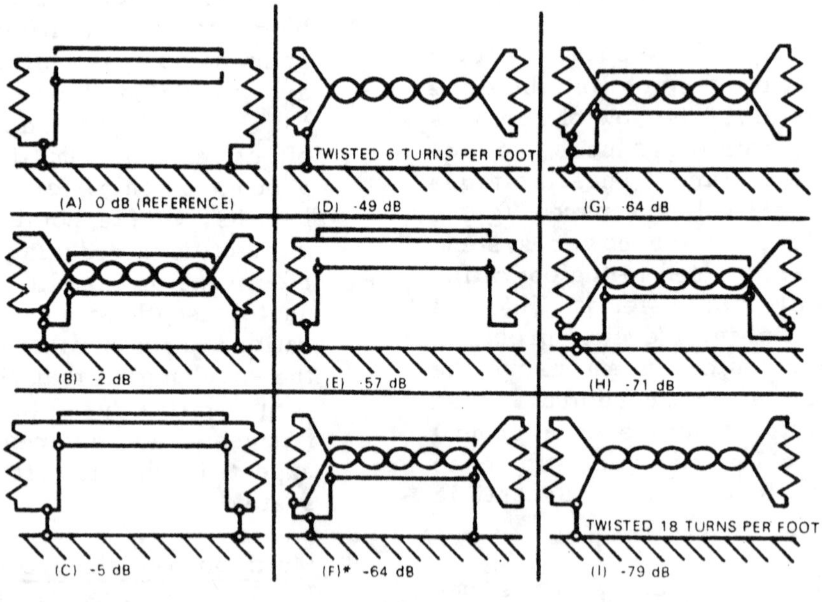

*PREFERRED CIRCUIT FOR HIGH FREQUENCIES

VALUES GIVEN ARE FOR CIRCUITS 1 INCH ABOVE GROUND PLANE BUT ARE ABOUT THE SAME FOR OTHER DISTANCES FROM GROUND PLANE.

Figure 7-7

Relative Susceptibility of Circuits to Magnetic Interference

b. Sending and receiving cable resistor termination grounded at the receiving end.

c. Sending cable resistor termination grounded and receiving cable resistor termination floating at the receiving end.

d. Sending cable resistor termination floating and receiving cable resistor termination grounded at the receiving end.

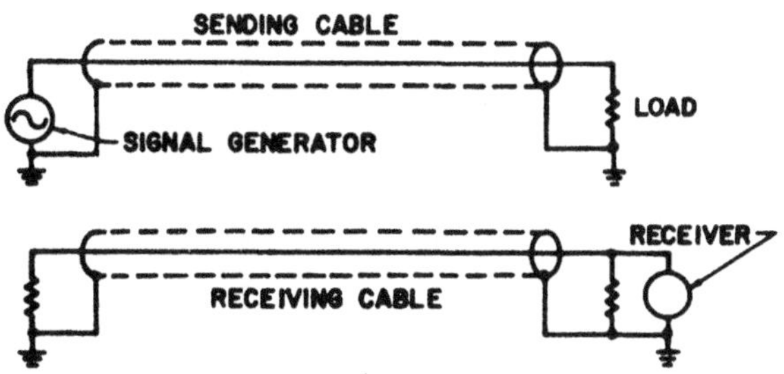

Figure 7-8

Test Setup for Coaxial Cable Crosstalk Measurements

Figure 7-9 shows some of the results of this test. It points out that specific rules on how to ground cannot be defined, and that the best grounding technique to employ is the one that works best for the particular circuitry involved. In this case, lowest crosstalk is obtained when the sending cable is single-point grounded and the receiving cable is multiplepoint grounded.

A comparison between two of the coaxial curves of Figure 7-9 and a shielded twisted pair configuration is shown in Figure 7-10. The sending and receiving cables of the twisted pair were grounded as shown in Figure 7-7 (F), that is, the cable shields were grounded at both ends, while the signal return leads were grounded only at the input ends. The advantages of the latter configuration can be seen in Figure 7-10.

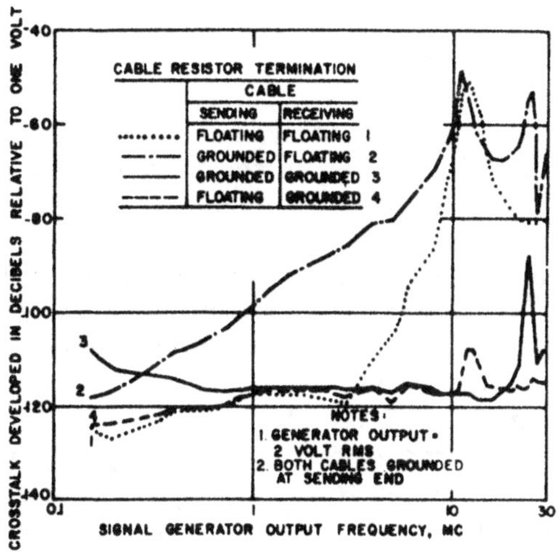

Figure 7-9

Measured RF Coupling Between Coaxial Cables

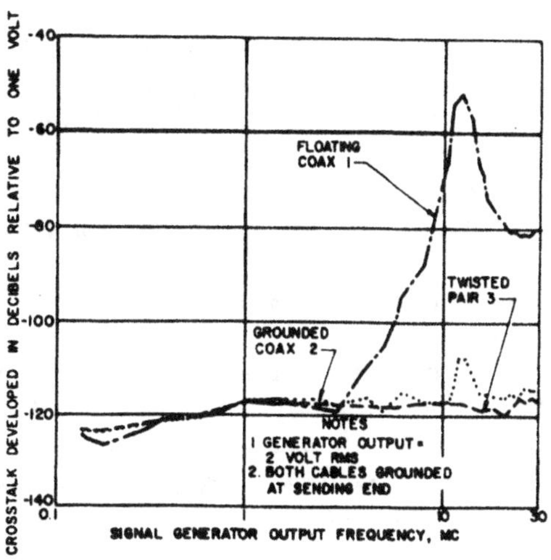

Figure 7-10

Measured RF Coupling in Shielded Cables

The performance of tests of this type indicate that, for low-level signals and low-impedance circuits where the distance between input connector and circuit input is small, i.e., an inch or two, the use of a twisted-pair alone may prove adequate. For long runs, the use of twisted-pair shielded wire becomes mandatory; this is true with unbalanced as well as with balanced output devices. In the balanced case, there should be no grounding except for the shield; in the unbalanced case, it is often necessary to ground one of the conductors as well as the shield. The shield and conductor should be grounded at the same point. Single-point grounding of the shield should be used for short runs and multiple-point grounding for long runs. If a shielded twisted-pair is part of a cable bundle and the bundle must go through a connector, three pins should be made available on the connector to provide insulated passage of the twisted-pair leads and the shield when it is not desirable to ground the shield at the connector.

Signals of high level will, in general, not be bothered by susceptibility of circuits. Rather, it may be a source of interference to lower level signal lines. For this reason, and dependent on other characteristics of the signal, either a twisted-pair or a shielded lead should be used. Multiple shielding may be required if the signal is of sufficient level. Grounding should be applied at both ends to prevent electric field radiation from the cable.

When cable shields are grounded, good electrical contact to the shield must be established. If possible, the shield should be grounded completely around the periphery of the connector shell. The use of pigtail grounding should be avoided on all cables, particularly those carrying signals above 1 MHz. When it becomes necessary to use a pigtail, the length should be minimized.

Figure 7-11 provides an indication of how the length of these ground leads can affect the coupling between two shielded cables[7].

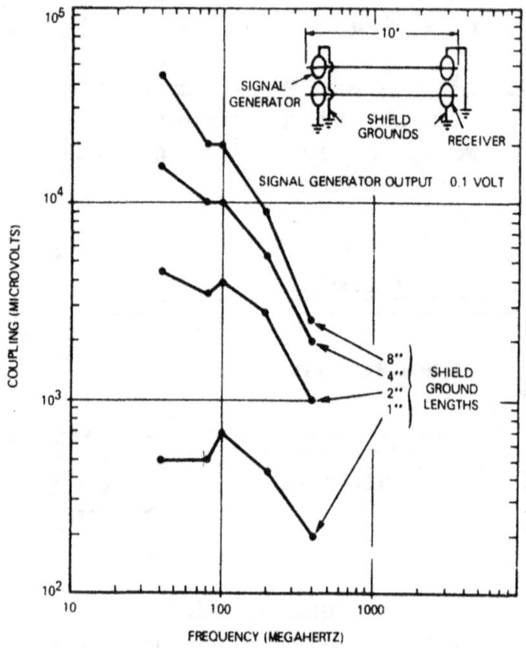

Figure 7-11

Cable Coupling Vs. Shield Ground Length

There may be situations when multiple-shielded cable usage is considered necessary. For example, one suggestion[3] has been made to use multiple shielded cable for unbalanced input circuits with sensitivity thresholds better than 50 millivolts and operating frequencies between 5 kHz and 1 MHz. When biaxial or triaxial cables are employed, single-point grounding should be used. An illustration of how to ground double-shielded cable is provided in Figure 7-12. Double or triple-shielded cable may be necessary for feeding high input or output impedance circuitry, particularly if the circuit is in a high electric field environment.

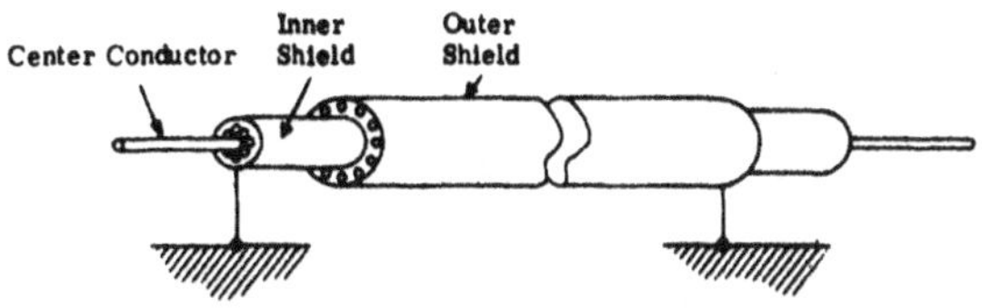

Figure 7-12

Example of Grounding a Double-Shielded Coaxial Cable

7.7 GROUNDING DESIGN GUIDELINES

Although the specific grounding philosophy to be employed on a GSE is very dependent on some of the detailed design objectives of that equipment, it is nevertheless possible to establish general grounding guidelines to follow. These are stated below, with the understanding that they should not be applied rigidly, and that alternate grounding methods should be tested before selecting a final design.

- Remember that the ground system of concern includes the combined effects of both the GSE and the equipment under test.
- Use single-point grounding when the dimensions of the circuit or component under consideration is small compared to the wavelength of concern (typically less than 0.15λ). Use multiple-point grounding when circuit or component dimensions exceed 0.15λ. When possible, ground large circuits or components at several locations, so that the separation between grounds is never greater than 0.15λ.
- There are occasions when transformer isolation and other isolation techniques can be used to prevent ground loops from occurring.
- Keep all ground leads as short and direct as possible. Avoid pigtails when terminating cables.
- It is advisable to maintain separate ground systems for signal returns, signal shield returns, power system returns, and chassis or case ground. They can be tied together at a single ground reference point.
- Ground reference planes should be designed so that they have high electrical conductivity, and so that they can be easily maintained to retain good conductivity.
- Circuits that produce large, abrupt current variations should have a separate grounding system, or should be provided with a separate return lead to the single-point ground. This will reduce transient pickup in other circuits.
- Grounds for low-level signals should be isolated from all other grounds.
- Never run supply and return leads separately, or in separate shields. A twisted pair is the best configuration for the supply bus and its return. Also, avoid carrying signal and power leads in the same bundle or in close proximity to one another. When signal and power leads must cross, make the crossing so that the wires are at right angles to each other.
- Use of differential or balanced circuitry can significantly reduce the effects of ground circuit interference.
- For circuits that operate below 1 MHz, tightly twisted pairs of wires (either shielded or unshielded, depending on application) that are single-point grounded offer the best approach to reduced equipment susceptibility.

- When coaxial cable is necessary for signal transmission, it appears that signal return through the shield and single-point grounding at the generator end (see Figure 7-7 (E)) offers certain advantages at the lower frequencies. However, other grounding arrangements should be considered. At high frequencies, multiple-point shield grounding is required.

Low-level sensitive transmission lines may require multiple shields. Single-point grounding of each shield is recommended.

GROUNDING BIBLIOGRAPHY

1. R.F. Ficchi, "The Grounding of Electronic Equipment," 8th Tri-Service Conference on Electromagnetic Compatibility, Oct. 1962, pp. 643-654.
2. U.S. Army Electronics Laboratories, Ft. Monmouth, N.J., "Interference Reduction Guide for Design Engineers," 1 August 1964.
3. National Aeronautics and Space Administration, "Radio Frequency Interference Handbook," NASA SP-3067, 1971.
4. C.B. Pearlston, "Considerations in Electromagnetic Interference," IEEE Transactions on Radio Frequency Interference, Vol. RFI-4, No. 3, Oct. 1962, pp. 1-16.
5. AFSC Design Handbook DHI-4, Electromagnetic Compatibility.
6. I.M. Newman and A.L. Albin, "An Integrated Approach to Bonding, Grounding and Cable Selection," 7th Conference on RFI Reduction and EMC, Nov. 1961.
7. Naval Air Systems Command, "Electromagnetic Compatibility Manual," NAVAIR 5335.

8.0 FILTERING

8.1 GENERAL

Even when a system has been well designed and incorporates proper shielding and grounding considerations, undesired energy can still be conducted through the system to degrade performance or cause malfunction. Filters can reduce this unwanted conducted energy to levels at which the system can function satisfactorily. Because of this role, filters are important in contributing to electromagnetic compatibility.

An electrical filter can be defined as a network of lumped or distributed constant resistors, inductors, and capacitors or their equivalent, or any combination thereof; that offers comparatively little opposition to certain frequencies or to direct current while blocking the passage of other frequencies[1]. The design of filters is an art as well as a science since much depends on the judgment and techniques used by the filter design engineer.

The purpose of this chapter is to provide the GSE design engineer with general filter design information that can be useful in the development of compatible avionics support units. However, some overall comments on EMI filtering are first considered appropriate.

First of all, it is important to point out that filters are often used only as stop-gap measures to problems that might have been resolved in a different way earlier in system design. For example, improved circuit linearity requirements might have obviated the need for a harmonic filter. Similarly, improved isolation of a relay circuit could have resolved a problem that later had to be attacked with transient suppression filters. An initial engineering effort should be made to design circuits that are inherently free of EMI. As part of this engineering effort filters should be considered to limit the magnitude of EMI currents and to confine the currents to the smallest practical physical area, but the decision to use a filter should be made at a stage of development that provides maximum choice in alternative EMC approaches.

The impetus to the establishment of equipment filtering requirements (or shielding or grounding requirements for that matter) are the formal and informal specifications imposed on the designer. Thus, formal interference specifications such as MIL-STD-461A limit the amount of conducted interference that may be introduced on a power line. Tolerable interference levels on critical internal equipment leads must be defined during an early stage of GSE design so that circuit designers know the conditions their subassemblies must meet. The ability to comply with these specification limits can then be continuously assessed in the breadboard stage, at the points when subassemblies are interfaced, in the prototype stage, etc. Discrepencies between design objectives and actual EMI levels can often be most easily resolved by adding or modifying filters, particularly as portions of the design are frozen.

While filters are necessary and should be placed where needed, care should be taken to avoid redundant filtering caused by uncoordinated efforts of separate design groups. Redundancy usually occurs when each "black box" is required to meet an interference control specification regardless of final location. Although trade-offs must be made, there is no substitute for a well thought-out system EMC control plan. If formulated well ahead of the system design, filter duplication can be avoided.

The effectiveness of any EMI filter is greatly influenced by the impedance of its source and load terminations. Manufacturers of EMI suppression filters normally specify the filter characteristics with fixed source and load impedances, usually 50 ohms. The actual characteristics may be different when used in a circuit that requires other terminations, or when used in circuits whose impedances are not fixed. This aspect must be taken into consideration when designing, specifying or using EMI filters.

The uncertainty in knowing the impedances into which a filter will work when it is installed in an equipment makes it difficult to evaluate what its effectiveness will be in a given situation. This is a particular problem with regard to power line filtering. Initial steps have been made to establish the statistics of filtering effectiveness based on measured values of line and load impedance.[20]

Certain guidelines are helpful in deciding what type of filter circuit to apply in any given instance. For example, if it is known that the filter will connect to relatively low impedances in both directions, then a circuit containing more series filter elements is indicated (a T-circuit, for instance). Conversely, a high-impedance system calls for a π-filter. If the filter is connected between two severely mismatched impedances, then an asymmetric filter circuit such as two L-section elements can be used. The series element faces the low-impedance side of the system.

The basic characteristic used to describe filter performance is its insertion loss. <u>Insertion loss</u> is defined as the ratio of voltages appearing across the system terminals immediately beyond the point of insertion of a filter, before and after insertion. It is represented as the ratio of input voltage required to obtain constant output voltage, with and without the filter in the system[2]. This ratio is expressed in decibels (dB) as follows:

$$\text{Insertion loss} = 20 \log \frac{E_1}{E_2} \qquad (8.1)$$

where:

E_1 = the output voltage of the signal source with the filter in the circuit.

E_2 = the output voltage of the signal source with the filter not in the circuit.

A number of other filter characteristics are important. The input and output impedance requirements have already been cited. Others include the attenuation in the pass-band, the skirt falloff characteristic (rate at which the filter insertion loss changes as a function of frequency), steady-state and transient voltage ratings, etc. These and other parameters will be discussed in more detail in Section 8.8 of this chapter.

8.2 FILTER DESIGN*

As indicated previously, filters are designed to attenuate at certain frequencies while permitting energy at other frequencies to pass unchanged. Reflective filters do this by using combinations of capacitances and inductances to set up a high series impedance or a low shunt impedance

* Except for ferrite filters, only lumped constant filters are considered in this design guide.

for the interfering currents . Lossy filters do this by absorbing the interference energy.

The passband of a filter is the frequency range in which there is little or no attentuation. The stopband is the frequency range in which attenuation is desired. The attenuation may vary in the stopband and is usually least near the cutoff frequency (the frequency at which a 3 dB insertion loss is obtained), rising to high values of attenuation at frequencies considerably removed from the cutoff frequency.

Filters can be grossly classified according to the relative positions of the passband and stopband in the frequency spectrum. There are four classes: low-pass, high-pass, band-pass, and band-reject, and the discussions to follow will deal with these classes. Attenuation as a function of frequency for each of the classes is shown in Figure 8-1.

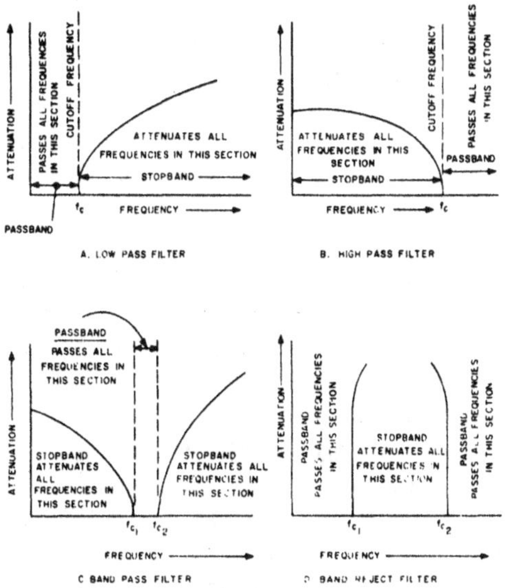

Figure 8-1

The Four Basic Classes of Filters

Design considerations for these four filter classes will next be presented. It should be pointed out that the Tables and Nomographs chapter of this design guide (Chapter 11) contains several normalized graphs and nomographs for convenience in making insertion loss calculations or defining filter parameters for selected filter types.

8.2.1 Low-Pass Filters

EMI control usually requires filters of the low-pass type. Power line filters are low-pass filters that pass DC or power frequency currents without significant power loss, while attenuating signals above these frequencies. Filters incorporated in amplifier circuits and transmitter output circuits are usually of the low-pass type so that the fundamental signal frequency can be passed while harmonics and other spurious signals are attenuated.

8.2.1.1 Shunt Capacitive Filters, and General Capacitor Characteristics

The simplest low-pass EMI filter is a shunt capacitor connected from the interference-carrying conductor to ground. It serves to bypass high-frequency energy, as indicated in the ideal representation of Figure 8-2. Under these circumstances the insertion loss of a shunt capacitor filter is defined[1, 3] by the relationship:

$$IL(dB) = 10 \log (1 + F^{\frac{1}{2}}) \qquad (8.2)$$

where

$F = \pi fRC$

f = frequency, in Hertz

R = driving or termination resistance, in ohms

C = filter capacitance, in farads

An actual capacitor incorporates both resistance and inductance. These effects are due to such factors as the foil inductance of the capacitor plates, lead inductance, foil resistance, and lead-to-foil contact resistance.

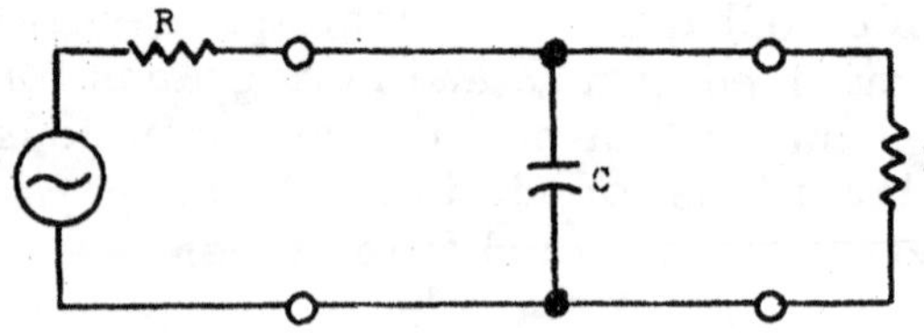

Figure 8-2

Capacitor Low-Pass Filter

The variations in these inductive and resistive effects depend upon the type of capacitor. Metalized paper capacitors, while small in physical size, offer poor RF bypass capabilities because of high resistance contact between the leads and the capacitor metal film. They are also a source of radio noise as the dielectric punctures and self-heals by burning away the metal film. This effect is indicated by the switch in the equivalent circuit shown in Figure 8-3. The standard wound aluminum foil capacitor may be employed as a radio frequency bypass in the frequency range up to 20 MHz. Its useful frequency range of operation is a function of capacitance and lead length. Its equivalent circuit is the same as that of the metallized paper capacitor, but R_S and S of Figure 8-3 are not in the circuit.

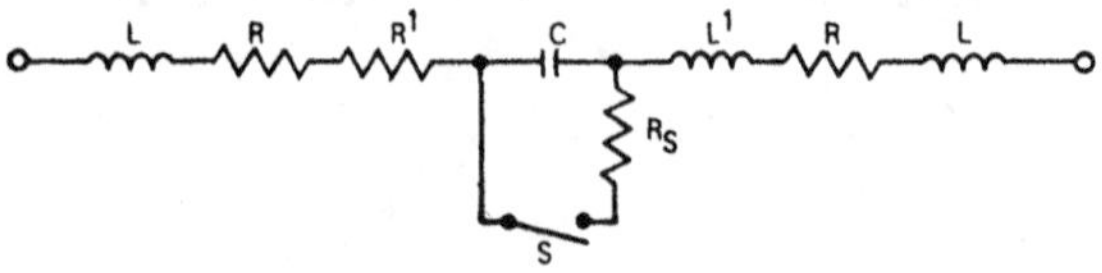

L = Lead Inductance

R = Lead-to-Foil Contact Resistance

R^1 = Resistance of Metallized Foil

C = Capacitance

L^1 = Foil Inductance

S = Short Circuit due to Voltage Puncture

R_S = Short Circuit Resistance

Figure 8-3

Metalized Capacitor Equivalent Circuit

As a result of these inductive effects, a capacitor will exhibit a capacitive reactance at low frequencies, and this situation will be maintained until its self-resonant frequency is reached. Above this frequency, the capacitor behaves like an inductive reactance. This effect is illustrated in Figure 8-4. Also note the effect of changing capacitor leadlength on this self-resonant frequency.

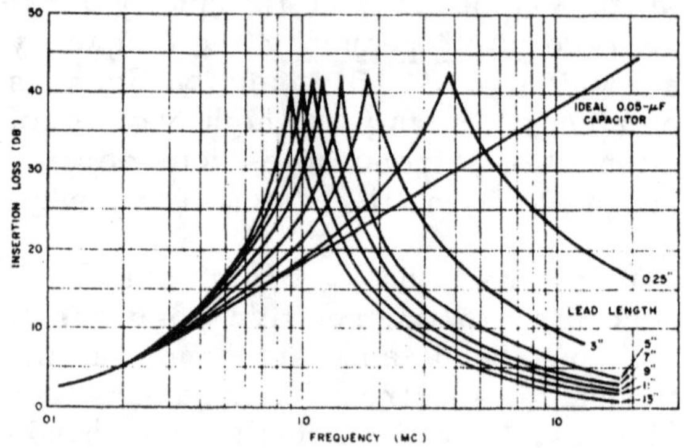

Figure 8-4

Insertion Loss of an 0.05-uf Aluminum Foil Shunt Capacitor

Mica and ceramic capacitors of small values are useful up to about 200 MHz. A capacitor of flat construction, if the capacitor plates are round as in a ceramic disc capacitor, will remain effective to higher frequencies than one of square or rectangular construction.

Other factors must be considered in selecting ceramic filter capacitors. A ceramic capacitor element is affected by operating voltage, current, frequency, age, and ambient temperature. The amount the capacity varies from its nominal value is determined by the composition of the ceramic dielectric. The dielectric composition can be adjusted to obtain a desired characteristic such as negative temperature or zero temperature coefficient, or minimum size. In obtaining one characteristic, other characteristics may become undesirable for certain applications. For example, when the dielectric composition is adjusted to produce minimum size capacitors, the voltage characteristic may become negative to the extent that 50 percent capacity exists at full operating voltage, and full ambient temperature may cause an additional sizeable reduction in capacity. Also, from the time of firing of the ceramic, the dielectric constant of the materials used may decrease; after 1000 hours, the capacitance may be as low as 75 percent of the original value. The designer should make ceramic capacitor selection based on required capacity under the most adverse operating conditions, and taking into account aging effects.

Capacitors of short-lead construction, and feed-through capacitors, are three-terminal capacitors designed to reduce inherent end lead inductances. Figure 8-5 shows the construction of these three-terminal types[4]. In each

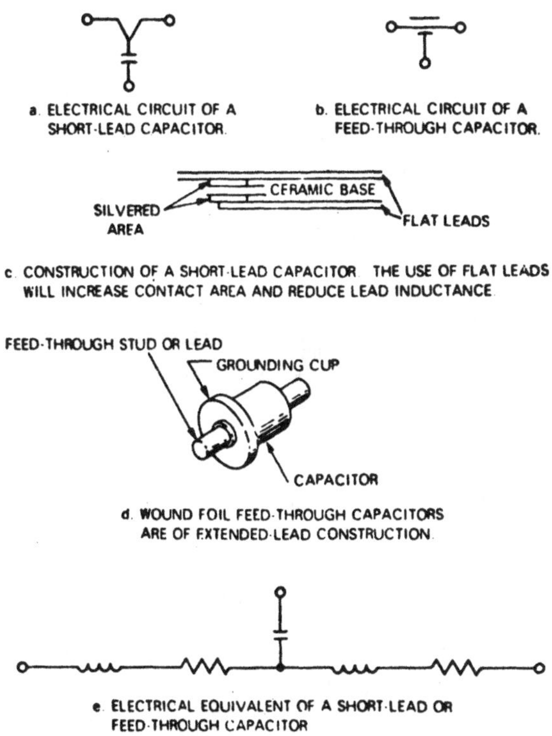

Figure 8-5

Three-Terminal Capacitor Construction

case, the inductance of the lead is not included in the shunt circuit. The wound foil short-lead capacitor is made with an extended foil type construction so that each plate of the capacitor can be soldered to a washer-shaped terminal. One washer is, in turn, soldered to the center lead, while the other is soldered to the case that is the ground terminal.

Theoretical insertion loss of three-terminal capacitors is the same as for an ideal two-terminal capacitor. However, the insertion loss of a practical three-terminal capacitor follows the ideal curve much more closely than does a two-terminal capacitor. The useful frequency range of a feed-through capacitor is improved further by its case construction, enabling a bulkhead or shield to isolate the input and output terminals from each other.

While the short-lead construction capacitor is ideally suited for EMI suppression in the frequency range of 1 to 1000 MHz, feed-through capacitors are available with a resonant frequency well above 1 GHz. The feed-through current rating is determined by the stud diameter. Figure 8-6 shows the insertion loss characteristics of typical feed-through capacitors, while Figure 8-7 indicates the construction details of a feed-through unit.

Capacitor selection for shunt capacitive filters, or any other filter application, is determined in part by the voltage, temperature, and frequency range in which the filter must operate. For 28 VDC applications, capacitors rated at 100 working volts DC (WVDC) are adequate. Metallized mylar capacitors offer the most compact design and good reliability. Their dissipation factor is very low, and lead length can generally be kept short to improve high frequency performance.

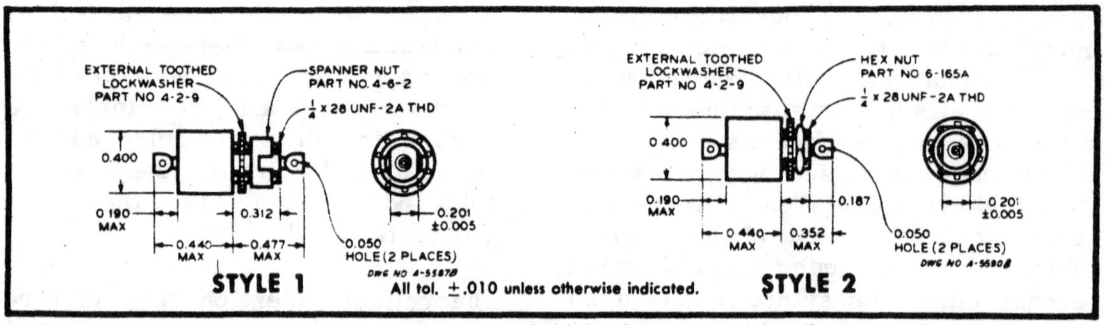

NOTE: Style 1 units are supplied with spanner nut and external toothed lockwasher, as shown above. If hex nut is preferred, add the letter "X" to the basic catalog number when ordering; i.e., 7JX2103X.

SPECIFICATIONS

Catalog No. Style 1	Catalog No. Style 2	Capacitance GMV (μF)	Max. D-C Ohms	Current (Amperes)	Voltage Rating
7JX2103	7JX2503	0.75	0.01	10.0	100VDC @ 85°C 50VDC @ 125°C
7JX2105	7JX2505	0.25	0.01	10.0	250VDC @ 85°C 150VDC @ 125°C
5JX3102	5JX3502	0.25	0.01	5.0 7.0	125VAC, 400Hz @ 125°C 125VAC, 60Hz @ 125°C

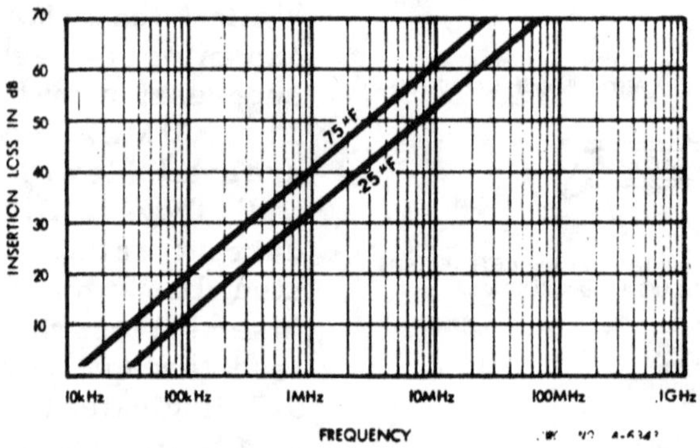

Courtesy of Sprague Electric

Figure 8-6

Typical Feed-Through Capacitor Characteristics

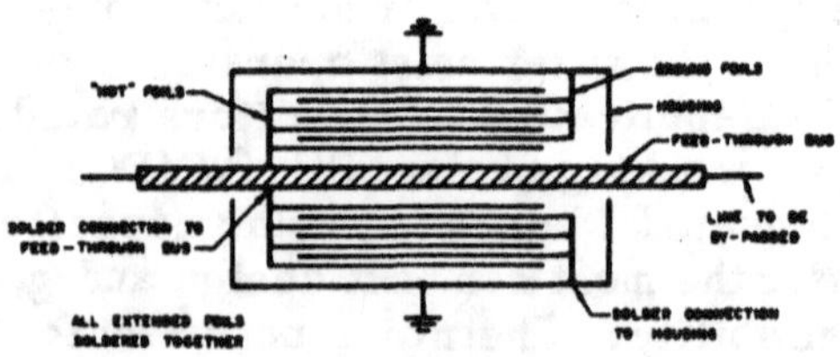

Figure 8-7

Construction of Feed-Through Capacitor

Wet-type electrolytic capacitors are used for dc filtering and sometimes used in EMI filters. They are single polarity devices, and their high dissipation factor or series resistance make them poor rf filters. An rf bypass capacitor should be placed across the output of DC supplies using electrolytics. The dissipation factors of electrolytic capacitors increase, and their capacitances decrease with age.

If a large value of capacitance is required in a small space, tantalum capacitors may be considered. Because tantalum capacitors are electrolytics, they are more sensitive to over-voltages, and are damaged by reverse polarity. The dissipation factor is considerably higher than for mylar or paper capacitors, and high frequency characteristics are poor. A large tantalum capacitor reaches its minimum impedance at 2 to 5 MHz or less, depending upon construction and capacitance value.

Capacitors for 120 VAC applications should be rated at 400 WVDC and be suitable for AC use. A unit of mylar and foil or of paper-mylar and foil is recommended. Dissipation factor is low and high frequency performance is good. For 240 VAC applications, an oil-impregnated paper and foil unit is recommended.

If good capacitor performance is to be expected above about 50 MHz, it is necessary to use designs incorporating feed-through techniques. As noted previously, lead inductance in a feed-through capacitor is not part of the shunt circuit, so that, compared to capacitors with leads, its insertion loss is not degraded as rapidly with increase in frequency.

8.2.1.2 Series Inductive Filters, and General Inductor Characteristics

Another simple form of low-pass filter is an inductor connected in series with the interference carrying conductor. It is ideally represented in Figure 8-8. In practice, its insertion loss can be defined[1, 3] by the relationship:

$$IL(dB) = 10 \log (1 + F^{\frac{1}{2}}) \qquad (8.3)$$

where

$$F = \pi f \frac{L}{R}$$

L = filter inductance, in henries

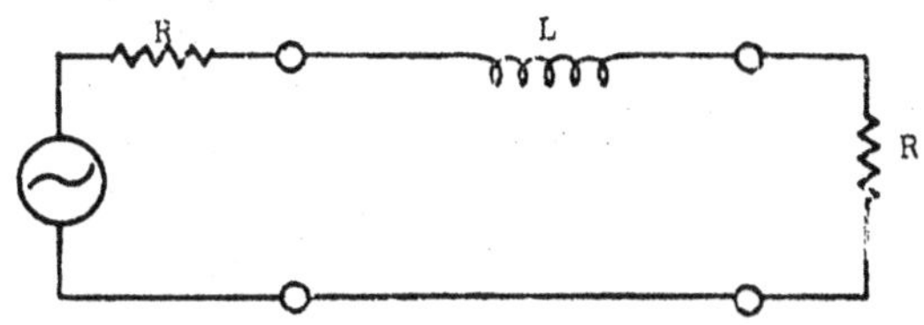

Figure 8-8

Inductor Low-Pass Filter

In practice, an inductor exhibits inductive reactance only until its self-resonant frequency is reached. Above self-resonance, it appears as a capacitive reactance, with the inter-winding capacitance becoming dominant.

Filter inductors are usually toroidal, wound on cores of powdered iron, molybdenum permalloy, or ferrite material. The size of the core is determined by required inductance and current rating. The magnetic flux (number of turns multiplied by the peak current) should not drive the core to more than 50 percent of magnetic saturation.

The choice of core materials is determined by operating frequency and current rating. Powdered iron cores can be used for all DC applications and for most 60 Hz applications. For high current 60 Hz devices, and for all 400 Hz applications, molybdenum permalloy cores should be used. Ferrite materials can be considered when the current that will flow through the inductor will not saturate the core.

Stray or distributed capacitance in a filter inductor has two detrimental effects: EMI may be coupled from input to output of the filter via the capacitance; and the capacitance may cause the filter to become self-resonant at one or more critical frequencies. Windings should be placed on the coil so that input and output turns are separated as much as possible to keep stray capacitance low. Distributed capacitance effects may be reduced by a careful arrangement of turns. In some cases, two or more coils wound on separate cores are connected in series to raise the self-resonant frequency.

Coil loss resistance is a measure of all power losses, hysteresis losses, and frequency-dependent absorption losses in the core. Loss resistance increases with frequency because of skin effect in the conductor, and due to changes in core loss with frequency. An increase in loss resistance represents an increase in attenuation in the filter passband. Losses in the core are not particularly detrimental, except that the insertion loss in the passband must be kept low.

8.2.1.3 Low-Pass "L" Section Filters

A primary disadvantage of single element filters is that their out-of-band falloff rate is only 6 dB per frequency octave (20 dB per decade). By combining both a shunt capacitor and a series inductance single element filter into an "L" configuration, a falloff rate of 12 dB/octave can be obtained.

The two possible representations of low-pass "L" section filters are shown in Figure 8-9. In one representation the capacitor shunts the source impedance, while in the other the capacitor shunts the load impedance. The insertion loss for the "L" section filter is independent of the direction of inserting the "L" section into the line, if source and load impedances are equal. When source and load impedance are not equal, the greatest insertion loss will usually be achieved when the capacitor shunts the higher impedance.

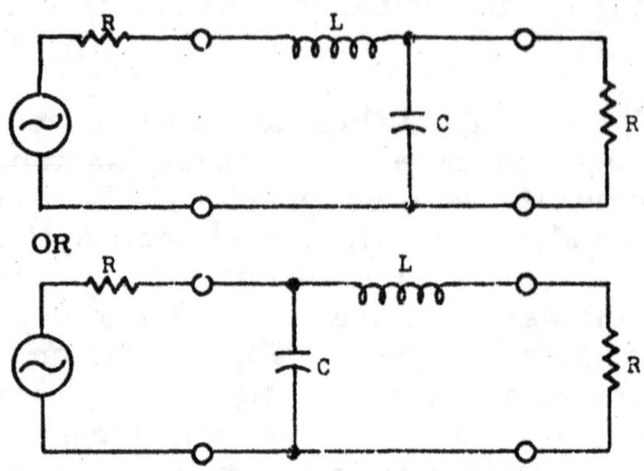

Figure 8-9
Low-Pass "L" Section Filter

The insertion loss of an "L" section lumped-constant network into 50 ohm resistance source and load impedances is[3]:

$$IL(dB) = 10 \log (1 + F^2D^2/2 + F^4) \qquad (8.4)$$

where

$$D = \frac{1 - d}{\sqrt{d}}$$

$$d = L/CR^2 = \text{damping ratio}$$

$$w_o = \sqrt{2}R/L = \sqrt{2}/RC \text{ (if } d = 1)$$

$$w_o = \sqrt{2/LC} \text{ (if } d \neq 1)$$

$$F = \frac{w}{w_o}$$

$$w = 2\pi f = \text{angular frequency}$$

The "damping ratio", d, relates the magnitudes of the filter elements to the magnitude of the source and load impedance. It is defined so that setting d equal to one (ideal damping) results in the elimination of the squared frequency term from the insertion loss equation and produces an abrupt transition from the pass-band to the stop region. The equations for Butterworth filter designs are obtained when d is set equal to one.

Values of d less than one result in insertion loss curves identical to those obtained when d is greater than one. That is:

$$\text{IL (for } d = n) = \text{IL (for } d = 1/n) \quad (8.5)$$

for two element filters. The insertion loss of a two-element filter is not changed when it is "turned around" so that the source and load terminals are transposed, so long as the source and load impedances are equal.

The physical size of an "L" section filter depends upon insertion loss requirement, current rating, and voltage rating, with the first two usually predominant. The "L" section type of filter may give poor high frequency attenuation because of stray inter-turn capacitance. In some cases, the "L" type may resonate and oscillate when excited by transients.

Figure 8-10 provides an example of a commercially available "L"-section low-pass filter. This type of configuration enables a unit to be manufactured that can maintain an adequate rejection level to 1 GHz.

BUTTON FILTERS

Courtesy of USCC/Centralab,
Electronics Div.
Globe-Union, Inc.

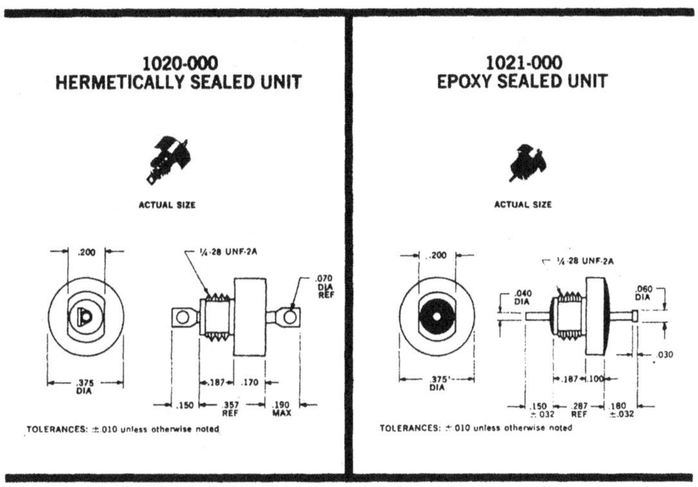

SPECIFICATIONS — Meet or exceed all applicable requirements of MIL-F-15733

Working Voltage — 50 VDC (−55°C to +125°C)

Current — 15 amps DC maximum

Dielectric Withstanding Voltage — 2 times rated voltage, with 50 ma maximum charging current

Insulation Resistance — 100 megohms minimum with 50 VDC, 50 ma maximum charging current

DC Resistance — .006 ohms maximum

Housing — Tin plated brass (standard)

Insertion Loss per MIL-STD-220, full load —

Frequency	db
30 kHz	15
150 kHz	28
300 kHz	33
1 MHz	44
10 MHz	60
1 GHz	70

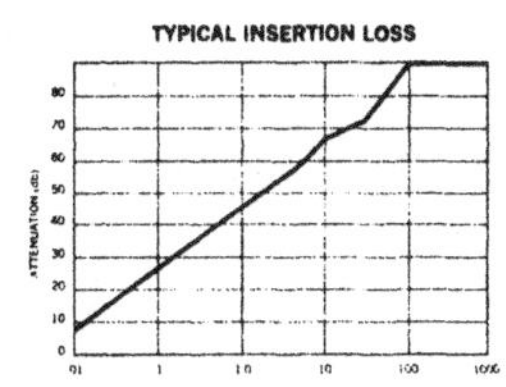

Schematic —

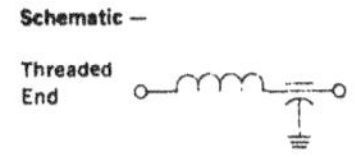

Figure 8-10

Example of Commercial "L" Section Filter

8.2.1.4 π-Section Filters

The "pi" section filter is the most common type of radio frequency interference suppression network. Figure 8-11 shows the circuit of the "pi" section filter. Advantages are ease of manufacture, high insertion loss over a wide frequency range, and moderate space requirements. Although voltage rating must be considered, current rating and attenuation are the most important factors in determining the size of the filter.

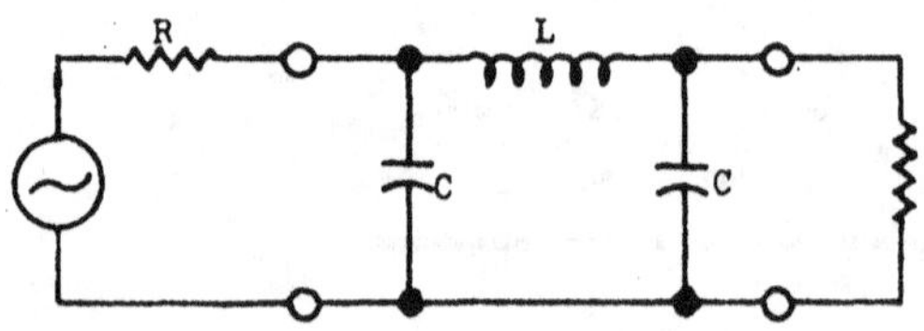

Figure 8-11

Low-Pass π Filter

The insertion loss of a lossless "pi" section network operating with equal source and load impedances is[3]:

$$IL\ (dB) = 10 \log (1 + F^2D^2 - 2F^4D + F^6) \quad (8.6)$$

where

$$D = \frac{1 - d}{\sqrt[3]{d}}$$

$d = L/2CR^2$ = damping factor

$$w_o = \sqrt{2/LC} = 2R/L = \frac{1}{RC} \text{ (if } d = 1)$$

$$w_o = \sqrt[3]{2/RLC^2} \text{ (if } d \neq 1)$$

$$F = \frac{w}{w_o}$$

$w = 2\pi f$ = angular frequency

Unlike the "L" section filter case, overdamping or underdamping of π-section (and T-section) filters result in entirely different affects. This is discussed further in Section 8.2.1.6 of this handbook.

A typical attenuation curve of a "pi" section filter has a slope of approximately 18 dB per octave; the high frequency performance can be improved by internal shielding within the filter case. However, the "pi" circuit is very susceptible to oscillatory ringing when excited by a transient. Representative data on a series of commercial low-pass "π" filters are shown in Figure 8-12.

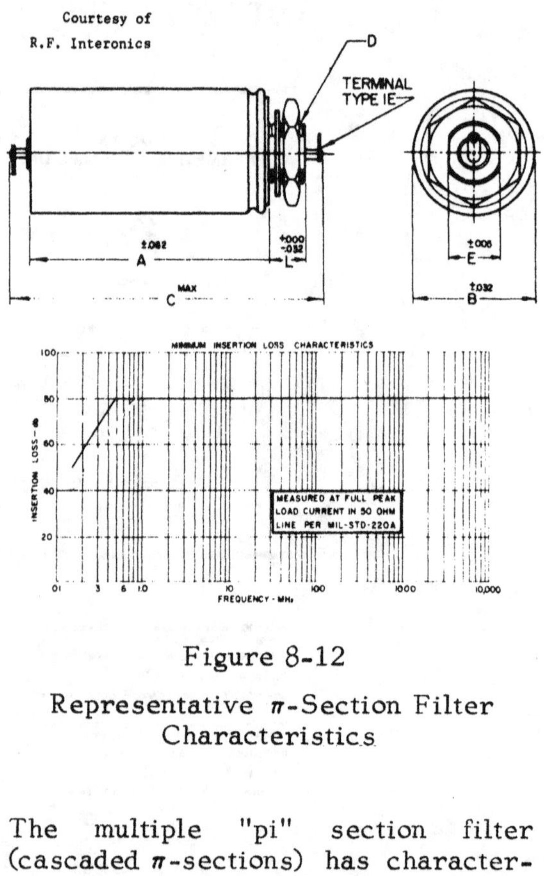

Figure 8-12

Representative π-Section Filter Characteristics

The multiple "pi" section filter (cascaded π-sections) has characteristics identical to those of the multiple "L" section filter. The attenuation curve of the theoretical multiple "pi" section filter rises at a rate of 20 dB more per decade of frequency than does a multiple "L" filter of the same number of sections. Although this may not be a significant factor when three or more sections are used, it does provide a capacitive input at both ends of the filter that is sometimes advantageous. An extensive use for this type of network is as a power-line filter in large installations, and for shielded enclosures where high attenuation is needed at very low frequencies.

8.2.1.5 "T" Section Filters

The "L" type low-pass filter can also be improved by the introduction of another series inductor. This addition forms a "T" section filter, which consists of two inductors in series with a shunt capacitor connected from the junction of the two inductors to ground (see Figure 8-13). Insertion loss is given by:

$$IL\ (dB) = 10 \log (1 + F^2D^2 - 2F^4D + F^6) \quad (8.7)$$

where

$$D = \frac{1 - d}{\sqrt[3]{d}}$$

$$d = R^2C/2L = \text{damping factor}$$

$$w_o = \sqrt{\frac{2}{LC}} = \frac{R}{L} = \frac{2}{RC} \quad (\text{if } d = 1)$$

$$w_o = \sqrt[3]{2R/L^2C} \quad (\text{if } d \neq 1)$$

$$F = \frac{w}{w_o}$$

$$w = 2\pi f$$

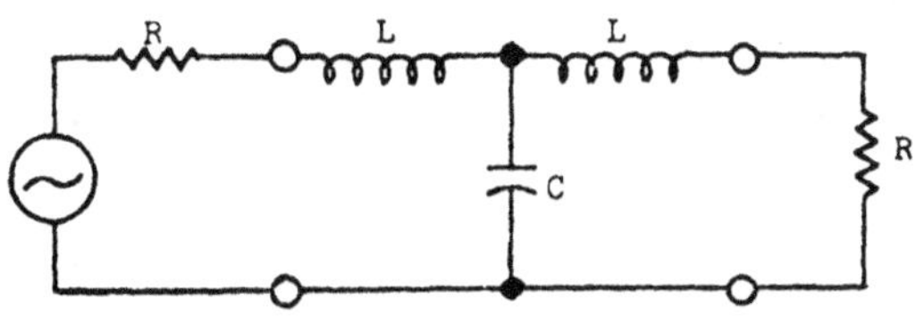

Figure 8-13

Low-Pass "T" Filter

The "T" type of filter is a very effective form of the lumped-constant type of filter for reducing transient interference. Its major disadvantage is the requirement for two inductors, which under some circumstances may present a size penalty. It provides the same out-of-band falloff rate as the π-section filter, that is, 18 dB/octave (60 dB/decade) for a single section.

Figure 8-14 provides representative information on commercially available T-section low-pass filters.

8.2.1.6 Insertion Loss Calculations for "π" and "T" Section Filters

The equations for the insertion loss of a T-circuit and a pi-circuit as given by Equations (8.6) and (8.7) are of the same form, but differ with respect to the definition of the equation parameters. The equation has three modes of response. When d equals one, the response is optimally damped and is the ideal (Butterworth) response curve. When d is greater than one, the response is in an overdamped mode. When d is less than one, the response is in an underdamped mode. In the underdamped case, the curve has a maximum in band loss of:

$$IL = 10 \log \left(1 + \frac{4D^3}{27}\right) \quad (8.8)$$

at the frequency where $F = D/3$. A minimum loss point will also occur at the frequency where $F = D$.

8.2.1.7 Lossy Line Filters

While the input and output impedances of some filters can be expected to match their intended source and load impedances over a fairly broad frequency range, it is more often the case that such matches will not occur. For

RATINGS

PART NO.	CURRENT AMPS	DCR AT 25°C OHMS	INSERTION LOSS CURVE	L1 MAX INCHES	L2 MAX INCHES
GF51F3A	0.10	3.20	A	.99	.94
GF51F3B	0.50	.66	B	.99	.94
GF51F3C	1	.36	C	.99	.94
GF51F3D	5	.022	D	.99	.94

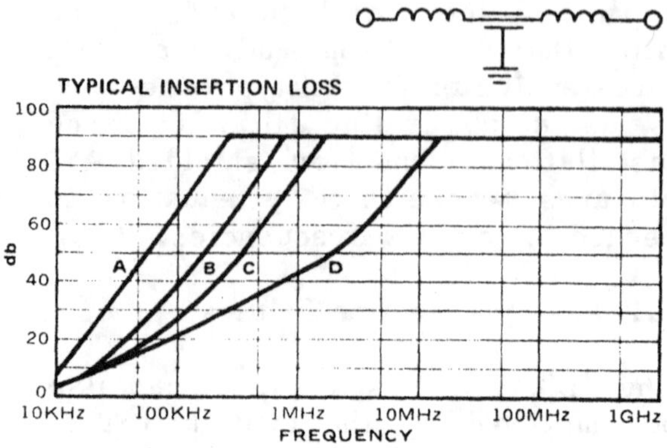

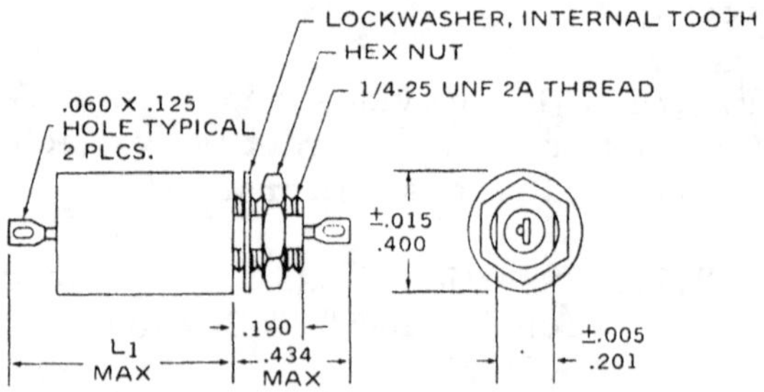

1. Tolerance: ±0.010 unless otherwise specified.
2. Case material and finish: steel, gold plated (optional).
3. Case marked with Genisco G and part number.
4. Typical weight: 5 to 12 grams.
5. Recommended torque during installation: 40 in. oz.
6. Each filter is supplied complete with lock washer and nut.

Courtesy of Genisco Technology Corp.

Figure 8-14

Representative Commercial Low-Pass T-Section Filter Characteristics

example, the input impedance of a powerline filter almost never achieves a match with the impedance of its associated power line. As another example, a transmitter harmonic filter is generally designed to match the transmitter output stage over the fundamental frequency range, but not necessarily at its harmonic frequencies.

Because of such mismatch situations, there have been many cases when the insertion of a filter into a line carrying interference has actually resulted in more, rather than less, interference voltage appearing on the line beyond the point of its application[3]. This deficiency in all filters composed of low loss elements has led to the development of dissipative filters that take advantage of the loss-versus-frequency characteristics of magnetic materials such as ferrites.

One form of dissipative filter uses a short length of ferrite tube with conducting silver coatings deposited on the inner and outer surfaces to form the conductors of a coaxial transmission line. The line becomes extremely lossy at radio frequencies, that is, it has high attenuation per unit length in the frequency range where either electric or magnetic losses, or both, become large. An example of the performance of a lossy line filter of this type is shown in Figure 8-15.

Dissipative filters of this type are necessarily low-pass. One of the large uses of such filters is in general-purpose power-line filtering, in which the dissipative filter is combined with conventional low loss elements to obtain the necessary low cutoff frequency.

Another method of achieving a dissipative filter is by use of lossy beads. Tubular ferrite toroids offer a simple, economical method for attenuating unwanted high frequency noise or oscillations. One bead slipped over a wire produces a single-turn RF choke that possesses low impedance at low frequencies and moderately high impedance over a wide high frequency band.

Ferrites are inert ceramics containing granulated iron compounds. They are free of any organic substances, and are not degraded by most environments. Generally, their inductance is small. Because of the high resistivity of ferrite beads, they may be considered insulators for most applications.

The presence of a ferrite bead on the wire causes a local increase of series impedance (largely resistive) presented to currents in the wire. Figure 8-16

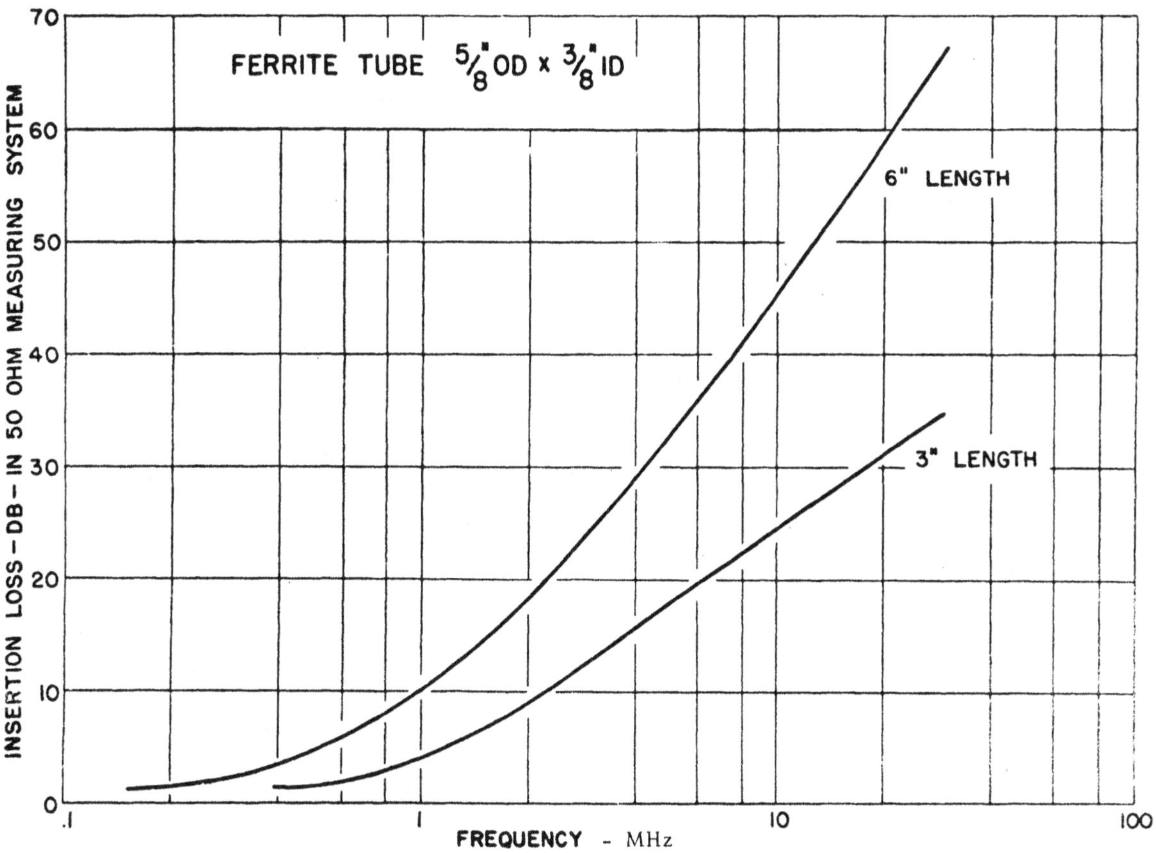

Figure 8-15

Insertion Loss of a Ferrite Tube Low-Pass EMI Filter

illustrates the effects of one ferrite bead on a length of wire[4]. Adding more or longer beads provides additional units of series inductance and resistance in direct proportion. Extra turns of wire can be passed through the bead, increasing both resistance and inductance in proportion to the square of the number of turns. Because of distributed winding capacitance, this technique is most effective at the lower frequencies.

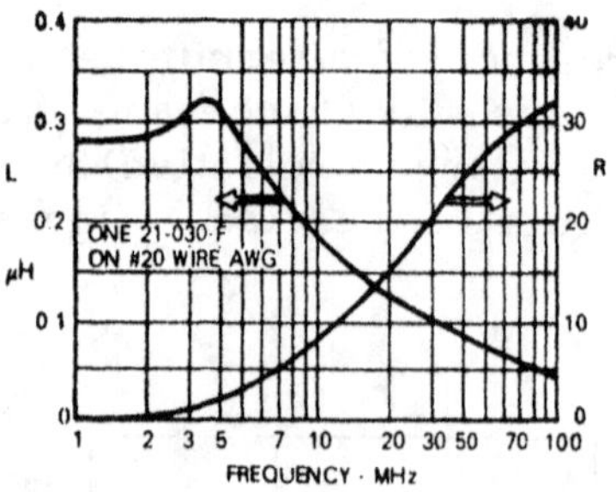

Typical values of equivalent series resistance and inductance attributable to ferrite beads.

TYPICAL PROPERTIES	
Flux Density (B) at 5 O_e	2400 G
Coercive Force (Hc)	0.56 O_e
Hysteresis Factor (h/μ^2)	22×10^{-6}
Initial Permeability (μo)	450
Permeability (μ) at 250G	900
Resistivity - Ohm - cm	$\geq 10^7$
Curie Temperature	$\geq$ 155 °C

Figure 8-16

Filter Characteristics of Ferrite Beads

High amplitude signals below 50 MHz may cause some reduction in the suppression effect due to ferrite saturation. However, as long as only one turn links the core, fairly high currents can be tolerated using representative materials before saturation is approached. At saturation, inductance and resistance will be low, but will return to normal values upon removal of the high field.

Lossy line suppressant tubing also provides an efficient means for suppressing undesired EMI and other spurious signals. The tubing can be slipped over standard wire and cable and suppresses both radiated and conducted energy. It can be used in environments from -55° to +250° C without electrical or mechanical degradation. The tubing will provide shielding from low frequency electrostatic interference and magnetic fields, and will not cause dc or low frequency ac losses. RF power handling capability is in excess of 10 watts/inch of tubing. Representative data on this type of tubing is shown in Figure 8-17.

SIZES AVAILABLE

Lundy Part No.	Inside Diameter (Inches)	Outside Diameter (Inches)	Tolerance on I.D. & O.D. (Inches)
LST-060	060	150	±.004
LST-080	080	170	±.005
LST-100	100	200	± 006
LST-125	.125	225	±.006
LST-150	.150	250	± 007
LST-175	175	300	±.008
LST-200	.200	.325	± 008
LST-225	225	.350	± 010
LST-250	250	375	± 010
LST-275	.275	400	± 010
LST-300	300	.450	± 010
LST-325	325	475	± 010
LST-350	350	500	± 010
LST-375	375	525	± 010
LST-400	400	550	± 010
LST-425	425	575	± 010
LST-450	450	.600	± 010
LST-475	470	650	± 010
LST-500	500	700	± 010

ATTENUATION CURVES

ATTENUATION VS LENGTH
MEASURED IN A 50-OHM SYSTEM PER MIL-STD-220A

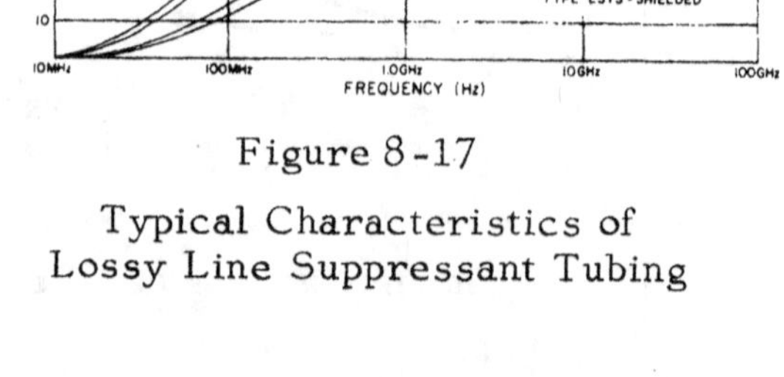

Figure 8-17

Typical Characteristics of Lossy Line Suppressant Tubing

Lossy line coaxial cable can also serve as a dissipative low pass filter. Data on such cables are provided in Figure 8-18. The lossy media have high rf attenuation and low Q. Short lengths of this cable (perhaps only a few inches long) may be adequate for achieving the desired suppression.

Still another form of ferrite filter that extends the ferrite bead concept is the filtering connector. Lossy filters are built directly into a male connector assembly, and offer low-pass filter performance as shown in Figure 8-19.

Improvement of high frequency rejection characteristics of a conventional low-pass filter may be obtained by employing a conventional reactive filter in cascade with a lossy line section. This arrangement can provide

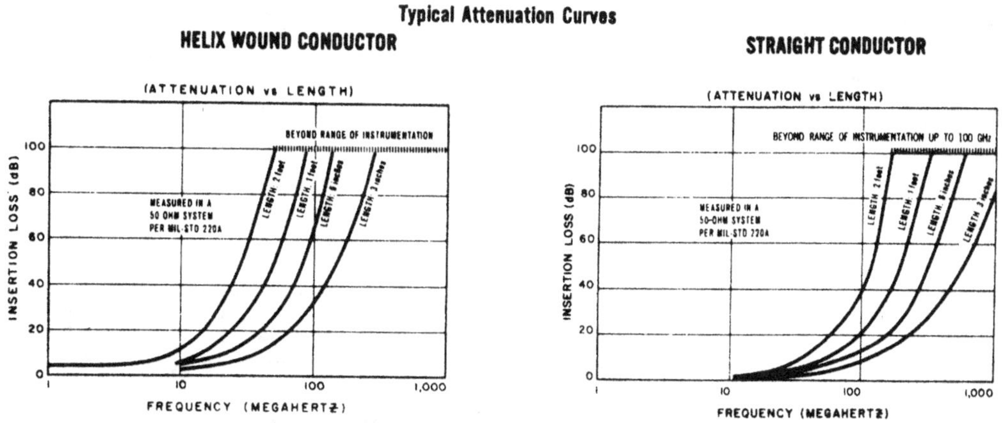

* Note: Attenuation characteristics vary slightly with different voltage ratings and shielding.

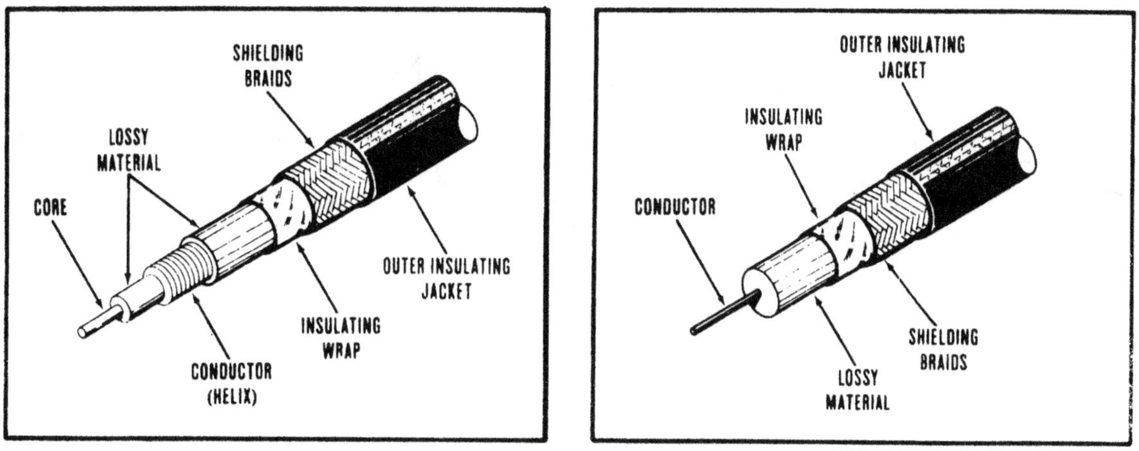

Basic Construction Features of Helix Conductor Flexible Filter

Basic Construction Features of Single Conductor Flexible Filter

Courtesy of Lundy Electronics & Systems

Figure 8-18

Typical Characteristics of Lossy Coaxial Cable

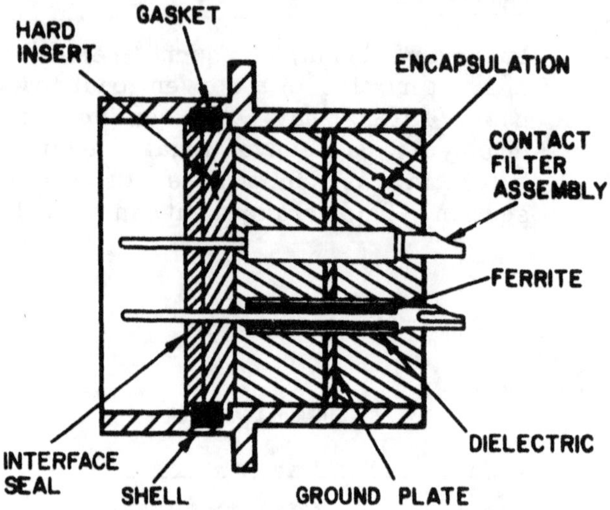

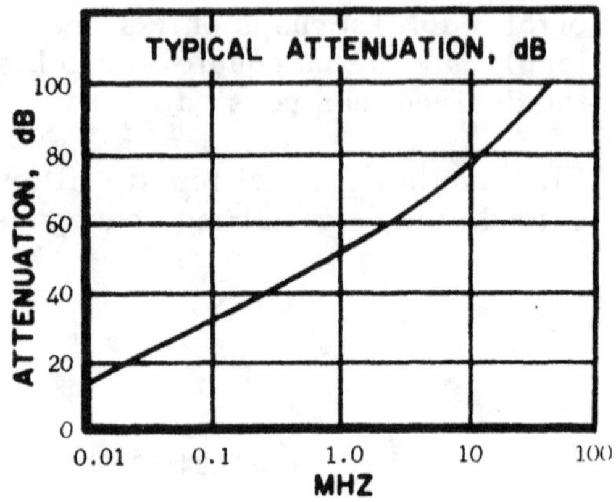

Minimum Attenuation from −55°C to +125°C and 100 MHz to 10 GHz	80 dB
D.C. Working Voltage (includes summation of the D.C. and low level A.C. superimposed peak voltages)	50 VDC
Dielectric Strength (for 5 sec. with charging current of 50 milliamperes maximum)	100 VDC
Feed Through Current (Nominal) D.C. and/or audio frequency RMS	7.5 Amperes
Insulation Resistance	250 Megohms
R.F. Current	0.25 Amperes
Operating Temperature Range	-55 C to + 125 C
Capacitance (μf)	1μf Nominal

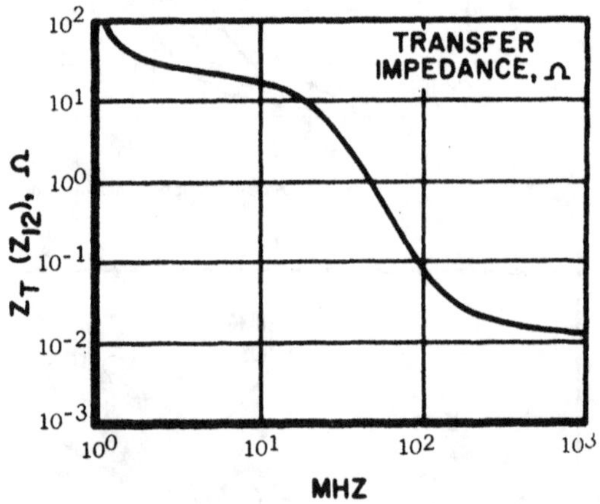

Figure 8-19

Typical Characteristics of Lossy Connector

an overall characteristic having both a rapid cutoff slope and a high-stop band attenuation. An example of the improvement in stop-band attenuation that can be gained by preceding a reactive filter with a lossy line section is illustrated in Figure 8-20.

Figure 8-20a shows the performance of a reactive low-pass filter constructed with lumped constant elements. The rapid cutoff at 400 MHz is followed by a high attenuation region between 400 MHz and 3 GHz, but at frequencies above 3 GHz the attenuation is greatly reduced. If the same low-pass filter is preceded by a section of coaxial line whose dielectric space is filled with a 6:1 ratio of iron-to-epoxy dielectric material, the attenuation characteristic is altered to that shown in Figure 8-20b. The addition of the lossy section has increased the passband attenuation only slightly, but the stop band attenuation has been increased to greater than 60 dB.

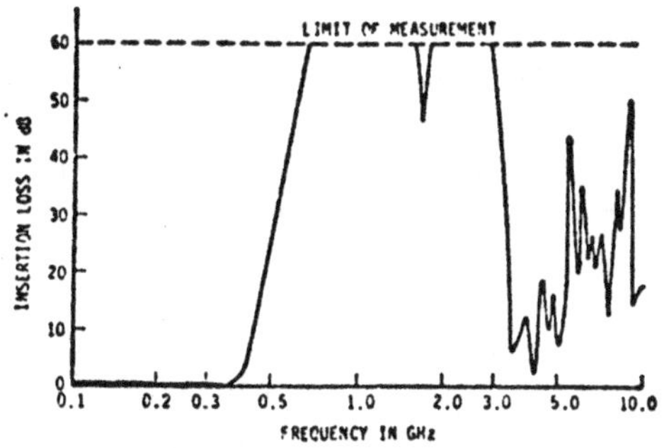

Figure 8-20a

Typical Low-Pass Filter Loss Characteristics, Low-Pass Filter Only

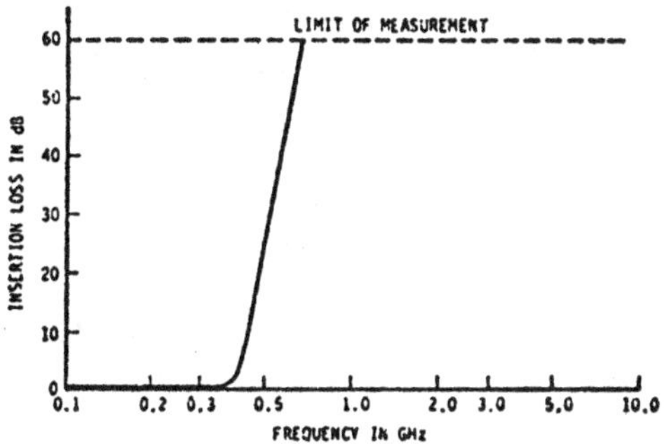

Figure 8-20b

Typical Low-Pass Filter Loss Characteristics, Plus Lossy Filter Section

When a lossy line section is used in cascade with a conventional low-pass filter, the passband insertion loss can be minimized by the proper choice of the dielectric material. However, there is always some passband loss introduced by the lossy dielectric. Such passband losses can be reduced by designing the reactive filter to have as wide a region as possible between the low-pass cutoff frequency and the first spurious passband, so that a minimum of lossy material is needed to provide the required stopband attenuation.

8.2.2 High Pass Filters

Although not as common as the low-pass type, high-pass filters also have an application in EMI reduction. In particular, such filters have been used to remove ac power line frequencies from signal channels and to reject particular lower frequency environmental signals.

Highpass filters can be designed by inverting the high pass filter response requirements, so that they become requirements on a low-pass filter[5]. Low-pass filters meeting this new requirement can be readily transformed back into the highpass filter of interest.

The low-pass filter transforms into a high-pass filter with the same impedance terminations and cutoff frequencies by replacing each coil with a capacitor, and vice versa, and by replacing the element values by their reciprocals. Thus, 2 Henries become 0.5 Farad, 10 Farads become 0.1 Henry, etc. The attenuation given by the low-pass filter at w, is now given by the high-pass filter at 1/w.

For example, a Butterworth low-pass, π-section has the element values shown in Figure 8-21. The cutoff frequency is 10 kHz. The filter is shown transformed into a high-pass filter with the same input and output impedances, and the same cutoff frequency. A similar transform relative to a T-section filter is also provided.

A transformation can be made from a low-pass filter to a high-pass filter having different impedance terminations and/or cutoff frequencies by applying scaling factors during the process of making the transformation. The scaling operations, if used before taking the reciprocals, are as follows:

$$L(\text{scaled}) = L(\text{original}) \frac{R(\text{scaled})}{R(\text{original})} \cdot \frac{W(\text{original})}{W(\text{scaled})}$$

$$C(\text{scaled}) = C(\text{original}) \frac{R(\text{original})}{R(\text{scaled})} \cdot \frac{W(\text{original})}{W(\text{scaled})}$$

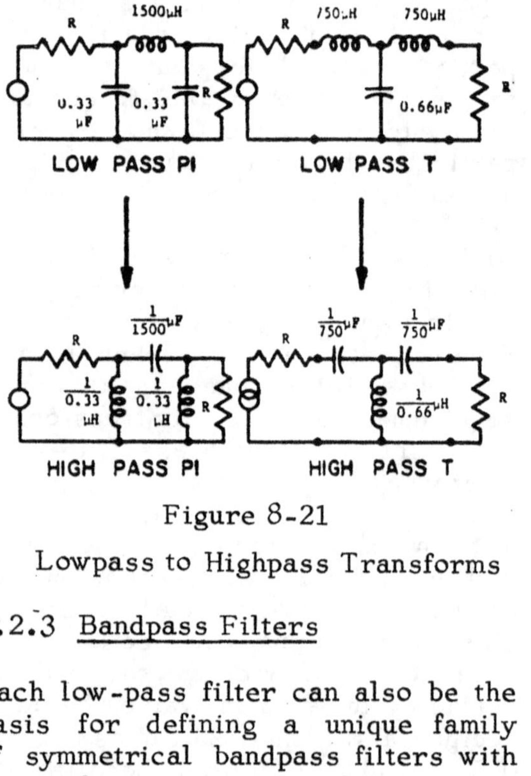

Figure 8-21

Lowpass to Highpass Transforms

8.2.3 Bandpass Filters

Each low-pass filter can also be the basis for defining a unique family of symmetrical bandpass filters with known characteristics, and vice versa. Thus, the requirements of a bandpass filter (a filter designed to pass an arbitrary frequency band and reject signals outside that band) can be readily established by use of the following transformation procedure:

a. Convert the desired bandpass filter requirements into low-pass filter requirements. The low-pass prototype has the same 3 dB bandwidth and insertion loss as the bandpass filter.

b. In the case of single section L, T and π filters having 50 ohm input and output impedances, select a low-pass filter with the required attenuation using the two and three element filter design equations discussed in this handbook.

c. Establish the filter element values in the manner previously described using the rf 3 dB bandwidth value.

d. Resonate each L and C at the required bandpass center frequency.

As an example of this procedure, consider a bandpass filter requirement of a center-tuned frequency at 1.0 MHz, and a skirt roll-off rate of at least 15 dB/octave. The required impedance level is 50 ohms input and output. There is to be no ripple in the pass-band; that is, response should at all points be monotonic. Bandwidth is to be 100 kHz between the -3 dB points.

A three element, Butterworth, low-pass pi-network is selected as the prototype low-pass filter. From Equation (8.6), it is found that the L and C values for such a filter are 160 μH and 0.03 μF, using a cutoff frequency of 100 kHz and a damping factor of unity.

Each of the above components are next resonated at 1.0 MHz using the relationship

$$f = \frac{1}{2\pi\sqrt{LC}} \tag{8.9}$$

The result is a 150 pF capacitor in parallel with L, and a 0.8 μH inductor in series with C. The final filter configuration is shown in Figure 8-22.

Note that the rf filter response is log-frequency symmetrical. That means that the response on a logarithmic frequency axis at some displacement above f_o is a mirror image of the equivalent displacement below f_o. Alternatively if the attenuation at a frequency xf_o, is N dB, then the attenuation will be the same at f_o/x. For the example above, the band-pass filter cutoff frequencies are shown in Figure 8-22 as 0.95 and 1.05 MHz, since the logarithmic effect is not evident at frequencies close to the tuned frequency of the filter.

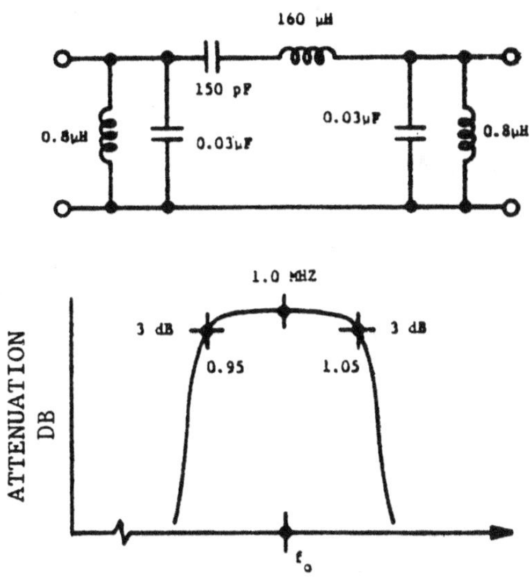

Figure 8-22

Example of Band-Pass Filter Design

Transformation is thus the essential principle in the design of bandpass filters. It reduces the design to a procedure of specifying the element values of a low-pass filter section and transforming this section into a bandpass filter. In this transformation, a capacitor is added across each coil of a size to resonate the coil at f_o. A coil is added in series with each capacitor of the low-pass filter of a size to resonate the capacitor at f_o.

The above approach is conceptually the same for multiple-section bandpass filters, or for filters whose input and output impedances differ, but the process becomes much more complicated. Tabular techniques are available to simplify the process under these circumstances. They are based on first converting the filter requirements to per-unit values (per cycle, per ohm of input impedance, etc.) designing the low-pass filter on that basis, and then converting back to the bandpass equivalent.

Butterworth filters have a maximally flat bandpass response. If some ripple within the pass-band can be tolerated, then a steeper descent into the attenuation band can be obtained. Tchebyscheff filters have a greater roll-off rate than do Butterworth filters for the same number of components, and are generally used in bandpass designs where bandpass ripple can be tolerated. Tabular approaches to the design of Tchebyscheff bandpass filters are also available[5,6].

8.2.4 Band-Rejection Filters

Band-rejection and notch filters are networks that, from an EMC standpoint, are designed to attenuate a specific narrow band of frequencies that may be causing interference problems. This type of device is normally used as a series rejection device between the interference source and the load. An alternative is to use a bandpass configuration that shunts the interference to ground.

Typical applications and locations of band-rejection notch filters include the following:

- At input terminals to reject strong out-of-band interference that would otherwise cause overload.
- At receiver input terminals to reject troublesome image frequencies.
- At receiver input terminals to reject

IF-feedthrough signals.

- At transmitter output or interstage terminals to reject harmonics.
- In ac or dc power distribution leads to reject radar PRF, computer-clock surges, or rectifier ripple.
- At audio amplifier input or interstage terminals to reject IF or BFO feedthrough, unwanted heterodynes, signal tones, radar PRF.

A notch filter or wavetrap may take the form of a lumped-constant inductor-capacitor circuit, or it may be a shorted quarter wave coaxial or waveguide stub, or a crystal or ceramic filter lattice. The inductive characteristics of capacitor leads and foil can be planned so that the capacitor acts as a self-contained wavetrap. For frequencies below about 1 MHz, a twin-T resistor-capacitor filter is often found to be an acceptable configuration.

The simplest types of wavetrap are a parallel or series resonant circuit such as those shown in (A) and (B) of Figure 8-23. The configuration of Figure 8-23(A) will give a high impedance at the resonant frequency, while the configuration of Figure 8-23(B) provides a low impedance at resonance. The disadvantages of these circuits are that their skirt falloff rates are low (6 dB/octave), and they do not present a good impedance match to either the source or load. Band reject performance can be improved by using parallel and series tuned elements in L, π or T configurations, as also illustrated in Figure 8-23.

The details on the design of the above types of band-reject filters are available in many filter textbooks and handbooks, and will not be discussed here. However, some additional comments are considered appropriate on one particular type of notch filter, because of its wide use in this type of application.

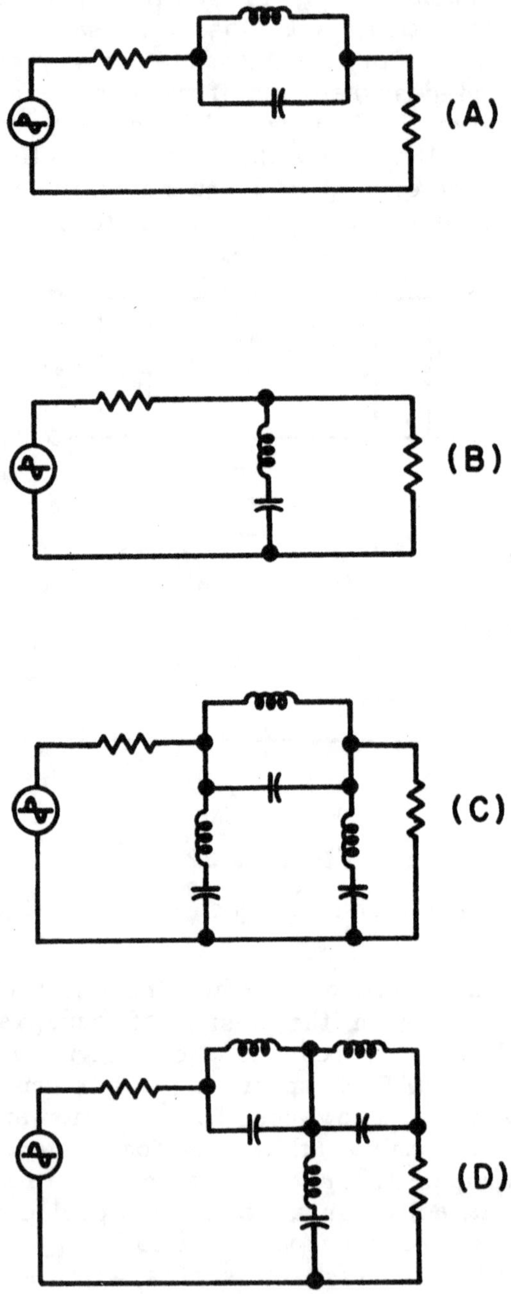

Figure 8-23

Band-Reject Filter Configurations

The twin-T notch filter shown in Figure 8-24 is useful as a band-reject filter in the lower frequency ranges. At low frequencies, the twin-T filter can achieve a circuit Q on the order of 100, which would not be economically feasible for a wavetrap or inductance-capacitance type filter at the same frequency. Shunting effects reduce its usefulness at high frequencies. The notch frequency is determined by[7]:

$$f_n = \frac{0.1592\sqrt{K}}{\sqrt{(R_1R_2C_1C_2)}} \qquad (8.10a)$$

where

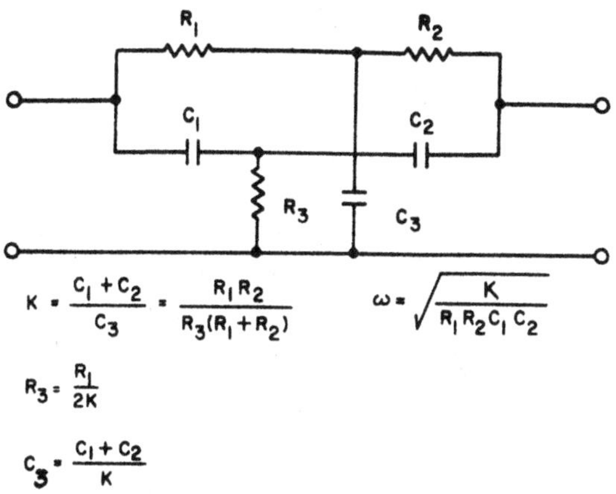

Figure 8-24

Twin-T Notch Filter

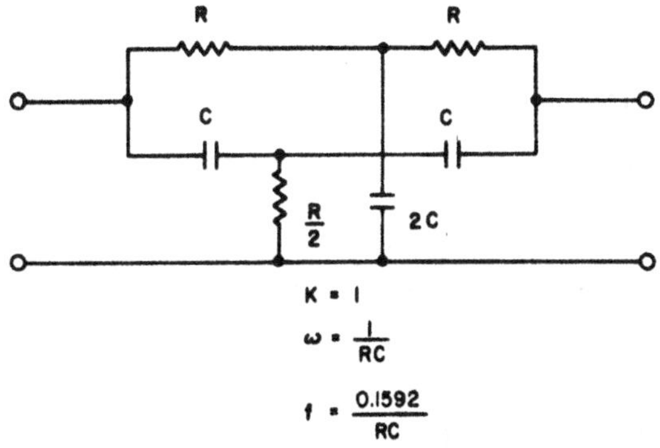

Figure 8-25

Twin-T Network with K = 1

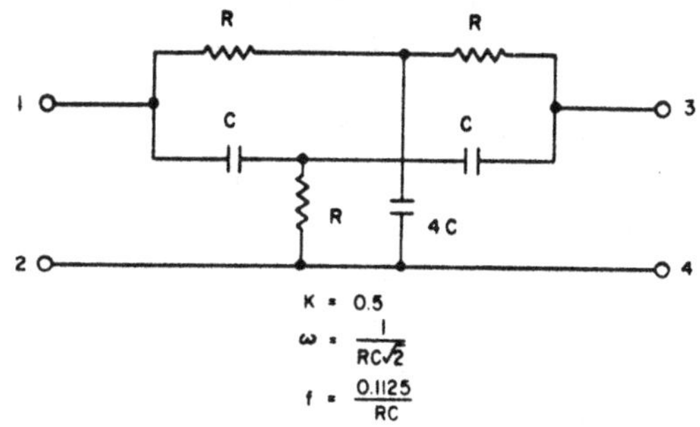

Figure 8-26

Twin-T Network with K = 0.5

Three special cases are of interest. The case when K = 1 gives the symmetrical form of Figure 8-25. With K = 1/2, a circuit with three equal resistances as shown in Figure 8-26 is obtained. In Figure 8-27, with K = 2, three equal capacitances result.

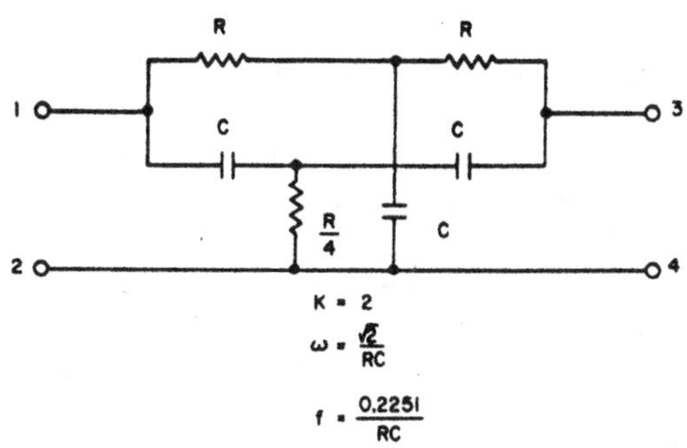

Figure 8-27

Twin-T Network with K = 2

Various transfer characteristics can be obtained with twin-T filters by varying the parameter K. For example, for the unloaded network and $R_1 = R_2$ and $C_1 = C_2$, the transfer function at $w = w_o$

can be shown to be

$$\frac{E_o}{E_i} = \frac{K(K-1)}{2+K(K+1)} \qquad (8.10b)$$

Thus for $K = 1$, a complete null is obtained, while for $K + 2$, $E_o/E_i = 0.25$. For $K = 0.5$, a $180°$ phase reversal is obtained, since $E_o/E_i = -0.091$.

A wide variety of characteristics can be obtained by removing the restrictions $R_1 = R_2$ and $C_1 = C_2$, as may be required to handle differing source and load impedances.

It should be pointed out that the twin-T notch filter parameters must be accurately selected to obtain attenuation at the null frequency. Getting the best possible null requires careful balancing in network tuning; a convenient way to do this is by use of trim capacitors or potentiometers.

8.3 TRANSIENT SUPPRESSION

The making and breaking of current, either mechanically or electronically, can introduce transient radiated and conducted effects. The transient may be generated within a GSE as a result of a switching function; or external to the test set, such as from another equipment tied to the same power line; or from the weapon system being tested.

If the source of the transient is external to the GSE and cannot be reduced at its point of origin, then the shielding, bonding, grounding and filtering techniques already discussed in this publication must be applied by the test equipment designer to prevent the transient energy from affecting GSE performance. If the source of the transient is within the GSE, the designer should be aware of particular switch transient filter techniques he can apply.

8.3.1 Inductive Loads

When an inductive load is opened by a switch contact, such as when a relay coil is opened, a reverse voltage is produced by the collapsing magnetic field. This reverse voltage increases until an arc occurs across the contact. The arc produces a wide frequency band of interference which is conducted and radiated away from the switch. Shielding and filtering will be necessary to reduce noise conduction and re-radiation from the switch contacts and wiring.

Figure 8-28 shows a number of suppression techniques that can be used to minimize transients in circuits that switch inductive loads[8] These techniques are as follows:

a. RESISTANCE DAMPING - Use of a non-inductive resistor across the load is the least expensive approach but it has drawbacks because the resistor will continuously dissipate power, and because the suppressor will increase relay dropout time. Maximum protection is obtained when the value of the suppression resistor, R, is equal to the coil resistance, R_L, but this will generally result in severe steady state power consumption requirements. Practically, R is kept as low as possible consistent with power capabilities and dropout time. A typical resistance value chosen is 10 times R_L, but this size may cause the dropout time of the relay to increase by as much as a factor of ten.

b. CAPACITANCE SUPPRESSION - A very popular relay suppression circuit consists of a series resistor capacitor combination placed across the coil. Care must be taken in this arrangement to avoid circuit resonance effects that will cause contact chatter, and to avoid overdamping so that dropout time does not become excessive. A typical arrangement is for R to be $\frac{1}{4}$ to $\frac{1}{2}$ of R_L. The value of the capacitance, C, can then be obtained from the relationship:

$$C = \frac{L}{RR_L} \qquad (8.11)$$

where

L is the inductance of the load

If the computed capacitance is very large, use the closest practical smaller capacitor. Capacitance between .01 to 1 uf is usually sufficient to minimize most transient effects.

The RC suppressor can create EMI problems when the relay is actuated, since there is an increase in current flow until the capacitor is charged. It is often preferable to place the suppressor across the relay contacts rather than across the coil to reduce this problem.

c. SINGLE DIODE SUPPRESSION - A series diode-resistor combination across the coil eliminates the power dissipation problem of the Resistor Suppressor. When the relay is energized from a dc source, the diode is back-biased so that the suppression circuit has no effect on relay operation. When current to the relay coil is opened, a current flows through the suppression circuit; this current is generally a function of the value of the suppression resistor R,

TYPES OF INDUCTIVE SUPPRESSION	VOLTAGE INPUT	RELAY CONTACTS		REMARKS
		CLOSING	DROPOUT	
A 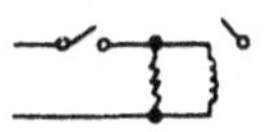 RESISTANCE DAMPING	AC or DC	NO EFFECT	FUNCTION OF RESISTANCE	Increase in power consumption. Resistance should be as low as practicable. Observe power rating, E^2/R, and heat dissipation.
B 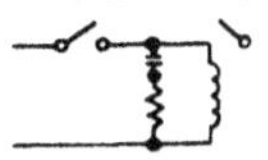 CAPACITANCE SUPPRESSION	AC or DC	SLIGHT EFFECT	SLIGHT EFFECT	Need series resistance of a few ohms. Capacitance value around .01 to 1 uF. Capacitance rated 10 times the input voltage.
C 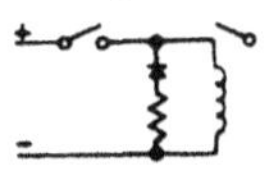 SINGLE DIODE SUPPRESSION	DC ONLY	NO EFFECT	SLIGHT EFFECT	Polarity is critical; diode is connected in backward or nonconductive direction. PIV should be higher than any transient voltage plus safety factor. Series resistance of a few ohms might be needed to increase inductance life.
D 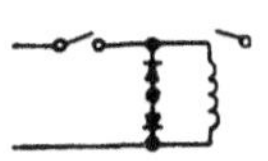 BACK-TO-BACK DIODE SUPPRESSION	AC	NO EFFECT	NO EFFECT	Avalanche voltage should be above input voltage. Power dissipation should be sufficient for transient current. Cost of device becoming a significant factor.
E 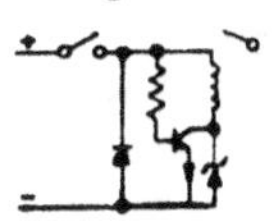 DIODE-TRANSISTOR SUPPRESSION	DC	NO EFFECT	NO EFFECT	Most effective in transient suppression. Voltage transient and dropout time negligible. Most expensive technique.

Figure 8-28

Comparison of Various Suppression Devices Across an Inductor (since the value of R is usually significantly larger than the forward resistance of the diode). The resistor value should be kept as small as possible consistent with relay dropout requirements.

Silicon diodes are used for this application because of their low cost, adequate current ratings, and ability to meet peak inverse voltage (PIV) requirements. Zener diodes have also been employed for this purpose.

d. DUAL DIODE BACK-TO-BACK SUPPRESSION - An extension to the single diode suppression approach is the use of back-to-back diodes across the relay coil. This arrangement is insensitive to the polarity of the relay control signal and can therefore be employed with AC relays. The component that conducts when power to the relay coil is removed may be a Zener diode, to clamp the transient at a specific level. Use of a Zener diode can help the dropout time problem as well.

e. DIODE-TRANSISTOR SUPPRESSION - A very effective transient suppression approach is to use two diodes, a resistor, and a transistor in an arrangement as shown in Figure 8-28E. When the relay switch is closed, the transistor saturates, allowing the coil to become energized. When the switch is opened, the transistor is cut off, forcing the transient energy to dissipate through the Zener diode Z and the conventional diode D. While this suppression circuit results in good transient control, it is a relatively expensive alternative, and should only be considered when the previously discussed approaches are not suitable.

8.3.2 Mechanical Switches

As indicated previously, any switching device causes transients during opening and closing. The arc generated across the switch contacts is extinguished when the energy stored in the inductance of the circuit is dissipated and the voltage drops below the value required to maintain the arc. To prevent or reduce the arc, the current, instead of being interrupted, can be channeled into another branch containing a series capacitance and resistor of a few ohms, as was suggested for reducing relay coil transients. In this way, the energy is partly stored in the capacitor and partly dissipated in the resistor, which also serves to damp any oscillations that may occur as a result of the added capacitance.

RC arc-suppression networks should be used whenever switches or relays are employed and equipment transient susceptibility is of concern. A capacitor should never be connected across contacts without including a series resistance, because the capacitor discharge through the switch contacts when the contacts are closed can cause a heavy current surge if not controlled by the resistor.

8.3.3 Transformer Switching

Transformer cores for GSE are often made from grain-oriented silicon steel. The high permeability of this material results in low exciting volt-amperes and low core loss. Because the cores operate at very high flux densities, the high inrush currents are of particular concern. The maximum inrush current occurs when the switch circuit is closed at the instant the voltage is zero.

The butt-joint air gap of the core can

be increased to control the inrush magnetizing current. In typical large cores, a spacer 0.005 inch thicker than the gap, will reduce inrush magnetizing currents by a factor of two or three; however, the exciting current is increased by about 2%. Another solution is to control the timing of the energizing switch or relay so that the transformer is connected when the voltage is at a maximum. The switch used in this application must be able to close the circuit at or near voltage maximum, when the flux change is approximately zero.

Both electrostatic and electromagnetic transformer shielding can be effective in reducing transformer-generated interference. The purpose of electrostatic shielding is to significantly lower the capacitance between primary and secondary transformer windings, and thus stop the direct flow of high-frequency currents through the transformer. This can be accomplished by enclosing one or both windings in a thin metal conducting shield, usually of copper or aluminum, with the ends of each shield insulated from one another. Either the shields are internally grounded to the core and case of the transformer, or a connection is provided so that external grounding can be accomplished.

Electromagnetic shielding is necessary to reduce stray magnetic fields outside the transformer. One method of doing this is to wrap and solder a copper band of .005 to .025 inch thick material around the winding and core (see Figure 8-28A). This band, called a shading ring, is a shorted turn of low resistance, and has the effect of creating a magnetic field that cancels much of the original transformer field. Another more expensive approach is to enclose the transformer in a high permeability case, or to line its regular steel case with high permeability material such as silicon steel.

8.3.4 Semiconductor Transients

Semiconductors that are used in power conversion or certain digital processing circuits operate as switches. The rapid current changes occurring in these devices when they are either turned on or turned off can give rise to EMI containing significant high-frequency spectral components. The spectral effects generally associated with pulsed signals are discussed in Section 2.2.2. of this design guide. However, some of the unique characteristics of transistors and other semiconductors that further contribute to spurious signal emissions are noted here.[17]

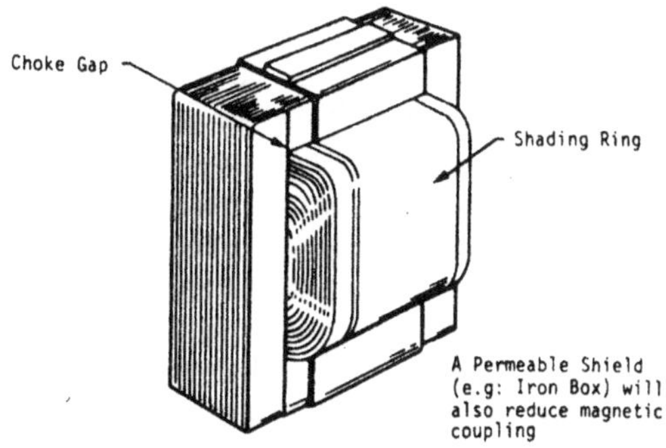

Figure 8-28A

Shading Ring in a Transformer or Choke[8]

8.3.4.1 Diode Recovery Time

Reverse recovery is a phenomenon common to all p-n junction diodes. Reverse recovery time (T_{rr}) occurs immediately after forward bias is removed and is a measure of the time required to remove ("sweep out") the minority carriers from the n and p regions and reunite their covalent bonds. T_{rr} for switching diodes ranges from

200 nsec to 12 nsec and is related to the magnitude of peak reverse current (I_{pr}). It has been shown that I_{pr} can be five times greater than the forward current for a 200 nsec diode and decreases to a value less than the forward current for a 12 nsec diode. In power supply applications, the reverse current pulses, once produced, are conducted back to the power source. The physical characteristics of the device, such as size and diffused impurities, can reduce reverse recovery time. Circuit conditions which reduce T_{rr} are low forward current, large reverse voltages, low impedances, and taking advantage of parasitic circuit capacitance.

8.3.4.2 Forward Recovery

Forward recovery transients have a similar interference generating effect as reverse recovery transients. Forward recovery time occurs immediately after a forward drive pulse is applied, and is a measure of the settling time of the p-n junction to a steady state conduction level. The most common methods of reducing forward transients and the resulting EMI are: (1) shape the driving pulse for minimum tolerable rise time, (2) keep the junction slightly forward biased, (3) reduce the drive pulse amplitude to the lowest level to maintain reliable switching action, and (4) select a device with fast recovery characteristics.

8.3.4.3 SCR Recovery

Silicon controlled rectifiers (SCRs) can produce reverse recovery transient currents and induce EMI similar to p-n junction diodes. The reverse recovery transients in SCR applications are intensified by increased forward current, predominantly inductive loads, and the large junction areas associated with these rectifiers. A T_{rr} for a typical SCR is on the order of microseconds. In full-wave SCR applications, the reverse recovery current has to be carried by the complementary SCRs. The reverse conducting SCR can produce an excessive forward current rise time in the conducting SCR. When SCRs are connected in series, mismatch of reverse recovery times can alter the reverse voltage distribution enough to exceed the maximum reverse voltage rating of the SCRs. Consideration of load current and surge current sharing is necessary when parallel SCR operation is required.

8.4 ACTIVE POWER LINE FILTERS

EMI filters made of passive elements are sometimes bulky and heavy. Active filters, using transistors, can provide large values of equivalent L and C without excessive size and weight. Moreover, the low impedance levels existing at low frequencies in power lines can be more easily accommodated with active devices.

Active filters for a dc line may contain capacitors as storage elements, modified series regulators to create a high impedance path, and modified shunt regulators in combination with high gain feedback systems for cancellation of interference. In the case of ac lines, cancellation is a most effective way to minimize interference. In contrast to conventional regulators used in regulated power supplies, these filters must not regulate the amplitude of the power to be passed.

Active ac power line interference filters pass, with high efficiency, only a narrow band about the power frequency. Voltage attenuation values of approximately 30 dB can be obtained even at very low load and source impedance levels; two filters may be cascaded for higher attenuation values.

The power line filter scheme shown in

Figure 8-29 uses phase cancellation and operates as follows: The input signal is fed into an ac-coupled amplifier through a notch filter, which is tuned to the fundamental of the power line frequency. The amplified interference signal, without the power line fundamental frequency, and with opposite polarity, is returned in series to the source through the transformer. All signals except the fundamental are attenuated by the gain of the amplifier. Within a limited range, a separate control circuit can provide automatic tuning of the notch filter to power line frequency and correct any change in the filter tuning itself. The voltage attenuation curve for a 220 volt, 20 ampere unit is also shown in Figure 8-29.

When inserted in the line, the filter introduces the equivalent of a small inductance of 700 μH at the pass frequency. The change in output voltage at full current rating caused by this inductance is negligible, i.e., 1 volt for a load power factor approaching unity and up to 6 volts in the worst case for zero power factor. This type of filter can handle large interference voltages, but with some

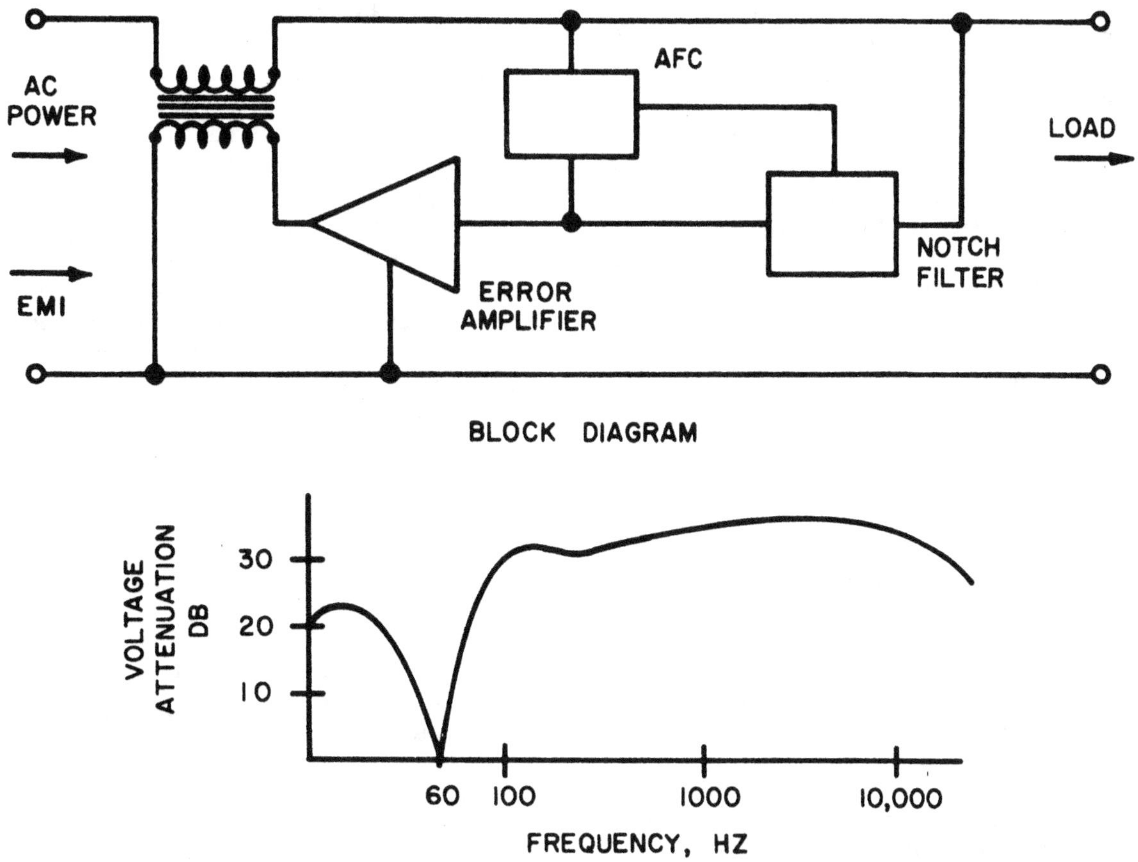

Figure 8-29
Active Filter for Powerline Interference

loss of efficiency. Efficiencies of 90 percent have been realized for 200 volt, 20 ampere filters.

8.5 NOISE BLANKING, CANCELLING, AND LIMITING CIRCUITS

8.5.1 Noise Blanking

If the interfering signal is a pulse, and it is not possible to prevent the pulse from getting into the system being developed, a blanking circuit might be used. The function of a blanker in the EMC sense is to render the system inoperative for the duration of the interference.

The blanking circuit can be triggered by the interfering pulse, or by an independent signal arriving before the interference pulse. When the interference pulse triggers the blanking circuit, delay lines must be provided to delay the desired signal long enough to allow the blanking circuit to turn off the system before the interference pulse reaches it. An independent pulse can be used if the arrival of the interference pulse is known beforehand by way of a trigger pulse, and the trigger can be hardwired to the GSE. Blanking action is usually provided by a simple gate circuit that can be actuated by the trigger pulse, and a timing circuit to control the duration of the gate.

Blanking circuits lack the simplicity of wavetraps and limiters, and can also cause interference. The periodic gating of the signal is a form of modulation, which will appear in the output as noise. The major disadvantage of blankers is that they introduce dead-time in a signal channel. For some applications, this dead-time cannot be tolerated.

8.5.2 Cancellation

An interference cancelling circuit suppresses interference by allowing the interference signal to travel along two paths. One of the paths carries only the interference signal. The interference signal is shifted 180 degrees in phase, adjusted in amplitude, and added to the channel containing the signal and interference. The result is the cancellation of the interference while leaving the desired signal.

This method can only be used when the path of entry and the nature of the interferring signal are known, and when the source of interference is accessible. Interference cancelling circuits are useful in suppressing interference from equipment when the interference signal has unique properties relative to a desired signal, such as radar transmitter interference to an analog device.

Figure 8-30 illustrates one arrangement for eliminating interference by cancellation. Two directional couplers are shown. One, indicated as DC-1, samples the interfering signal at the offending transmitter. The signal is routed through delay lines to introduce the required phase shift, attenuated to set the power level, and then coupled to the receiver via another directional coupler, indicated as DC-2.

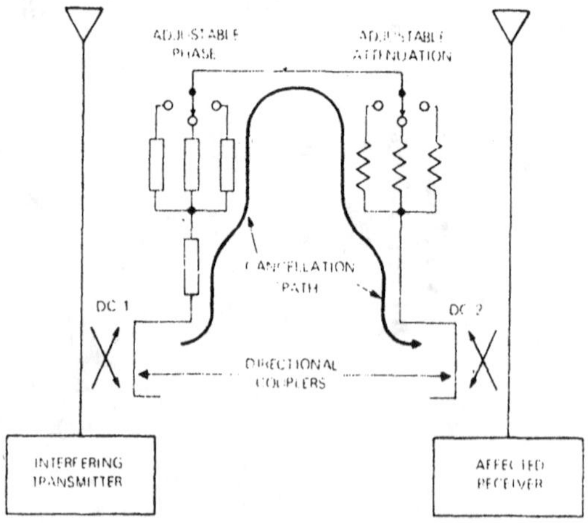

Figure 8-30

Block Diagram of an Interference-Cancelling Network

For cancelling interference sources not a part of the same system being developed, a different pickup arrangement will be needed, and the phase and attenuation control methods will be more complex. One pickup arrangement that has been employed is a separate receiver to pick up the interferring signal. For this configuration, it is necessary to include automatic phase (or frequency) and gain control circuitry to control the characteristics of the signal in the cancellation path. This type of EMI reduction technique should be considered only when the type and source of interference is known, and when system operation and outside influences can be tightly controlled.

As an example of the expected capabilities of cancellation circuitry, consider a prototype canceller designed to suppress analog signals in the 225-400 MHz range[9]. Such a device typically provided at least 35-40 dB of rejection to the interference signal at desired-to-interference signal frequency differences of as small as 0.025 MHz, so long as the interference-to-signal ratio in the interference channel was 20 dB or more.

8.5.3 Limiting

For interfering signals consisting of short impulses of large amplitude, noise limiters can prevent the peak amplitude of the interference from rising above the desired signal and overloading the system. Noise limiters are not effective against thermal noise, or steady-state EMI. There are two types of noise limiters: peak amplitude limiters and gated noise limiters.

The peak amplitude limiter reduces the effects of interference by clipping all impulses above a preestablished threshold level. To be most effective, the clipping level should be as close as possible to the peak value of the desired signal, but should not be adjusted to clip so heavily as to distort the desired signal. Diode devices are commonly used as limiters, although tube or transistor amplifiers biased to swing into a nonlinear region on strong signals may also be employed. The disadvantage of peak amplitude limiters is that they introduce nonlinear effects, with the resultant possibility of cross-modulation on strong signals.

The gated noise limiter is a low frequency device that might also be classed as a blanker. Basically the circuit consists of a diode or combination of diodes having a fast time constant network at the input, and a slower time constant network at the output. As long as the charge-discharge period at the output can follow the desired input signal, the diode gate continues to conduct and the input signal is delivered to the output. However, an impulse will possess a rise-decay rate greater than that provided for at the noise gate output. This will cause the gate to be back-biased and cut off momentarily until the slower time constant at the output can catch up with the input. Noise impulse steep wavefronts are thus gated out. The gated noise limiter is generally effective only when used with near-sinusoid waveforms.

The usefulness of both types of noise limiters depends on the width of the system bandpass. The bandpass must be broad enough to avoid "ringing" on noise impulses. Also, bandwidth-limiting can alter the noise impulse envelope so that subsequent limiter action is ineffective.

8.6 FILTER TESTS

8.6.1 General

There exist several techniques for measuring filter performance, based on the type of filter and its intended application. The tests to be described are those which appear to provide the most meaningful results of such performance. They include:

- Insertion Loss Test in accordance with MIL-STD-220A
- Filter Admittance Transfer Test
- Parallel Signal Injection Test
- Series Signal Injection Test

These test methods are applicable to both signal line and power line filters. In all cases, the test procedures are designed to measure the filter insertion loss while the filter is carrying rated load current.

The MIL-STD-220A Insertion Loss Test[2] and the Filter Admittance Transfer Test are very similar. They differ mainly in that the former test is restricted to the situation of 50 ohm input and output impedance terminations. This case may not be representative of the actual filter operating condition. The Filter Admittance Transfer Test provides the flexibility of varying the output load impedance.

In determining power-line filter attenuation characteristics, it is also important to use measurement techniques which give results that demonstrate installed performance. For example, Figure 8-31 provides one illustration of the variation of ship power line impedance. Not shown in the figure are the very low values of expected line impedance (well below one ohm) below 15 kHz. Variations of impedance as a function of location on a power-line also occur.

A meaningful measurement should indicate minimum attenuation that the installed filter will exhibit under such impedance variations. A worst-case filter test method is therefore a desirable technique for evaluation of filters, when the filters will be used in a variety of installations where impedance conditions can vary widely. That is the objective of the Parallel and Series Signal Injection Tests.

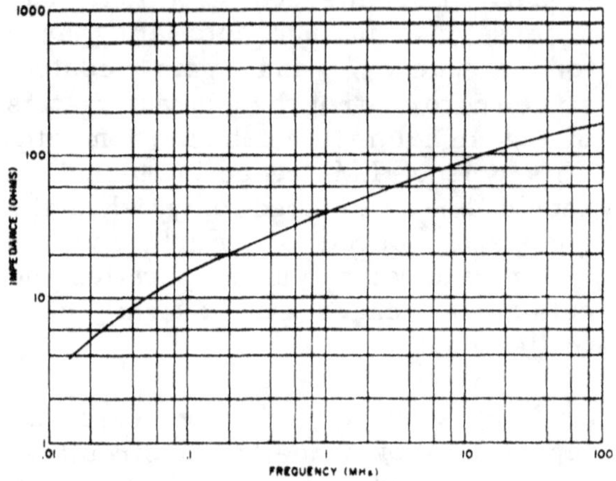

Figure 8-31

Sample of Power Line Impedance

For simulated installation testing, the current injection probe technique used in the Series Signal Injection Tests will furnish meaningful results, provided that the current injection probe design does not cause changes in loading conditions by inserting large values of impedance into the filter circuit; and provided that it does not cause coupling between filter input and output. However, the limited frequency range offered by this test necessitates use of a different test method at high frequencies, usually that defined by MIL-STD-220A.

Before discussing the individual filter test procedures, it is important that a few unique terms to be used in this section of the design guide be defined:

- Buffer Network: An isolation network, consisting of (a) an inductor designed to have a very high self-resonant frequency and capable of carrying the full rated load current of the filter under test, and (b) a source-end shunt by-pass capacitor.

- Current probe: A calibrated toroidal winding capable of being placed around a current-carrying conductor. This arrangement then becomes a 1:N transformer, where N is number of turns on the toroidal core. Probes of this type are used for both current detection and current injection. In the latter case, the probe windings must be heavy enough to carry the required current for specified test power levels.

- Power bias: Direct or low-frequency current passed through a filter under test. The presence of such current enables filter performance to be tested under load.

- Substitution measurement: A technique for finding the amplitude of an unknown voltage or current by using a detector to indicate an arbitrary value at some combination of detector device control settings and, without changing the detector system, replacing the unknown source with a known calibrated voltage or current that can be adjusted to give the same detector indication.

- Transfer admittance: The ratio of network signal output current divided by the network input signal voltage.

- Transfer impedance: The ratio of network signal output voltage divided by the network input signal current.

The test procedures provided in this chapter are for guidance purposes only; refer to the cited references for further test details.

8.6.2 MIL-STD-220A Insertion Loss Test

8.6.2.1 Objective

MIL-STD-220A covers a method of measuring, in a 50 ohm system, the insertion loss of feed-through suppression capacitors, and of single- and multiple-circuit radio-frequency filters. MIL-STD-220A defines insertion loss in accordance with Equation (8-1).

8.6.2.2 Test Set-Up

Figure 8-32 is a general test set-up for filter insertion loss measurements as given in MIL-STD-220A. The isolation attenuators in this set-up have input and output impedances of 50-ohms to provide matching for the signal generator and the receiver. These networks also present 50-ohm loading at both the input and output of the filter. Using the insertion-loss measuring equipment specified, the test set-up is capable of measuring insertion loss within $\pm$ 1.0 dB over the required frequency range.

All test equipment must be shielded and filtered so that leakage from the signal generator or any portion of the signal-source circuitry does not affect the output of the receiver when the generator and receiver are operating at the output level and sensitivity needed to make the required maximum insertion-loss measurement. The direct-current source used in making these measurements must be a floating dc source and not connected to ground.

It is essential that cable with the general characteristics of type RG-214/U be used to connect the isolation attenuators together for the filter-out condition, and to connect the filter under test to the isolation attenuators for the filter-in condition. The length of cable connecting the isolation attenuators in the filter-out condition

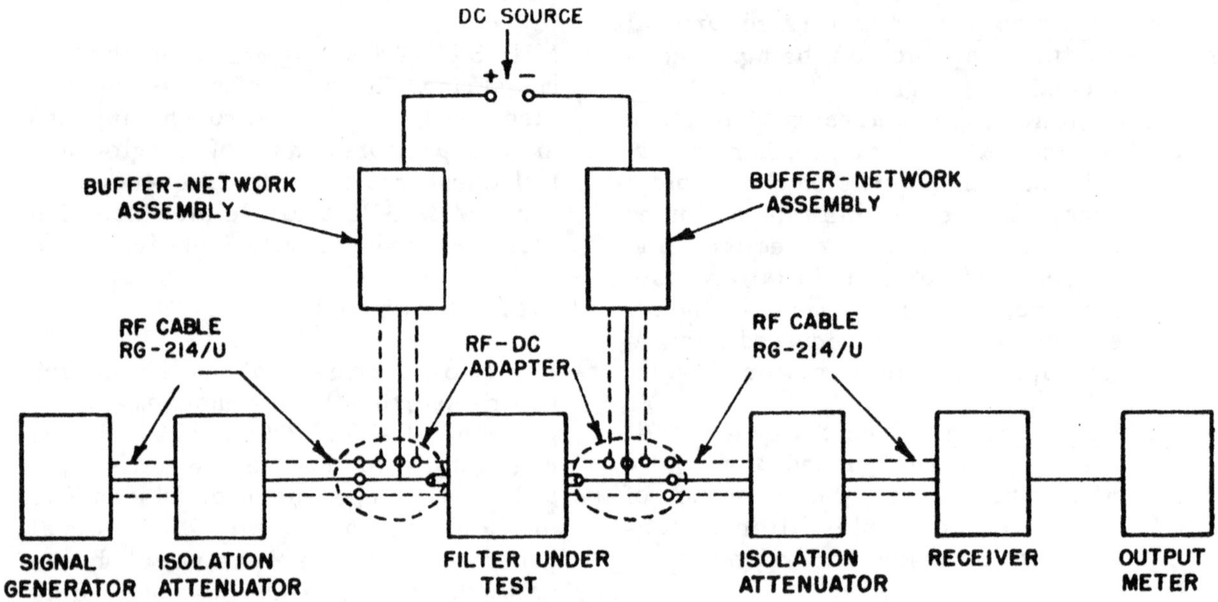

Figure 8-32

Test Arrangement for MIL-STD-220A Filter Test

should be within 6 inches of the combined length of the two cables connecting the filter under test to the isolation attenuators in the filter-in condition. Type N 50-ohm coaxial connectors conforming to specification MIL-C-71 should be used where applicable. When coaxial switches are used, they should have a 50-ohm characteristic impedance, and a maximum voltage standing wave ratio (VSWR) of 1.1 to 1 at the frequency of measurement.

The buffer-network assembly and RF-dc adapter assemblies shown in Figure 8-33 and in Figure 8-34 are specified by MIL-STD-220A when performing insertion-loss measurements with rated current applied. The buffer-network assembly provides isolation between the rf test signal and the dc power supply. The complete measurements system can be used for filter tests over the frequency range of 100 kHz to 20 MHz. The indicated buffer-network is suitable for continuous use with currents up to 100 amperes.

8.6.2.3 Procedure

The test equipment is set up as shown in Figure 8-32 except that a cable of the type and length specified for the filter-out condition may be inserted between the attenuators when applicable.

- Adjust the signal generator to the desired test frequency, with the attenuator set for the lowest convenient value of output voltage.

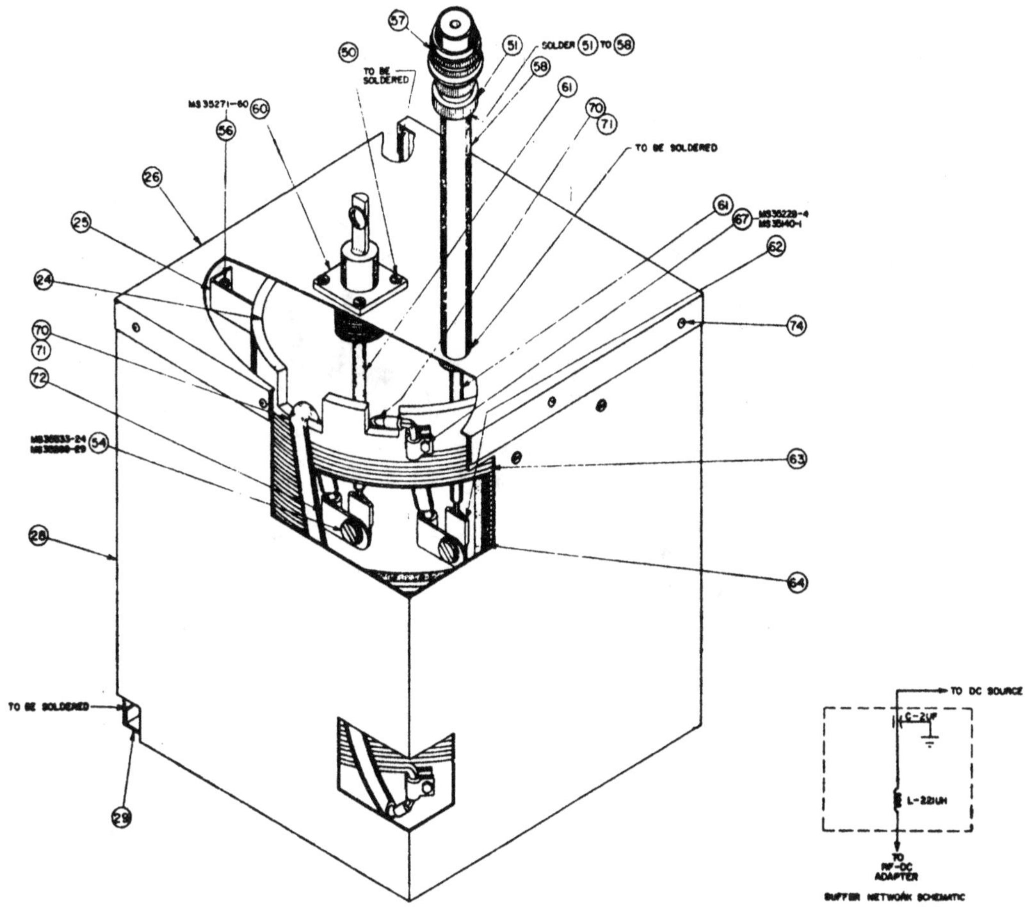

Figure 8-33
MIL-STD-220A Buffer Network

- In the filter-out condition, tune the receiver to resonance at the frequency of the generator, and set the gain controls so that receiver sensitivity will be high enough and the level of circuit noise low enough to allow clear reception of the signal required for the measurements.

- Adjust the output of the generator to give the lowest possible stable and readable indication on the output meter, care being taken not to saturate or overload the receiver. Record the generator output level (E_2).

- Remove the cable used for the filter-out condition, and insert the filter under test and its connecting cables between the isolation attenuators.

- Readjust the receiver to resonance, and adjust the output of the generator until the output meter gives the same indication as that obtained for the filter-out condition. Record the generator level (E_1).

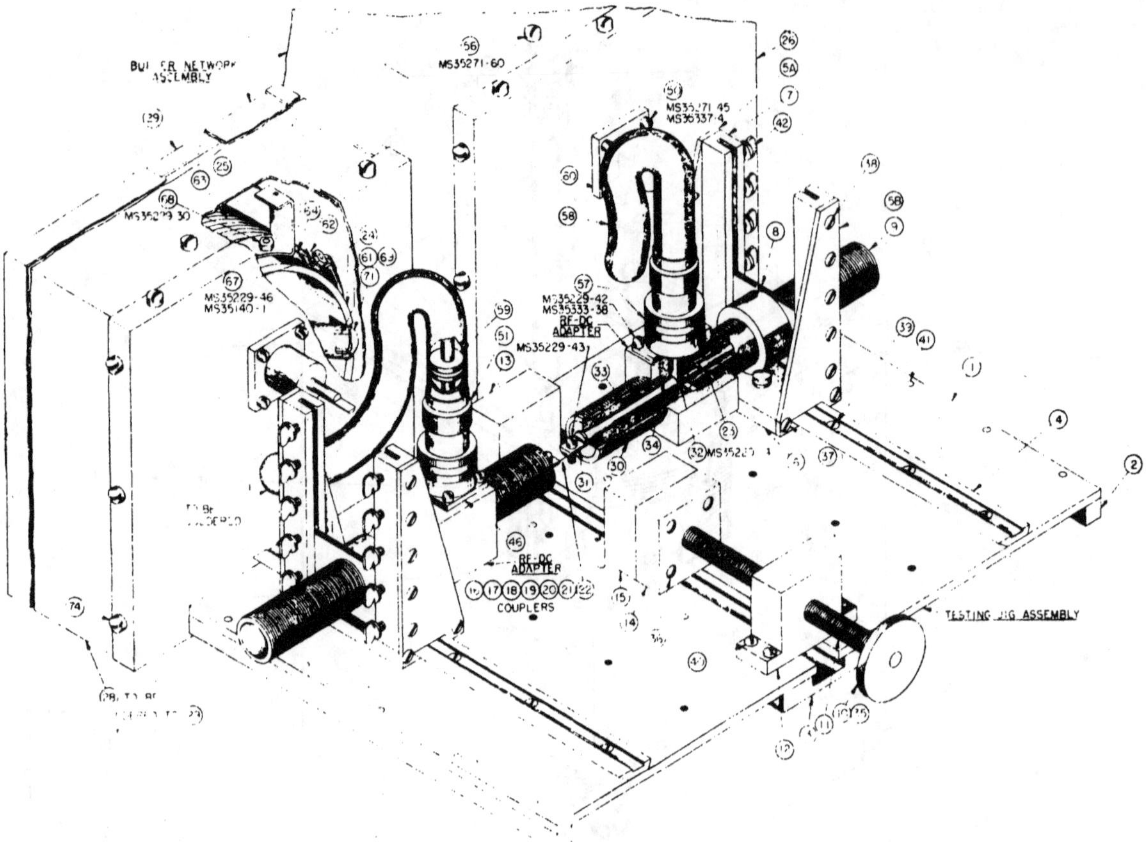

Figure 8-34

MIL-STD-220A Filter Test Fixture

- The insertion loss of the filter being tested under the specified conditions and at the frequency of measurement may then be expressed in dB as

$$\text{Insertion Loss} = 20 \log \frac{E_1}{E_2} \tag{8.12}$$

- If a multiple-circuit filter is under test, measure the insertion loss of each circuit of that filter with each of the other circuits open-circuited, and also with each of the other circuits short-circuited. By definition, the insertion loss of the filter at the frequency of the test is considered to be equal to the lesser of these two measurements.

- If full-load insertion-loss measurements are being performed, the nominal dc rated current of dc components, or the dc equivalent of the peak alternating-current (ac) rated current of ac components should be applied during the above test.

8.6.2.4 Limitations and Precautions

- Measurements using MIL-STD-220A can be extended to 1000 MHz, but caution should be exercised in judging the accuracy of the test results at frequencies above 30 MHz.

The MIL-STD-220A buffer-network limits measurement accuracy below 100 kHz and above 20-30 MHz for rated load tests.

- Be sure the dc power supply is an ungrounded supply.
- As a precautionary measure, apply the power bias starting with zero voltage from the power supply.

8.6.2.5 Advantages/Disadvantages

- The 50-ohm system used allows the performance of different filters to be measured against a common standard.
- Power bias load currents of up to 100 amperes may be used in the filter tests.
- The buffer-networks, as defined by MIL-STD-220A, limit low frequency measurement down to 100 kHz and high frequency measurements up to approximately 30 MHz, for full power load tests. For no load tests, the test range may be extended up to 1000 MHz. Figure 8-35 provides a comparison of load and no-load tests taken in accordance with this procedure at frequencies below 100 kHz[10].
- This test is based on a 50-ohm matched impedance condition at both input and output of the filter under test. Impedance conditions of other than 50-ohms will result in different insertion loss values. It should be recognized that GSE terminations are not nessarily 50 ohms.

8.6.3 Filter Admittance Transfer Test[11]

8.6.3.1 Objective

The objective of this test is to measure filter performance under extremes of

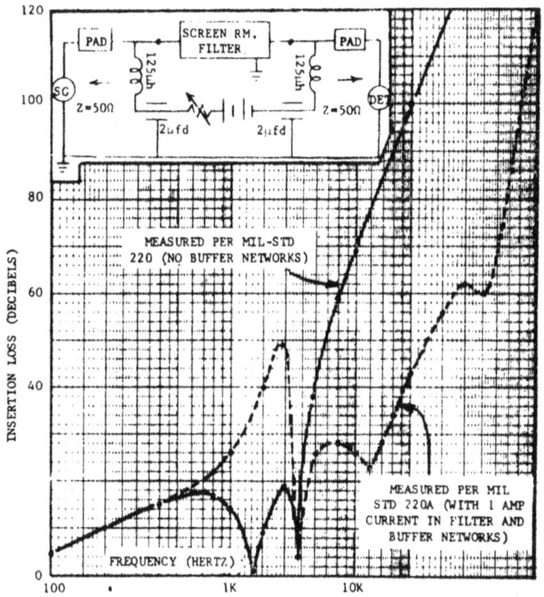

Figure 8-35a

Insertion Loss of 100 Ampere Screen Room Filter

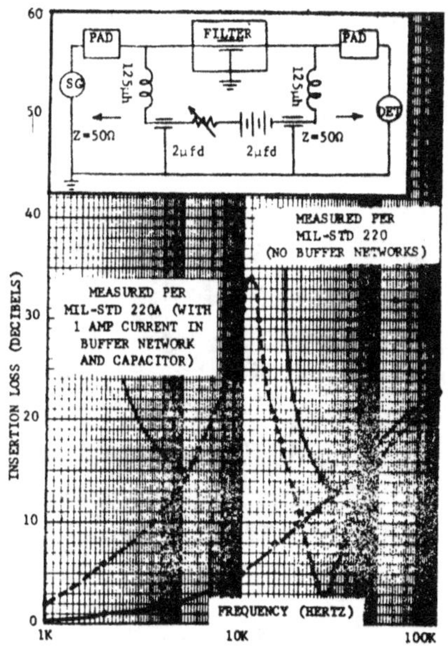

Figure 8-35b

Insertion Loss of 1.0 uF Feed Through Capacitor

load impedance with readily available instruments.

The measurement set-up for the admittance transfer test is shown in Figure 8-36. The method uses a current probe to measure the filter network output current, and a direct measurement of input test voltage to the filter. The insertion loss of the network is calculated from:

$$\text{Insertion Loss} = 20 \log_{10} \left[\frac{V_i + i_o R_L}{i_o R_L} \right] \text{, dB} \quad (8.13)$$

where:

V_i = Input test signal voltage

R_L = Filter Load Impedance

i_o = Filter output current into load = E_o/Z_T

Z_T = Current probe transfer impedance

E_o = Current probe output voltage

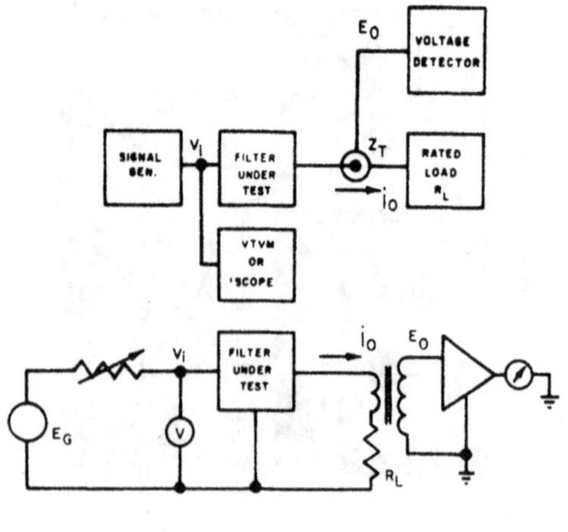

Figure 8-36

Transfer Admittance Test Block Diagram and Schematic Diagram

This relation is obtained by assuming that the filter is a two-terminal device, and is applicable for absorptive-type filters with any load R_L. However, for lumped parameter L-C type filters, which are essentially four-terminal devices, the formula is applicable only when R_L is much smaller than the series transfer resistance, R_T, of the filter, i.e., V_i/i_o.

In many respects, this test is similar to MIL-STD-220A and results can be correlated when R_L = 50 ohms, if R_T is much greater than 50 ohms. The test has the advantages over the MIL-STD-220A procedure in that measurements can be made with different filter output load impedance terminations. In this respect, the transfer admittance test can more closely approximate actual usage conditions, since the output load resistance can be made very low. However, it should be pointed out that at frequencies below filter cut-off, the series resistance of a lumped parameter low-pass L-C filter will be small and it will be difficult to satisfy the condition that R_L be much smaller than R_T.

8.6.3.2 Test Set-Up

The test block diagram is shown in Figure 8-36. The signal generator should have a power output level, in dB, at least equal to the detector sensitivity in dB, plus the attenuation of the filter, plus the probe gain. For example, assume the generator has a +10 dBm maximum output (600,000 μV, or 117 dBμV), the detector is a receiver with 0.1 μV sensitivity (-20 dBμV), and the anticipated filter specification at the test frequency is 100 dB. The probe transfer impedance is 0.31 ohms, or -10 dB re 1 ohm. The calculation is: (8.14)

Generator Output	-	Filter Loss	+	Probe Gain	-	Detector Noise	=	Measurement Margin
(117)	-	(100)	+	(-10)	-	(-20)	=	27 dB

Since the measurement margin is positive, the signal generator is adequate.

The output voltage should be measured with a detector having a high input impedance over the entire test frequency range. The current probe should have a transfer impedance between 0.1 and 10 ohms, and should be placed around the filter output conductor as close to the filter output as possible. Do not clamp the probe around the outside shield of a coaxial cable; clamp it around only the "high" side or center conductor of the coax. Use a ground plane for current return. Be sure a cable of correct characteristic impedance connects the probe to the detector.

8.6.3.3 Test Procedure

At each test frequency, perform the following steps and record the indicated data:

- Adjust the signal generator frequency and output level.
- Record the output frequency and generator output voltage level.
- Adjust the detector until the reading on the output indicator is greater than one-third full scale.
- Measure the voltage delivered by probe to the detector input, using either substitution or direct-reading techniques.
- Using the probe transfer impedance, the voltage delivered from the probe, and the input voltage to the filter, calculate the filter insertion loss for the test frequency and filter termination used.

8.6.3.4 Limitations and Precautions

- The series impedance of the test filter must be much greater than the load impedance.
- Insertion loss can be accurately measured by this method only when the frequencies are well into the attenuation band, e.g., when the filter insertion loss is greater than 20 dB. Thus, neither the shape of the pass-band nor the 3 dB point can be accurately determined in this way.
- The current probe turns ratio must be high enough not to reflect significant series impedance levels compared to the value of test load impedance. At the same time, the probe must not have too many turns, or the value of transfer impedance will be too low and the probe frequency response and sensitivity will be limited.
- For high values of insertion loss, care must be taken that the input and output terminals of the filter are electrically isolated, and that generator and detector are not coupled in any way to by-pass the filter under test. While this requirement applies to all tests, it is particularily important when using current probes.

8.6.3.5 Advantages/Disadvantages

- Insertion loss values for varying load conditions can readily be found using this procedure, subject to the constraints indicated above.
- High levels of attenuation can be measured, with proper isolation of filter terminals.
- A wide dynamic measurement range is possible.
- Test results can be correlated with MIL-STD-220A when R_L is 50 ohms and the filter series impedance is

much greater than 50 ohms.

- Low values of insertion loss cannot be accurately measured.
- The filter 3 dB point cannot be accurately measured.
- The test arrangement is not designed to measure filter attenuation under load, although the method could be adapted to incorporate this feature.

8.6.4 Parallel Signal Injection Test[11, 12]

8.6.4.1 Objective

This test is intended as a "worst-case" filter test technique, where the test filter is operated under rated power conditions, and the input and output capacitors of the filter are resonated at each test frequency. The test is designed for low pass pi-filters, having as the last element at each end a shunt capacitor. Under these conditions, the filter insertion loss is found, and will be the minimum value for any service load or source impedance to be encountered. This test uses a method of parallel test signal injection.

8.6.4.2 Test Set-Up

The test arrangement is shown in Figure 8-37.

A test signal current source is inserted in parallel with the filter input and a detector is inserted in parallel with the filter output. The input current level is monitored with a probe; the probe-inserted impedance is not critical since the impedance level of the current source is large in comparison to the signal source impedance of about 50 ohms.

The highpass filters shown in Figure 8-37 are needed to keep the low-frequency power-biasing voltage from reaching the signal source and detector. To avoid errors resulting from radiation, the oscillator, amplifier, and detector should be placed outside the shielded room used in performing the test, or the filter should be mounted within a separate enclosure in the shielded room.

The power bias source for testing the filter under rated load can be either the standard AC line voltage or a DC power supply of variable output voltage and required maximum test current output. The set-up shown in Figure 8-37 assumes a 115v, 60 Hz source, with amplitude control using an autotransformer and an ammeter to measure power load current. The variable frequency test signal current source is a standard signal generator and, if the generator output level is insufficient, a booster amplifier.

The variable inductors shown in Figure 8-37 are used to resonate the filter shunt capacitors at each test frequency.

8.6.4.3 Test Procedure

With the equipment arranged as shown in Figure 8-37, the following steps can be performed to find the lowest possible insertion loss of a particular pi-section filter:

- Determine the value of the pi-filter shunt capacitances, both input and output.
- Calculate the range of resonating series inductor values required for the range of test frequencies.
- Place the correct series inductors in the test circuit and adjust for resonance at the test frequency.
- Adjust the test signal generator at the test frequency, and adjust the detector for indication of the test signal.

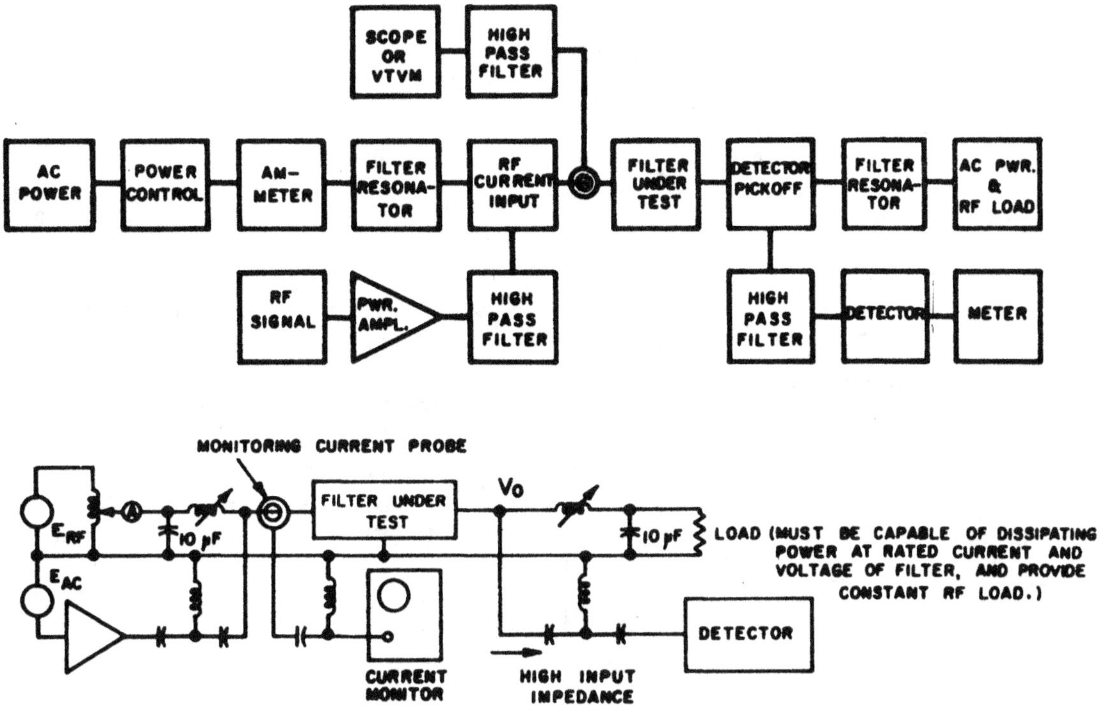

Figure 8-37

Parallel Signal Injection Block Diagram and Schematic Diagram

- With the filter input and output resonating, and with power bias of the desired level applied, record the detector indication of received test signal.
- Also record the injected test current level into the filter (I_2), and the power bias level from the AC or DC power supply.
- Record the frequency used in the test.
- Remove the filter under test, and connect the test signal source directly to the detector and load (filter-out condition).
- Decrease the test signal current level until the same detector indication is obtained as with the test filter in the circuit.
- Record the input test (I_1).
- Calculate the insertion loss of the filter from:

$$\text{Insertion Loss} = 20 \log_{10}\left(\frac{I_1}{I_2}\right) \tag{8.15}$$

8.6.4.4 Limitations and Precautions

- The upper frequency limit of the Parallel Signal Injection approach is dictated by inductance introduced into the test circuit by the circuit configuration, test leads, and components used. In general, an upper limit of 5 to 10 MHz can be expected.

- The lower frequency limit of this approach is affected by the size of the inductor needed for resonance and the feedthrough capacitor needed for isolation. The capacitor is shown as having a value of 10 μf in Figure 8-37. The lower frequency value may be approximated using Table 8-1:

Table 8-1

Theoretical Low-Frequency Limit for Parallel Injection Testing (Q = 10; R = 50 ohms)

Filter Capacitance (μF)	Lower Frequency Limit (kHz)	Value of Resonating Inductor (μH)
1	32	25
10	3.2	250
20	1.6	500
30	1.1	750
40	0.8	1000
50	0.64	1250

- Be sure that filter input and output terminals are electrically isolated.
- Apply power bias gradually, starting with zero input power voltage to the test set-up.
- Be sure that the resonating series inductors can carry the power bias current.
- Protect the power bias circuitry from test filter short circuit failure with fuses or circuit breakers.
- Values of source and load impedance will depend in part on the shunting effects of the test signal current source and the test signal detector.
- Be sure the ammeter used to measure power bias current has full scale values at least equal to or preferably greater than rated filter load current. Some pi-filters draw appreciable amounts of reactive current, and some allowance should be made for a meter safety factor.

8.6.4.5 Advantages/Disadvantages

- The method of resonating the input and output filter capacitors gives "worst-case" performance data for the filter.
- The filter is tested under rated load current conditions.
- Good low frequency performance data can be obtained, limited only by the required value of the series resonating inductors and shunt by-pass capacitors.
- Accurate measurements of worst-case insertion loss can be made up to about 10 MHz.
- Load impedance may be varied to simulate actual operating conditions.
- Measurements at VHF and UHF frequencies cannot be accurately performed.
- A wide range of series resonating inductor values is required for general purpose testing.
- For a wide dynamic range (large values of filter attenuation), power booster amplification may be needed for the injected test current signal.

8.6.5 Series Signal Injection Test[11,12]

8.6.5.1 Objective

The objective of this test is to perform a "worst-case" evaluation of a pi-section filter (resonating the filter input and output) without hard wire connection to the filter under test. As indicated previously, hard wired connections limit the range of values of source and load impedance under which the filter can be tested. In this test a signal injection current probe acts as a series current source for the test filter input. The test filter can be operated under rated power conditions during the test.

8.6.5.2 Test Set-Up

The instrumentation arrangement is given in Figure 8-38. Two current probes are used for injecting the test signal and for detecting the filter output.

The input test signal is connected through a highpass filter to a current injection probe; the highpass filter prevents the power bias signal picked up by the probe from reaching the test signal generator during power bias tests. A second probe couples current on the output side of the filter to a signal detector.

The large feedthrough capacitors are needed during power bias tests to bypass any external power-line impedance, forming essentially a short circuit to ground for the test signal. These capacitors should be at least five times as large as the test filter input capacitors, and should have very low self-inductance.

The power bias source for testing the filter under rated load can be either the standard AC line voltage or a DC power supply of variable output voltage and required maximum test current output. The equipment set-up shown in Figure 8-38 assumes a 115v., 60 Hz power bias source, with power bias level control using a variable autotransformer. An ammeter is used to measure power bias current level. The test signal current source is a standard signal generator; a power booster amplifier may be needed if the filter under test has high insertion loss

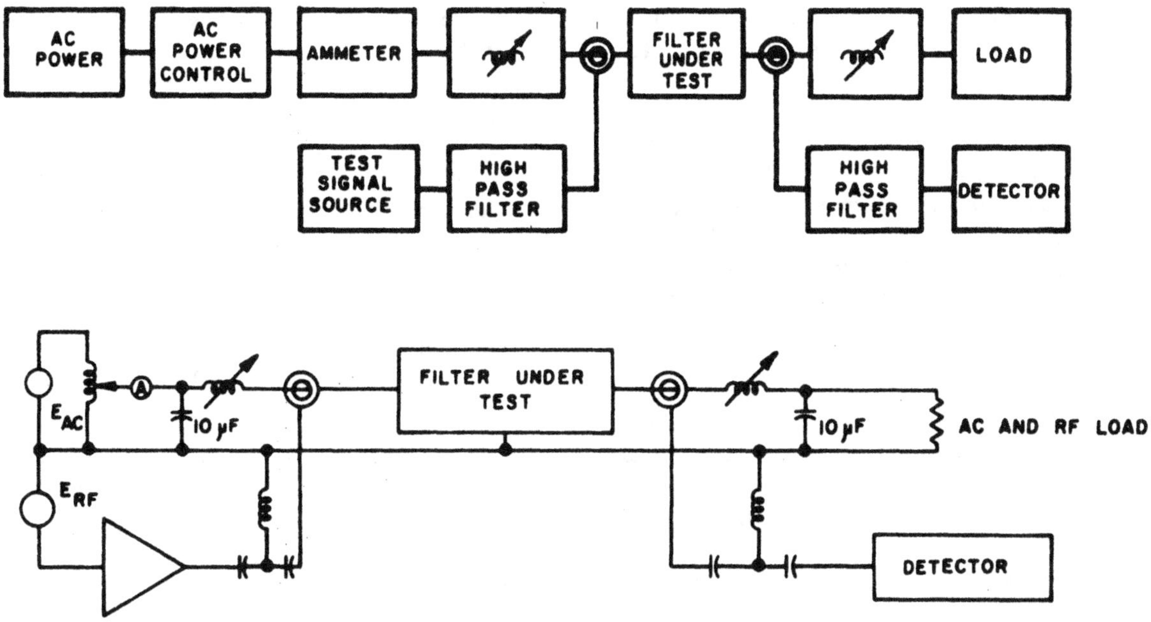

Figure 8-38

Series Signal Injection Block Diagram and Schematic Diagram

and the detector has low sensitivity.

The input current probe should be designed to minimize the resistance inserted into the test circuit, and should be capable of carrying the maximum current to be injected as a test signal. Neither the input nor the output current probe should inject appreciable series impedance at the filter input or output, respectively. The variable inductors shown in Figure 8-38 are used to resonate the input and the output capacitors in the pi-filter under test.

8.6.5.3 Test Procedure

With the test equipment arranged as shown in Figure 8-38, the following steps can be performed to find the lowest possible insertion loss of a pi-section filter:

- Determine the value of the input and output shunt capacitance in the pi-filter to be tested.
- Calculate the range of resonating series inductor values required for the range of test frequencies.
- Place the correct series inductors in the test circuit and adjust for resonance at the test frequency.
- Adjust the test signal generator at the test frequency, and adjust the detector for indication of the test signal.
- With the filter input and output resonating, and with power bias applied, record the detector indication of the received test signal.
- Record the power bias level from the AC or DC power supply, and the injected test voltage level (E_2).
- Record the frequency used in the test.
- Remove the filter under test, and connect the test signal source directly to the load (filter-out condition).
- Adjust input signal voltage level to obtain the same indication on the detector as was obtained with the test filter installed.
- Record the test signal (E_1).
- Calculate the insertion loss of the filter using Equation (8.1).

8.6.5.4 Limitations and Precautions

- The lower frequency limit of the Series Signal Injection Method is dictated by the high value of inductance needed to resonate the filter capacitance, the feedthrough capacitor impedance needed to provide isolation, and the roll off of the transfer impedance curve of the current probes.
- The high frequency limit of this method is determined by the amount of inductive reactance inserted into the test circuit by the current probes. The maximum frequency value, based on input and output filter capacitances and probe-inserted reactance, may be approximated using Table 8-2:

Table 8-2

Upper Frequency Limit for Worst-Case Testing for Series Injection

Filter Capacitance (μF)	Upper Frequency Limit (MHz)	Inserted Inductive Reactance (mΩ)
52	0.100	30
25	0.150	43
2.3	0.500	140
0.52	1.000	300

- Care must be taken to insure that the input and output sections of the filter under test are electronically isolated.

- Injected series impedance from the current probe transformation must be kept as low as possible, consistent with adequate probe transfer impedance. Refer to Section 8.6.6 of this publication for further details on the design parameters and characteristics of current injection probes.

- Be sure that the test load power bias dissipation capability is not exceeded.

- Apply power bias gradually starting with zero input power voltage to the test set-up.

- Be sure the series resonating inductors can carry the power bias current.

- Protect the power bias circuitry from test filter short circuit failure by fuses or circuit breakers.

- Be sure the ammeter used to measure power bias current has full scale values at least equal to or preferably greater than rated filter load current.

8.6.5.5 Advantages/Disadvantages

- Series signal injection using current probes eliminates hard-wired connections between the signal generator and the filter under test, and between the filter and the detector.

- Filter load impedances can be varied over a wider range.

- Possible damage to generation and detection equipment from the power bias source is practically eliminated.

- The series injection technique has a lower maximum frequency limit for measuring pi-filters than does the parallel injection method. The current probes inject a series impedance into the test circuit, limiting the high frequency measurement range.

- Low frequency measurement limits are established by the large values of series inductors needed to resonate the filter capacitors.

- For wide dynamic range, current probes able to handle appreciable amounts of power must be available.

- Large by-pass capacitors will be necessary if testing to very low frequencies is required. Additionally, probe low frequency transfer impedance performance may also establish the low frequency test limit.

8.6.6 Current Injection Probes

8.6.6.1 General

The Series Signal Injection Test method just discussed (Section 8.6.5) for performing filter evaluations under load requires that a test signal be coupled to the filter input through a current probe device. Since commercially available probes are not normally employed in this way, some discussion is needed of the design characteristics of a current probe that can be used for signal injection applications.

There are five major design objectives for a current injection probe. These objectives are as follows:

- Low insertion resistance - The effective resistance the probe couples to the filter input circuit must be kept as low as possible to assure that filter resonances are not damped out, and that other measurement errors are not introduced.

- Low insertion reactance - The reactance introduced into the filter input circuit by the probe must also be kept to a minimum, since the higher this inductance, the lower will be the maximum frequency at which the Series Signal Injection test can be performed.

- High transfer impedance - The current injection probe is actually

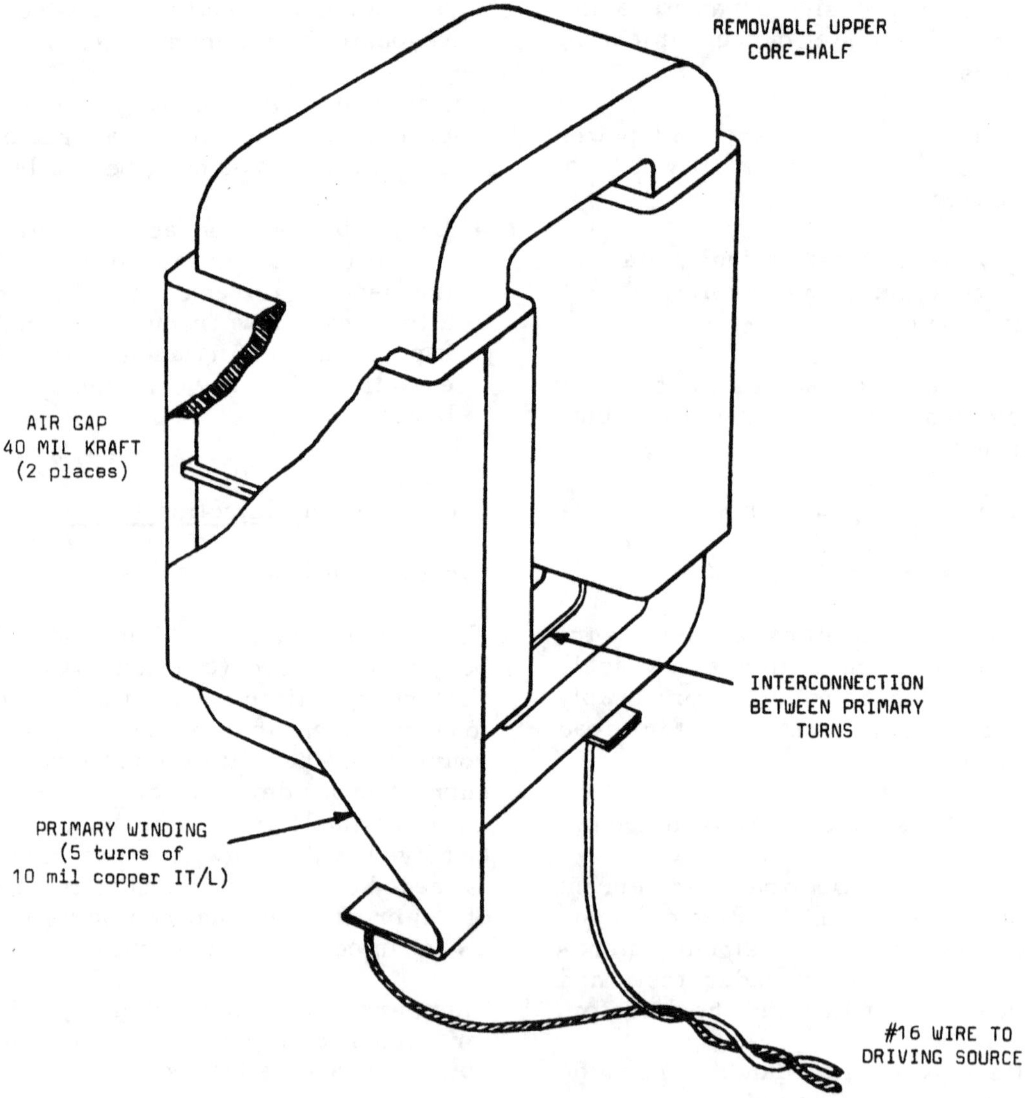

Figure 8-38A
SAMPLE CURRENT INJECTION PROBE

being used to inject a voltage in series with the filter under test and this voltage level is proportional to the probe transfer impedance.

- High current handling capability - The probe must be designed to handle the rated prime power load current (100 amperes or more in some cases). Also, the primary winding of the probe must be able to carry a current that will result in adequate dynamic range for the insertion loss test.

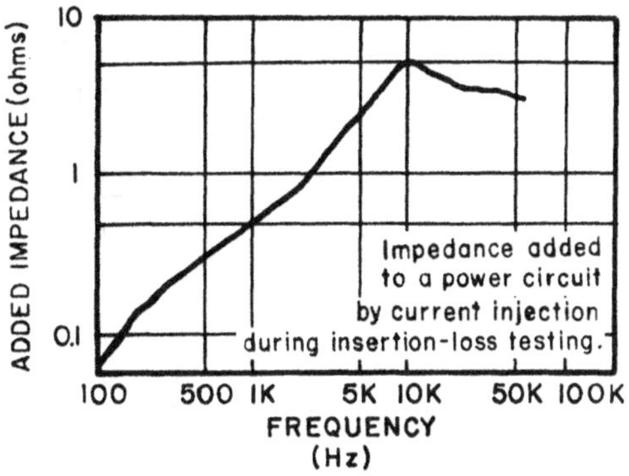

Figure 8-38B

Impedance Added by Sample Probe

- Good external shielding - Sufficient shielding must be provided to assure that the probe is not the cause of the test signal bypassing the filter.

As in most designs, the above requirements are not mutually exclusive. For example, the higher the number of primary turns on the probe, the lower the insertion resistance. However, the transfer impedance will decrease and the transfer impedance may increase with increased turns. Probe design must therefore be a compromise in performance objectives.

8.6.6.2 Sample Probe Design

One example of a current probe that has been used successfully in making Series Signal Injection filter tests is shown

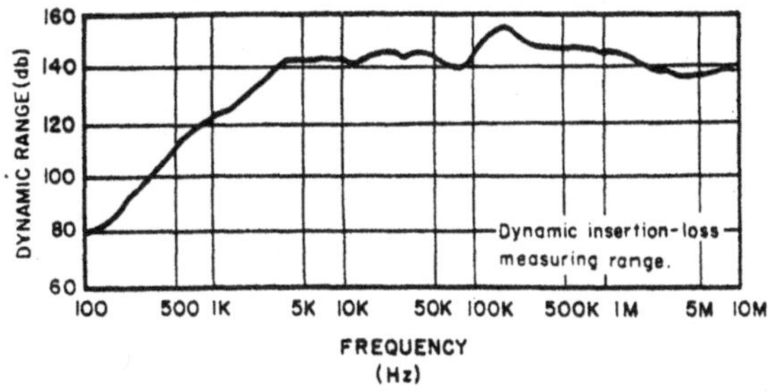

Figure 8-38C

Sample Probe Dynamic Range

in Figure 8-38A[18]. The probe is constructed for a U core and ribbon conductors. The core is 12.9 sq. cm. in area, and consists of 2 mil grain-oriented silicon iron weighing 5.8 pounds. The upper half of the core is removable in order to allow the placement of a conductor within the probe aperture.

An airgap of 80 mils total is provided by 40 mil spacers in each side of the core.

The primary windings of the probe consist of two turns of flat ribbon conductor that are as wide as the vertical sides of the core. The two-turn windings on each core side are connected together by a one-inch wide strip that also forms a fifth turn. The ends of the conductor windings are brought out for connection to the driving source.

The unit is encapsulated in a high temperature epoxy compound and mounted in a brass electrostatic shield case. Heavy copper terminals are provided in order to facilitate high current excitation at low frequencies. All materials used in the construction are rated at least for +200° C.

One test arrangement employed an audio oscillator and amplifier combination driving the probe. The amplifier was rated at 75 watts and had 4 and 8 ohm output terminals. The impedance reflected into the input circuit under test was found to be as shown in Figure 8-38B. This impedance level can be reduced by using a separate step-down transformer at the amplifier output, or by designing a probe having more primary turns.*

The level of current that can be injected into the filter input at frequencies between 4 and 300 kHz can be raised significantly by capacitively tuning the probe. This increases the impedance of the capacitor-probe combination so that it more closely matches that of the driving source. Approximately 100 uf are required at 4 kHz, decreasing to 0.1 uf at 300 kHz. The capacitor is adjusted for a voltage peak across the probe windings.

Figure 8-38C shows the representative dynamic range capability of the probe when it is tuned as indicated above. The advantage of this technique at frequencies higher than 4 kHz is evident from the figure.

A typical test setup using the current injection probe is illustrated in Figure 8-38D.[19] Note the shielding precautions taken to prevent the input signal energy from bypassing the filter under test.

It should be noted that the use of an input current probe is not restricted to the Series Signal Injection test, where the filter input and output are resonated to obtain a "worst-case" test result. Figure 8-38E shows the results of using an input signal probe to test a filter under different line conditions, for fixed terminating resistances. The partial saturation of the filter inductor cores under loaded conditions causes the reduced insertion loss above 10 kHz.

*See Reference 12 for an example of a current injection probe with a 30:1 turns ratio.

1. Signal Generator, such as HP-200-AB, AN/URM-25F
2. Power Amplifier, McIntosh 75 watts
3. RF Voltmeter, HP-400-E
4. Variable Capacitor, 0 to 110 uF
5. Current Injection Probe
6. Variable resistive load bank
7. Power filter being measured
8. Clamp-on power current probe meter
9. RF current probe, 10 Hz to 15 MHz, Genistron GCP-5120
10. Isolation pad, 10 dB
11. RF Field Strength meter

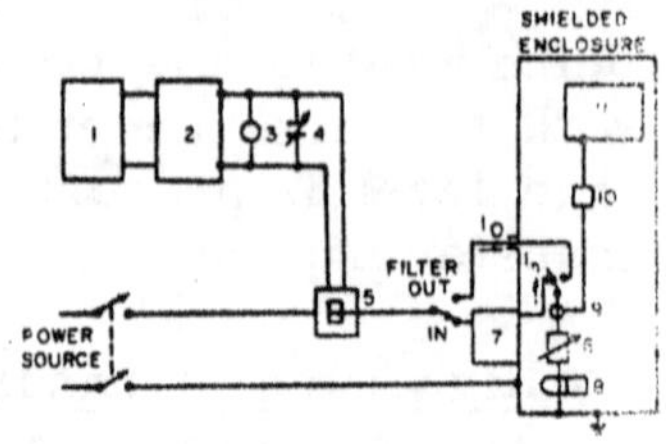

Figure 8-38D

Filter Test Block Diagram

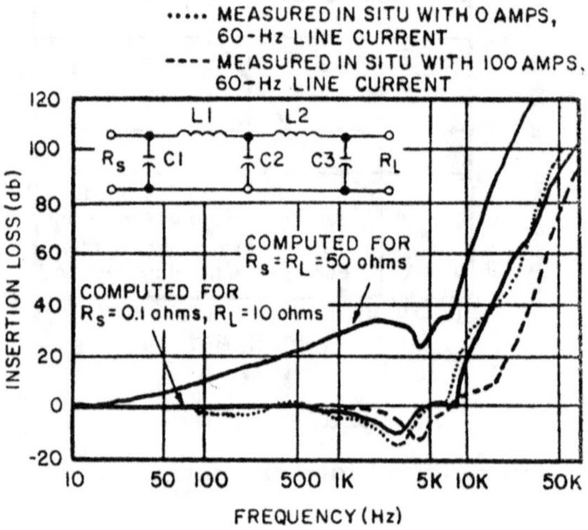

Figure 8-38E

Insertion Loss Data Using Current Injection Probe

8.6.7 Leakage Current Test

8.6.7.1 Objective

The objective of this test is to determine the leakage current per phase of a power line filter when driven by an AC power source. This may be done at 60 Hz or 400 Hz as appropriate, with the filter separate from or installed in a GSE. The leakage current test may be performed in conjunction with the insertion loss test (para. 8.6.2) or EMI tests (per MIL-STD-861). See paragraph 8.8.j for safety considerations.

8.6.7.2 Test Set-up

The test configuration is shown in Figure 8-38F. Three alternatives for measuring the leakage current are shown. One is to use a millimeter (with negligible internal impedance) inserted in the ground return to make the measurement directly. An alternative is to insert a known resistance (250 ohms or less) in the ground return, use a millivoltmeter to measure voltage drop across the resistor, and calculate the leakage current. Another alternative is to put a current probe clamp around the ground return, measure voltage out of the probe with a millivoltmeter, and calculate the leakage current from the probe calibration data. One phase of the power input is energized while the other two phases and the neutral are open circuited.

If the filter is tested separately, the power source should provide a waveform that has the characteristics specified in Table 8-3. The filter may be terminated in the GSE or may be short circuited during the test. True RMS voltmeters should be used for voltage measurements, or individual

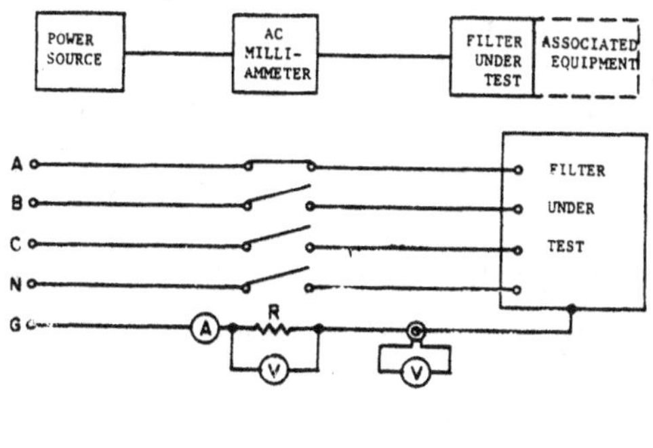

Figure 8-38F

Leakage Current Test Block Diagram and Schematic Diagram

frequency components can be recorded separately and combined on an RMS basis.

8.6.7.3 Test Procedure

With the equipment arranged as shown in Figure 8-38F, perform the following steps:

Energize one input to the filter and determine the appropriate ammeter or voltmeter reading.

Convert voltmeter reading to current value using resistance value or clamp calibration factor.

Repeat above steps for other inputs to the filter and add the results vectorially.

8.7 FILTER INSTALLATION AND MOUNTING

When filters are used, it is absolutely necessary to follow certain installation guidelines if good results are to be obtained. The rf impedance between case and ground must be kept as low as possible. Otherwise, the filter insertion loss will be seriously degraded at the higher frequencies. Effective separation of input and output wiring is mandatory because the radiation from wires carrying interference signals can couple directly to output wiring, thus circumventing and nullifying the effects of shielding and filtering. If complete isolation is affected between input and output, filter insertion loss will approach the design figure.

Both filter mounting guidelines given above can be readily satisfied by the use of bulkhead mounted feedthrough filters such as shown in Figure 8-39.

A wide variety of these filters is available as off the shelf items from a number of manufacturers that can satisfy most filtering requirements.

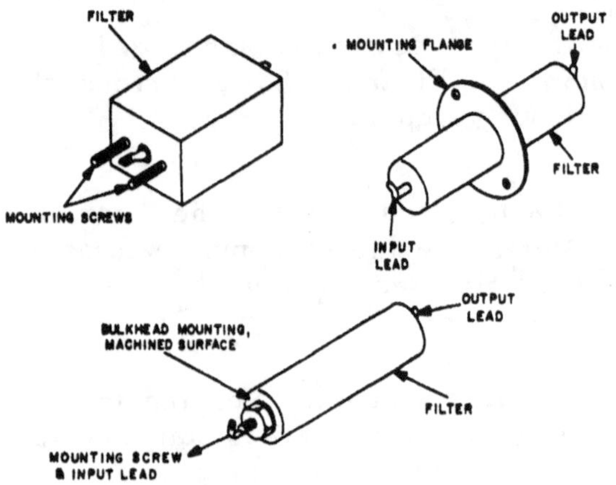

Figure 8-39

Typical Feed Through Filters for Bulkhead Mounting

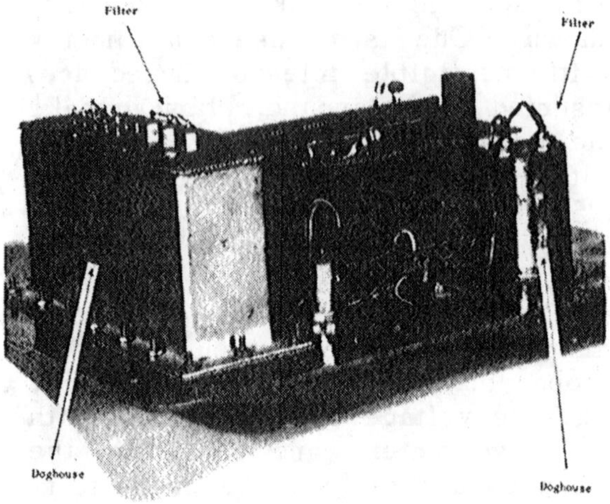

Figure 8-40

Example of Filter 'doghouse'

Figure 8-40 shows the preferred method of mounting feedthrough filters in equipment that must be used in a high level electromagnetic environment such as exists on an aircraft carrier flight deck. Here the filters are mounted in a metal enclosure behind the front panel known as a "doghouse" that provides excellent isolation of filter input and output terminals. This enclosure can be constructed with an access panel on one of the sides. However, the panel should be attached to the doghouse using an rf gasket following the manufacturers recommendations for obtaining an effective seal. (See 5.5.2)

Another effective filtering method uses a filtered pin connector. This arrangement, called a lossy connector, was shown in Figure 8-19. In this case the EMI filter is an integral part of the connector. The filtered pin connector will generally require less equipment space for mounting than will a connector/filter/doghouse assembly.

8.8 SPECIFYING FILTERS

In designing or selecting a filter for a particular application, many parameters must be taken into account if the filter is to be effective. Its insertion loss versus frequency curve is obviously the primary characteristic that determines the suitability of the filter for a particular EMI application. However, other electrical and mechanical requirements must also be designated. These include the following:

a. IMPEDANCE MATCHING - The elements of the filter must be chosen so that the impedance network matches the line into which it is inserted. This is especially true of transmission lines, so that the filter does not impair the normal function of the equipment at both ends of this line. When a filter must be installed in a circuit where its source impedance and/or load impedance is either not known or may vary over a relatively wide range, it may be desirable to terminate the filter into fixed impedances to stabilize its performance.

b. VOLTAGE RATING - Consider the voltage rating on the filter, particularly if used on power lines. Under some conditions, the voltage may deviate by a large amount from its normal value. In addition, short duration pulses whose amplitudes are well above rated line voltage may be on the power circuits, both AC and DC. The filter voltage ratings must be sufficient to provide reliable operation under the extreme conditions expected.

c. VOLTAGE DROP - Determine the maximum allowable voltage drop through the filter and design accordingly.

d. CURRENT RATING -Current rating should be for the maximum allowable continuous operation of the filter. Calculate the current rating for filter elements, such as capacitors, inductors, and resistors. Whenever possible, the current rating of filters should be consistent with the current rating of the wire, circuit breakers, or fuse with which the filter will be used. A filter with a higher current rating than the circuit in which it is installed will often add a weight and space penalty. A filter with a lower rating will have poor reliability and may be a safety hazard. The safety factor used in rating filters should also be consistent with those used for other circuit components.

e. FREQUENCY - Consider both the operating frequencies of the circuit and the frequencies to be attenuated. In general, the cost of a filter rises rapidly as the required rate of skirt falloff goes up, so be careful in identifying insertion loss versus frequency needs. Also, be sure to select the appropriate type of filter for your needs - you might be better off with a band-reject filter to reduce the level of a single close-in narrowband source, than to use a sharp falloff low-pass filter.

f. INSULATION RESISTANCE - The insulation resistance of the filter may vary during the life of the filter. Determine the maximum allowable variation of this resistance for proper filter operation, and design accordingly.

g. SIZE AND WEIGHT - Size and weight may be important in some GSE filter applications. When space is at a premium, adding or subtracting various filter elements may be traded against reduced size and weight of the filter. Filter manufacturers are fairly flexible in being able to provide a wide choice in shape of the filter unit, its method of mounting, and the methods of making connections.

h. TEMPERATURE - The filter must be able to withstand the environmental operating ranges of the equipment in which it will be used. In most cases, the GSE requirement will probably necessitate a -65 to +85°C temperature range.

i. RELIABILITY - Filter component reliability must be commensurate with the equipment reliability requirement, and should be high relative to other equipment components. This is primarily dictated by the fact that faults in EMI filters may be somewhat more difficult to locate than faults in other components.

As an aid to filter procurement, Table 8-3 is provided. This type of form is similar to that contained in the Appendix of MIL-F-18327C, but is applicable to both MIL-F-18327C[14] and MIL-F-15733E[15] requirements. It can be useful to the GSE designer in establishing his filter requirements, whether they are to be met in-house or by an outside source. The designer who must specify EMI filter requirements should familiarize himself with MIL-F-18327C and MIL-F-15733E (as well as MIL-STD-202D[16], not only as far as Table 8-3 is concerned, but to obtain a clear understanding of the requirements that must be imposed on filters.

j. LEAKAGE CURRENT (POWER LINE FILTERS) - The capacitance between individual inputs to a power line filter and the filter case should be minimized to control leakage currents on the ground return. Low leakage will prevent shock hazards to personnel if the filter case should become ungrounded, and will also reduce common-mode interference (hum) as well as prevent inadvertent tripping of ground fault indicating devices. Maximum filter leakage current in a single equipment should not exceed 30mA per phase (5mA per phase in test equipment) in order to comply with existing specifications. This translates to 0.6 μf and 0.1 μf (0.1 μf and 0.016 μf for test equipment) of input capacitance per phase at 60 Hz and 400 Hz, respectively. Where filtering requirements cannot be met without exceeding the input capacitance limits, an isolation transformer should be used to isolate the equipment from the power source. Other means such as grounding the power line neutral at the filter case, should be explored. Paragraph 8.6.7 outlines a test procedure to test for leakage current.

TABLE 8-3

Filter Information Sheet

Type Designation FR ________________ FL ________________ Specification Sheet (if applicable) ____________

Type of filter: High pass ☐ Low pass ☐ Band pass ☐ Band suppression ☐ Power line application ☐ , including the Power line frequency, ____Hz Dual functioning ☐ Other ☐ (specify) ____________.

Outline drawing (including case, terminal, and mounting dimension, and terminal identification):

Finish: ____________ Stamping/Marking Reqts: ____________ Equivalent to ______ except ________.

REQUIREMENTS:

CASE: Metal ☐ Magnetic ☐ Other ☐ (specify) __________. Maximum allowable weight ____________.

TERMINALS: Solder lug ☐ Pin ☐ Solid wire lead other than printed circuits ☐ Special Specify ____________.

Terminal Strength Test, Method 211, MIL-STD-202D:

Test condition A (pull test) ☐ Applied Force in lbs: 1 ☐ , 2 ☐ , 4½ ☐ , 9 ☐ .

Test condition B (bend test) ☐ Number of bending operations: 2 ☐ , 5 ☐ .

Test condition C (wire-lead bend test) ☐ Load: ½ lb ☐ , 2½ lb ☐ , 5 lb ☐ .

Test condition D (twist test) ☐ .

Test condition E (torque test) ☐ Torque applied to non-wire, rigid type terminals if greater than 1 inch ______ torque (ounce-inches).

Sealing: MIL-F-18327C: Paragraph 4.7.3.1 ☐ , 4.7.3.1.1 ☐ , 4.7.3.2 ☐ ; Method 112, Condition A, MIL-STD-202D ☐.

Dielectric withstanding voltage: Method 301, MIL-STD-202D.

At atmospheric pressure:

Magnitude of test voltage______volts AC.

Duration of test voltage (if other than 60 seconds) ______seconds.

Points of application of test voltage ______.

At reduced barometric pressure: Method 105, MIL-STD-202D.

Test condition B ☐ . Other (specify)______.

Insulation resistance: Method 302, MIL-STD-202D.

Test condition A ☐ Other (specify)______.

Points of measurement: Between terminal and mounting ☐ , between terminal and case ☐ , Other (specify) ________

Overload: Paragraph 4.6.10, MIL-F-15733E ☐.

Stability at temperature extremes: Paragraph 4.7.6, MIL-F-18327C ☐.

Life: Method 108, MIL-STD-202D.

Test condition B [ ], Test condition D [ ], Test condition F [ ], Duration of test if other than test condition B, D, & F ________ hrs.

Operating conditions: Test potential ______, Duty cycle______, Load______, Test Temp ______, Tolerance ______.

Temperature rise: Paragraph 4.7.9, MIL-F-18327C [ ]. Paragraph 4.6.4, MIL-F-15733E [ ].

Vibration: Low frequency [ ], Method 201, MIL-STD-202D. Load condition ________.

High frequency [ ], Method 204, MIL-STD-202D. Load condition ________.

Shock: Method 205, MIL-STD-202D, Paragraph 4.7.11.1, MIL-F-18327C and Paragraph 4.6.16, MIL-F-15733E [ ],

Test condition A [ ], B [ ], C [ ]; Method 207, MIL-STD-202D, Paragraph 4.7.11.2, MIL-F-18327C [ ].

Thermal shock: Method 107, MIL-STD-202D, Paragraph 4.7.12, MIL-F-18327C.

Temperature cycling: Method 104, Condition A, MIL-STD-202D, and Paragraph 4.5.15.2, MIL-F-15733E [ ].

Immersion: Method 104, MIL-STD-202D,

Test condition A [ ], B [ ], C [ ].

Moisture resistance: Method 106, MIL-STD-202D, Paragraph 4.7.14, MIL-F-18327C [ ].

Paragraph 4.6.18, MIL-F-15733E [ ].

Flammability (grades 5 and 7): Method 111, MIL-STD-202D.

Salt Spray: Method 101 of MIL-STD-202D [ ].

<u>Electrical characteristics</u>

Impedance:

Input ______ ohms. Output ______ ohms. Transfer ______ ± ______ ohms (if applicable). Load ______ ohms.

Resonating capacity ______ pf ± ______ pf (if applicable).

Capacitance to ground: Method 305, MIL-STD-202D and 4.6.3, MIL-F-15733E: ______ pf at ______ kHz

______ pf at ______ kHz

Signal input voltage ______ volts.

Current rating: AC, ______ amps DC, ______ amps

Voltage rating: AC, ______ volts DC, ______ volts

Duty Cycle (if intermittent): Time on ________ Time off ________

Maximum allowable voltage drop: ________ volts

Reference frequency ______ (Hz) [ ] (kHz) [ ] (MHz) [ ].

Insertion loss (at reference frequency) ________ dB (min.).

Discrimination Frequency Range [ ] (Hz) [ ] (kHz) [ ] (MHz)	Min (dB)	Max (dB)

FILTERING BIBLIOGRAPHY

1. HIASD, AFSCM 80-9, Handbook of Instructions for Aerospace Systems Design, Volume IV - Electromagnetic Compatibility, 20 April 1964.

2. Department of Defense, Washington, D.C., MIL-STD-220A, "Method of Insertion-Loss Measurement," 15 December 1959.

3. Air Force Systems Command Design Handbook, "Electromagnetic Compatibility," DH 1-4, Washington, D.C., Second Edition, 1972

4. Naval Air Systems Command, Washington, D.C., "Electromagnetic Compatibility Manual," NAVAIR 5335, 1968.

5. Phillip R. Geffe, "Simplified Modern Filter Design," John F. Rider Publisher, Inc., New York, N.Y. 1963

6. Department of the Navy, Washington, D.C., "Reliability Design Handbook," Section III.

7. Bertram Wellman, "Convenient Twin-T Circuits," 1958 Electrical Design Manual, Rogers Publishing Co., Englewood, Colo., 1958, pp. 26, 27.

8. Donald R.J. White, "Electromagnetic Interference and Compatibility," Vol. 3, EMI Control Methods and Techniques, First Ed., 1973.

9. Ronald J. Prehoda, "Evaluation of a Real-Time Interference Cancellation System," NWL Technical Report TR-2329, August 1969.

10. R.B. Cowdell, et. al., "Evaluating Filters in Situ Under Heavy Load Currents and Normal Working Impedances," 1967 IEEE Electromagnetic Compatibility Symposium Record, July 18-20, 1967, pp. 352-364.

11. S.H. Eisbruck and F.A. Giordano, "A Survey of Power-Line Filter Measurement Techniques," IEEE Transactions on Electromagnetic Compatibility, Vol. EMC-10, No. 2, June 1968, pp. 238-242.

12. H. Weidmann, and W.J. McMartin, "Two Worst-Case Insertion Loss Test Methods for Passive Power-Line Interference Filters," IEEE Transactions on Electromagnetic Compatibility, Vol. EMC-10, No. 2, June 1968, pp. 257-263.

13. Naval Ship Engineering Center, Electronics Installation and Maintenance Book, Electromagnetic Interference Reduction, NAVSHIPS 0967-000-0150, June 1970.

14. Department of Defense, Washington, D.C., MIL-F-18327C, "Filters: High Pass, Low Pass, Band Pass, Band Suppression, and Dual Functioning, General Specification For," 25 May 1966.

15. Department of Defense, Washington, D.C., MIL-F-15733E, "Filters, Radio Interference, General Specification For," 24 March 1965.

16. Department of Defense, Washington, D.C., MIL-STD-202D, "Test Methods for Electronic and Electrical Component Parts," 14 April 1969.

17. Naval Air Facility, Crane, Indiana, "Design Guide for EMI Reduction in Power Supplies," draft copy.

18. R.B. Cowdell and W.K. Grossmann, "Development of a Current Injection Probe (CIP) for High Power Level Filter Analysis," Genistron Division, Genisco Technology Corp., Final Report on Contract NBY-62222 for Naval Civil Engineering Laboratory, Port Hueneme, Cal., Nov. 1966.

19. D.B. Clark, et. al., "Power Filter Insertion Loss Evaluated in Operational-Type Circuits," IEEE Transactions on Electromagnetic Compatibility, Vol. EMC-10, No. 2, June, 1968, pp. 243-255.

20. H. M. Schlicke, "Assuredly Effective Filters," 1975 IEEE Electromagnetic Compatibility Symposium Record, page 5AIId.

21. E. E. Wetherhold, "Passive LC Filter Design," Home Study Course 69/51, Measurements & Control Journal, June 1978.

9.0 TESTING REQUIREMENTS AND TECHNIQUES

9.1 INTRODUCTION

Specialized equipment testing is necessary to insure that GSE will ultimately operate in their anticipated EM environments without performance degradation, and that they will not degrade other equipment in these environments. EMC testing of ground support equipment is necessary at both the "component" or sub-system level, as well as at the equipment and system level, and specifically in the following areas:

- Design testing of breadboard or engineering models, to obtain component and circuit engineering performance estimates that will assist in meeting intra-equipment and intra-system needs.
- Performance testing of equipment or system prototypes to evaluate operation in intended environments, or to evaluate compliance to applicable specifications.
- Acceptance testing of production units, to formally demonstrate or provide proof of operation as specified or required on the first article produced.

Planning to support these test areas should be initiated early in GSE design and development stages, so that sufficient consideration can be given to the EMI implications of the particular design employed, and meaningful tests can be defined and implemented. Such tests should not only provide input data from which basic EMC design decisions can be made, but should be able to offer essentially continuous information that can lead to design updates, reduce redundant EMC design, and avoid EMC retrofits.

Previous chapters of this design guide have already provided considerable inputs to the EMI testing area. Test techniques to evaluate shielding effectiveness (Section 5.6), bonding effectiveness (Section 6.5), and filtering effectiveness (Section 8.6) have been discussed. The purpose of this chapter is to provide additional technical information the GSE designer may need in planning and conducting EMI tests. This includes:

- Shielded enclosure requirements
- Testing guidelines
- Radiated and conducted equipment testing

9.2 SHIELDED ENCLOSURE REQUIREMENTS

9.2.1 Enclosure Limitations

A shielded enclosure for equipment test purposes is nothing more than a conductive "box" for enclosing all or part of a test setup. The practical requirements of power access, personnel access, light, heat, and air conditioning, result in discontinuities to this enclosure that must be considered a part of enclosure design.

The main advantage of the shielded enclosure for interference measurements is the rf isolation it provides to and from the outside world. Its use allows meaningful emission and susceptibility measurements to be made, both conducted and radiated, in locations where such testing would not ordinarily be possible because of the level of the ambient environment. It also prevents high level signals that must be generated during the course of testing from causing interference to other devices that may or may not be part of the test.

There are several difficulties in making measurements inside of a shielded enclosure. One is that many radiated tests employ antennas whose dimensions may be large compared to the size of the available enclosure. The result of antenna elements coming close to the conducting walls of the room is to load the antenna and cause changes in its calibration and radiation characteristics.

A second difficulty is that the multiple reflections of emitted signals that will occur in a shielded enclosure will result in standing waves in the room and potential measurement errors. This effect becomes particularly significant above the lowest resonant frequency of the room (see Figure 9-1)[1]. It can be aggravated by personnel within the enclosure, who can cause the standing-wave pattern in the room to change as they move about. There is no way to avoid this problem, but it may be alleviated to an extent in some measurements by moving the test antenna over a small volume to determine the worst-case susceptibility/interference radiation, by using microwave absorbing materials to reduce reflections, and by keeping the number of people and volume of extraneous equipment in the room to a minimum. A third difficulty is that practical room sizes result in many radiated tests being performed at distances for which $r \ll D^2/2\lambda$ (see Section 2.3.2), where levels of field strength are very difficult to establish and measure.

The alternative to testing in a shielded enclosure is open-site testing[2]. This requires a location several miles from

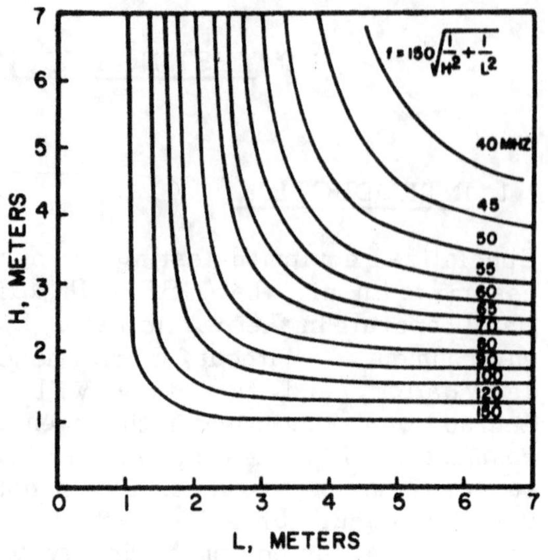

Figure 9-1

Lowest Natural Resonant Frequency Chart

any concentration of machinery or electronics and also some distance from overhead power lines. Such a site assures a reasonable ambient level, usually limited only by atmospheric noise and signals in known crowded frequency bands. The advantage of such a site is offset by inconvenience in having to go to a remote site for testing, and in the special precautions necessary to get power to the site. Also, use of an open-site does not eliminate the need for a shielded enclosure where it is necessary to isolate one part of a particular test employing a radiating signal source from another part of the test that uses a sensitive detection device.

9.2.2 Enclosure Design Considerations

There are three basic types of shielded enclosures. They include:

- Single wall
- Double wall, with the walls connected together at many points

- Double wall, with the walls connected together at only one point

The enclosures are assembled in panel sections. The sections may be bolted, clamped, or welded together to create the room. In the case of the multiply-connected double-wall enclosure, the inner and outer walls are connected around the periphery of each panel member. In the case of the single-point connected double-wall enclosure, the inner and outer walls are obviously kept separate electrically when the panel members are assembled.

Enclosures are made of either screen or solid metal. Typical screen materials are copper (22 x 22 - .015, measuring 22 x 22 wires per inch, each wire being 0.015" in diameter), bronze (18 x 20 - .011), and hot galvanized (.011 - 162). Typical solid sheet materials used include 2402. copper and 24 gage galvanized steel, or a combination of copper and steel sheets in double wall enclosures[3].

Room cost increases as one goes from a single-wall to a double-wall multiply-connected, to a double-wall single-point connected room. It also is higher for a solid wall room than for a screened room, and for a welded room versus a bolted or clamped room. Since room shielding effectiveness also goes in these same directions, it is important to carefully define the requirements needed in an enclosure before taking steps to purchase a room.

As far as power and signal feed-through requirements are concerned, a number of adequate connectors are available commercially that can provide the necessary continuity through the enclosure wall, and the isolation of signal leads that must be obtained for effective testing. The power lines are filtered for each phase separately and for the neutral at the entrance to the enclosure. Adequate filters are available commercially from a number of manufacturers[3].

The heating and air-conditioning ducts for both fresh air and return air are generally built into a single barrier panel. Their form is that of a honeycomb whose element dimensions act as waveguides operating below cutoff. The cutoff frequency of the honeycomb is many times that of the highest frequency used in the enclosure in order to protect the enclosure characteristics. The correct penetration of the barrier panel is generally specified by the manufacturer[3].

The opening of the enclosure that is the most critical and the one that can cause the most EMI problems is the entrance/exit door. There are many methods that will preserve the continuity of the room, and a wide variation is found amongst manufacturers.

The most common method of providing the barrier continuity is through the use of finger stock around the perimeter of the door. The finger stock can either be mounted around the door itself, or around the door frame. When the door is closed, the fingers are compressed, providing a good electrical contact throughout the door perimeter. One manufacturer employs a rather complex system that makes use of a set of pressurized air bags to force the door edges against a mating flange. This provides a good intimate contact so that the required uniformity of attenuation characteristics is protected.

There must be lighting in the enclosure. This should in all cases be incandescent, because other types of lighting usually involve ionization processes and subsequently produce rf noise.

Figure 9-2 provides an illustration of a commercially available solid-wall shielded enclosure. In Table 9-1, some of the general electrical characteristics of the three major room types are summarized. The 120 dB limit usually represents a testing limit, and not the limit of the room itself.

Courtesy of Erik A. Lindgren & Assoc., Inc.

Figure 9-2

Representative Solid Wall Shielded Enclosure

The characteristics of rooms designed to meet special performance requirements are not included in Table 9-1. For example, rooms have been designed and built that provide a magnetic shielding effectiveness of about 60 dB at 60 Hz, and reach the limit of typical measurements (120 dB) at 200 Hz.

9.2.3 Microwave Absorbers

A well-designed rf anechoic (echoless) chamber must first be a good shielded enclosure, and then be further modified or finished into an anechoic enclosure. The basic requirements of such a chamber are high attenuation through the walls and high absorption on the inside walls.

Microwave-absorbent materials such as polyurethane foam, vinyl, and epoxies are used in anechoic chambers. These materials, usually pyramidal in shape, are placed on the surfaces of the chamber to absorb microwave energy. Signals from the test equipment or equipment being tested that are not absorbed by the equipment can be absorbed by the walls so that reflections are prevented or minimized. Cone-shaped sections of the absorbent materials are also available for incorporation in standard shielded enclosures.

The size of the cones or wedges determines, to a great extent, the lowest frequency at which the absorbent material is effective. The lowest practical limt at the present time is about 200 MHz, and requires using very large wedges. This rather high frequency, from an interference test standpoint, reduces the usefulness of the rf anechoic chamber for general-purpose radiated energy measurements.

Figure 9-3 indicates a representative application of microwave-absorbent material. The material shown in the figure is effective about 300 MHz, and provides a reflection loss of about 40 dB at 1 GHz.

Table 9-1

Summary of Typical Shielded Enclosure Performance*

Enclosure Type	Wall Type	Material	Magnetic Fields		Electric & Plane Waves	
			60 Hz	15 kHz	1 GHz	10 GHz
Double Electrically Isolated	Screen	Copper	2 dB	68 dB	120 dB	77 dB
		Bronze	0 dB	40 dB	110 dB	57 dB
		Galvanized		50 dB	50 dB	
	Solid	24 Ga. Steel	15 dB	84 dB	120 dB	90 dB
		Cu & Steel	18 dB	86 dB	120 dB	106 dB
Double Non-Electrically Isolated	Screen	Copper		52 dB	90 dB	
		Bronze				
	Solid	24 Ga. Steel		68 dB	90 dB	
		Copper			90 dB	
Single Shield	Screen	Copper	6 dB	42 dB	60 dB	
		Bronze				
	Solid	Copper			75 dB	
		24 Ga. Steel		48 dB	80 dB	

*Measurements made in accordance with MIL-STD-285.

Courtesy of Emerson & Cuming, Inc.

Figure 9-3

Representative Anechoic Chamber

9.3 TESTING GUIDELINES

9.3.1 General

In performing EMI testing of GSE, both standard and specialized test equipment may have to be employed. In either case, general guidelines can be established regarding the required characteristics of this equipment, and the test setups in which they might be used. The following paragraphs are intended to document some of these guidelines.

9.3.2 Signal Sources

General laboratory signal sources and generators are usually suitable for EMC applications as well. The generator should have good frequency and amplitude stability, with an output level calibrated to within at least 2 dB. Its output VSWR should not exceed 1.3:1 (<0.1 dB mismatch loss) when terminated in the design output impedance.

Many signal generators exhibit some case or power line leakage that becomes particularly noticeable when the generator attenuator is near its maximum setting. This leakage may create a problem under low-level testing conditions. Particular attention should be given to keeping this leakage level down, through addition of a power-line filter to the generator, by improving the generator case shielding or enclosing the generator in a small shielded enclosure, or by using a lower leakage generator.

Care should be taken that signal generator harmonics are attenuated sufficiently to prevent false GSE responses from these signals. Low-pass or bandpass filters of known insertion loss at the desired frequency may be used for this purpose.

Many EMI tests necessitate a large dynamic measurement range. This would be the case in order to evaluate shielding or filtering effectiveness. Under these circumstances, high generator or source output levels are desirable. The particular level required depends on the type of test being performed, the sensitivity of the detection device being used, and other factors. Signal generator/power amplifier combinations can provide an adequate power level. Pulse generators can often be used to advantage when high peak power levels are needed.

Some pulse signal generators of the master-oscillator, keyed-amplifier type generate a cw output during the interval between pulses. This residual cw energy should be well below the peak pulse level (perhaps 30 dB down), and data collected using this type of generator should be carefully monitored to be sure a false response is not being obtained from the cw component of the signal.

9.3.3 Sweep Generators and Oscillators

Much EMI testing must cover measurements over a wide frequency range. Under these circumstances, it is convenient to use a frequency sweeping technique, and automatically record the detected output signal through a broadband receiver.

Commercial sweep oscillators are available that can be used for this purpose. They typically sweep an octave range at microwave frequencies (and wider at lower frequencies), have a wide range of sweep rates, have a fairly flat output level over the sweep range (within 0.1-0.2 dB), and put out ten to 100 milliwatts of power[3].

9.3.4 Attenuators

Fixed and variable attenuators are used extensively in EMI testing. The units should be accurate to within 0.5 dB when terminated in their characteristic impedance. As noted elsewhere in this design guide, proper termination of attenuators, as well as other test equipment, is important in maintaining test accuracy.

Variable or stepped attenuator assemblies are also frequently built into signal sources and detection devices. When attenuators are incorporated into a detector, it is important to know where they are located relative to the detector input. For example, it is often assumed that the attenuator in a frequency-selective voltmeter is at the input terminals of the voltmeter, and insertion of attenuation will prevent voltmeter front-end saturation; this is not always the case for low attenuation values. The initial attenuation may be added at IF, and therefore have no effect on the front-end situation.

When attenuators are clustered in a switch arrangement, particular care should be taken to be sure that energy does not get bypassed through the switch assembly and result in erroneous readings. About 60 dB of isolation appears to be a practical level of switch decoupling to be expected[4].

9.3.5 Detectors

A wide variety of detection systems are employed for EMC testing. These devices might include power meters, crystal-video detectors, wide-band rf voltmeters, oscilloscopes, spectrum analyzers, and various other standard indicators. However, there is one detection system that is unique to interference analysis; that is an Electromagnetic Interference (EMI) meter.

The EMI meter is a two-terminal tunable voltmeter. It is essentially a calibrated super-heterodyne receiver having a wide dynamic range. Calibration is usually built into the EMI meter, using cw, random noise, and impulse noise sources. An impulse noise source operates by generating a very narrow (less than a nanosecond) repetitive pulse by alternately charging and discharging a short length of transmission line.

For performing radiated field measurements, a series of calibrated antennas are used. EMI meters are available that cover the frequency range of 30 Hz to 40 GHz. Peak and average detection modes are available.

Although current EMI meters are fairly well shielded, the dynamic range requirements of some tests may necessitate additional meter shielding. One way of doing this is to enclose the meter (or other type of detector) in a small shielded enclosure.

9.4 RADIATED AND CONDUCTED EQUIPMENT TESTING

9.4.1 General

In addition to the shielding, bonding, and filtering tests already discussed in this design guide, a number of other test procedures can be identified that are concerned with the EMI performance of a total equipment or system being developed. These procedures are for the purpose of exploring the radiated and conducted emission and susceptibility characteristics of an electrical/electronics device over a wide frequency range.

The generally accepted standard for these types of tests is MIL-STD-462[5]. The tests contained in MIL-STD-462 are listed in Table 9-2. The reader who will be concerned with equipment-oriented EMC measurements, either to obtain an indication of overall GSE performance in the course of equipment design, or to evaluate a contract compliance requirement, is urged to familiarize himself with this specification.

Table 9-2

INDEX OF MIL-STD-462 EMI MEASUREMENT TESTS

Test	Title
	CONDUCTED EMISSION (CE)
CE01	30 Hz to 20 kHz, Power Leads
CE02	30 Hz to 20 kHz, Control and Signal Leads
CE03	20 kHz to 50 MHz, Power Leads
CE04	20 kHz to 50 MHz Control and Signal Leads
CE05	30 Hz to 50 MHz, Inverse Filter Method
CE06	10 kHz to 12.4 GHz, Antenna Terminal
	CONDUCTED SUSCEPTIBILITY (CS)
CS01	30 Hz to 50 kHz, Power Lead
CS02	50 kHz to 400 MHz, Power Lead
CS03	30 Hz to 10 GHz, Intermodulation, Two Signal
CS04	30 Hz to 10 GHz, Rejection of Undesired Signals At Input Terminals (2-Signal Generator Method)
CS05	30 Hz to 10 GHz, Cross-Modulation
CS06	Spike, Power Leads
CS07	Squelch Circuits
CS08	30 Hz to 10 GHz, Rejection of Undesired Signals At Input Terminals (1-Signal Generator Method)
	RADIATED EMISSION (RE)
RE01	30 Hz to 30 kHz, Magnetic Field
RE02	14 kHz to 10 GHz, Electric Field
RE03	Spurious and Harmonic Emissions 10 kHz to 40 GHz
RE04	20 Hz to 50 kHz, Magnetic Field
RE05	150 kHz to 1 GHz, Vehicles and Engine-Driven Equipment
RE06	14 kHz to 16 Hz, Overhead Power Lines
	RADIATED SUSCEPTIBILITY (RS)
RS01	30 Hz to 30 kHz, Magnetic Field
RS02	Magnetic Induction Fields
RS03	14 kHz to 10 GHz, Electric Field
RS04	14 kHz to 30 MHz, Electric Field

9.4.2 Emission Tests

Most GSE do not constitute EMI emission hazards. However, equipment that employ digital clocks, or that generate switching transients, or that incorporate coherent or incoherent signal sources, may be potential interferers over either radiated or conducted paths.

Conducted emission tests at one time were performed by direct electrical coupling to the power or control lead under test through an impedance transformation device. Today, these tests are almost exclusively executed using a magnetically coupled probe device, calibrated in terms of the current flowing in the line under test. If the test is on a power line, a 10 microfarad capacitor is used across each line to establish a low impedance between the line and ground.

Radiated emission tests use an EMI meter or equivalent detector coupled to either a magnetic loop or an electric field sensor. The antennas probe the field around the unit under test and measure the maximum level of radiation. The MIL-STD-462 test distances are 7 centimeters for the magnetic field sensor and 1 meter for the electric field sensor. Sensor calibration is in terms of far-field equivalent levels, even though it is recognized that a significant measurement error can result in using this calibration procedure.

9.4.3 Susceptibility Tests

Equipment-oriented EMI susceptibility tests are very similar in most respects to tests already discussed in this publication. The procedures employed to make radiated susceptibility measurements are the same as the shielding effectiveness tests, except that the indicator is no longer a detection system whose sensor is located inside of a box, but is the system under evaluation. Similarly, conducted susceptibility tests are very much like the filter effectiveness tests, but with device performance being the susceptibility indicator. The MIL-STD-462 conducted procedures couple the test signal onto the line under test through a special impedance matching transformer for tests below 50 kHz, and through a capacitor whose reactance is less than 5 ohms for tests above 50 kHz.

The devices employed for radiated susceptibility measurements in MIL-STD-462 include the following:

- Magnetic loop, with current through the loop being the indicator of input level, and the loop located 5 cm from the equipment under test.
- Antennas, located 1 meter from the equipment under test
- Parallel Plate Transmission Line, of different construction than the one discussed in Section 5.6.7, since the specification requires testing using this device to only 30 MHz.

The direct application of MIL-STD-462 susceptibility testing under some circumstances may not provide a complete picture of GSE operational capabilities and limitations. Differences in test procedures that the GSE test engineer should consider relative to MIL-STD-462 susceptibility testing procedures include (a) the use of modulated input signals for the tests, and (b) the use of levels more representative of the expected GSE operational environment. Both of these modifications can result in more realistic indications of GSE performance.

9.4.4 Transient Testing

Both conducted and radiated transient response tests are designated in

MIL-STD-462. The unique nature of this type of test is worthy of comment here.

A standardized pulse is employed to test equipment transient response. The characteristics of this pulse are shown in Figure 9-4. Transient or spike generators are commercially available that generate this pulse, with peak amplitudes as high as 600 volts[3].

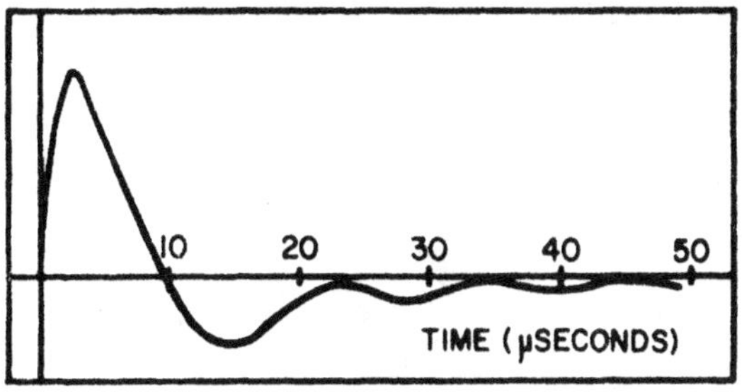

CHARACTERISTICS

(1) Pulse Width = 10 microseconds to first zero crossing, 5.5 microseconds at half-voltage point

(2) Pulse Repetition Rate = 3 to 10 p.p.s.

(3) Pulse Rise Time = approximately 1.5 microseconds, 10 to 90% voltage levels

(4) Voltage Output = Not less than 200 V. peak

(5) Output Control = Adjustable from 0 to 200 V. peak

(6) Output Spectrum = 150 dB uv/MHz at 25 kHz decreasing to 115 dB v/MHz at 30 MHz

(7) Phase Positioning = 0 to 360 degrees

(8) Source Impedance (with injection transformer) = 0.06 ohms

(9) Transformer (current capacity) = 30 amperes

(10) External Synch = 50 to 800 Hz

(11) External Trigger = 0 to 20 p.p.s.

Figure 9-4

MIL-STD-462
Transient Pulse Characteristics

The pulse may not be representative of many signals that might be encountered in a GSE environment, and the test engineer should consider modifying the pulse shape to more realistically represent his needs. In particular, the pulse rise time of the test signal may be significantly longer than pulses created by flourescent lighting, ignition systems, and rotating machinery[3].

The conducted transient susceptibility test is performed on AC or DC power lines. Figures 9-5 and 9-6 show alternative series and parallel test arrangements. Tests should be performed at different spike repetition rates (including those coincident with any clock rates or gating operations of the equipment under test), at different phase positions relative to the normal power-line signal, and at both positive and negative polarities.

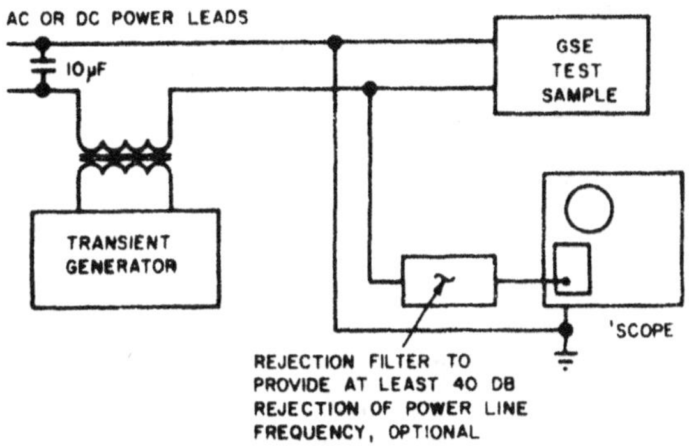

Figure 9-5

Conducted Transient Susceptibility, Series Injection

The radiated transient susceptibility test is an induction field test. It can be performed on input cable bundles or on equipment cases. To measure equipment transient susceptibility via cable bundles, the output of the transient generator is fed to a coil that is wrapped around the cable (see Figure 9-7). To measure transient

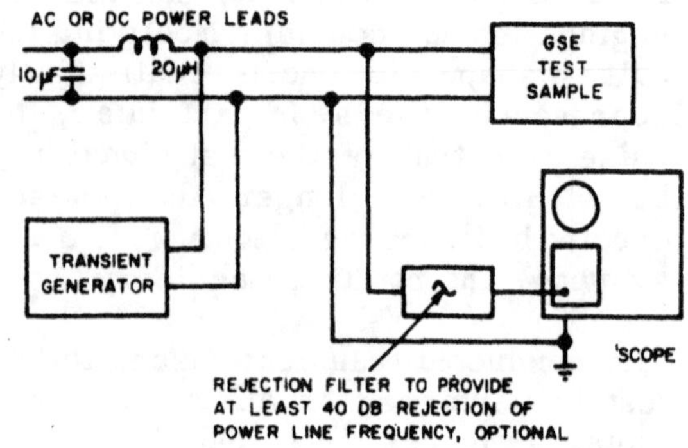

Figure 9-6

Conducted Transient Susceptibility, Parallel Injection

susceptibility through the case, three turns of wire are wrapped around the case and excited by the generator (see Figure 9-8). The generator parameters are then varied as for the conducted transient susceptibility case, to establish the susceptibility threshold of the unit under test.

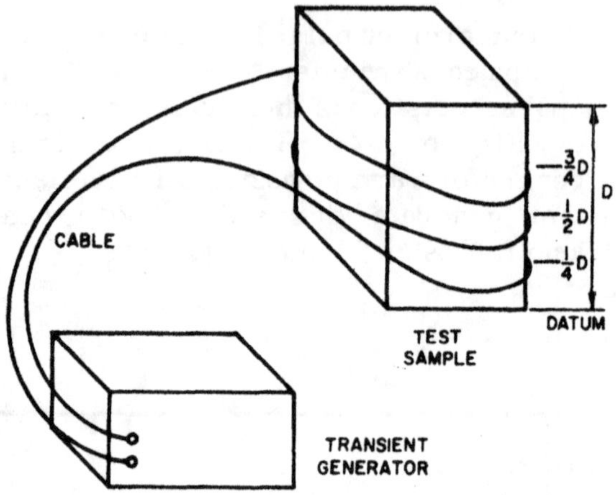

Figure 9-8

Radiated Transient Susceptibility, Case Test

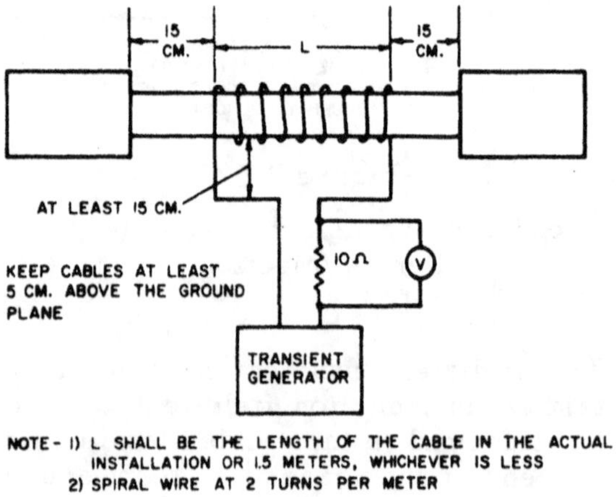

Figure 9-7

Radiated Transient Susceptibility, Cable Test

BIBLIOGRAPHY

1. J.E. Bridges, "Proposed Recommended Practices for the Measurement of Shielding Effectiveness of High-Performance Shielding Enclosures," IEEE Transactions on Electromagnetic Compatibility, Vol. EMC-10, No. 1, March 1968, pp. 82-93.

2. NAVAIR 5335, Naval Air Systems Command Electromagnetic Compatibility Manual.

3. D.R.J. White, "Electromagnetic Interference and Compatibility, Vol.4, EMI Test Instrumentation and Systems," 1971.

4. H.W. Denny, "Considerations in the Design of a Radio Frequency Compatibility (RFC) Test Set," Proceedings of the Tenth Tri-Service Conference on Electromagnetic Compatibility, Nov. 1964, pp. 112-130.

5. Department of Defense, Washington, D.C., MIL-STD-462, "Electromagnetic Interference Characteristics, Measurement of," 31 July 1967.

10.0 SPECIFICATIONS & STANDARDS

10.1 GENERAL

The previous chapters of this handbook have dealt with the objectives of EMC design, and possible methods for achieving the desired end result of EM compatibility. This chapter summarizes the appropriate government design and test specifications that can be employed to quantitatively control or influence GSE EMC design and performance. These specifications identify performance requirements to be met, specific design practices to be employed, and other guidance pertinent to meeting desired EMC design goals.

The specifications selected for inclusion in this design guide have been chosen for appropriateness to Navy avionics GSE. The list is not intended to be all inclusive as regards the entire EMI/EMC discipline. Specifications have been omitted from the list relating to more specialized EMC problem areas or to types of situations not generally expected to be encountered during GSE design. The basis for specification selection were as follows:

- Direct relationship to EMC design of Navy Avionics GSE
- General EMC design guidance for GSE
- Checklist for general design to prevent inadvertent oversight or omission of specific types of problems
- General information regarding system engineering and performance requirements

Information is also provided in this chapter on how to obtain copies of required government specifications.

10.2 APPLICABLE SPECIFICATIONS

The specifications considered appropriate to EMC design of GSE are contained in Table 10-1. The table lists the title and number of the specification, its scope of application and coverage, and comments regarding usage, content, and any unique features.

The data in Table 10-1 reflect three general categories of specifications:

- Criteria and tests for measuring system performance
- Sub-system test and design techniques and requirements
- Component and hardware selection, design and application

For those special cases or situations where the data of Table 10-1 are not sufficient, reference should be made to the Department of Defense Index of Specifications and Standards for all current related documents.

10.3 OBTAINING SPECIFICATIONS

The Department of Defense single stock point (DOD-SSP) for unclassified specifications and standards is:

Naval Publications and Forms Center
5801 Tabor Avenue
Philadelphia, Pensylvania 19120
Phone: 215/697-3321

The DOD-SSP has the responsibility to index, issue, and control the current specifications and standards listed in the Department of Defense Index of Specifications and Standards (DOD-ISS).

Types of documents stocked and distributed are:

- Military Specifications
- Military Standards
- Federal Specifications
- Federal Standards
- Qualified Products Lists
- Industry Documents
- Military Handbooks
- Air Force-Navy Aeronautical Standards
- Air Force-Navy Aeronautical Design Standards
- Air Force-Navy Aeronautical Specifications
- U.S. Air Force Specifications
- Other Departmental Documents
- Air Force-Navy Aeronautical Bulletins
- U.S. Air Force Specification Bulletins

The Department of Defense Index of Specifications and Standards (DODISS) is prepared at NPFC and is published in three parts annually with bi-monthly supplements. Parts I and II of the DODISS and the Federal Supply Classification Listing are available to private industry on request from the Superintendent of Documents, Government Printing Office, Washington, D.C. 20402. Both include the basic index and the cumulative bi-monthly supplements for each part as they are published.

A sample of Form DD 1425, Specifications and Standards Requisition, to be used when requesting documents from NPFC, is shown as Figure 10-1.

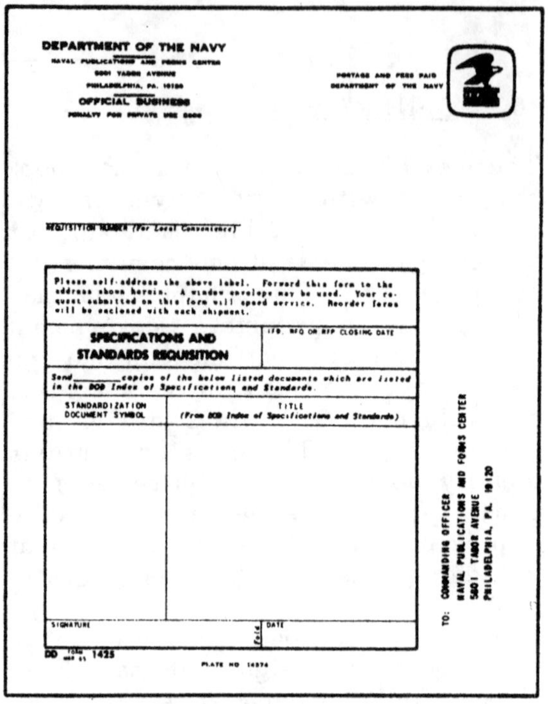

DEPARTMENT OF THE NAVY
NAVAL PUBLICATIONS AND FORMS CENTER
5801 TABOR AVENUE
PHILADELPHIA, PA. 19120
OFFICIAL BUSINESS
PENALTY FOR PRIVATE USE $300

POSTAGE AND FEES PAID
DEPARTMENT OF THE NAVY

REQUISITION NUMBER (For Local Convenience)

Please self-address the above label. Forward this form to the address shown herein. A window envelope may be used. Your request submitted on this form will speed service. Reorder forms will be enclosed with each shipment.

SPECIFICATIONS AND STANDARDS REQUISITION	IFB, RFQ OR RFP CLOSING DATE

Send ________ copies of the below listed documents which are listed in the DOD Index of Specifications and Standards.

STANDARDIZATION DOCUMENT SYMBOL	TITLE (From DOD Index of Specifications and Standards)

SIGNATURE	DATE

Fold

DD FORM 1425

PLATE NO 14574

TO: COMMANDING OFFICER
NAVAL PUBLICATIONS AND FORMS CENTER
5801 TABOR AVENUE
PHILADELPHIA, PA. 19120

Figure 10-1

Standard Order Form, DD 1425, Specifications and Standards Requisition

Table 10-1
Specifications for EMC Design

Specification	Title	Scope	Comments
MIL-B-5087B (ASG) 15 October 1964 Superseding MIL-B-5087 (ASG) 30 July 1954	MILITARY SPECIFICATION BONDING, ELECTRICAL AND LIGHTNING PROTECTION, FOR AEROSPACE SYSTEMS This specification has been approved by the Department of the Air Force and by the Bureau of Naval Weapons.	This specification covers the characteristics, application, and testing of electrical bonding for aerospace systems, as well as bonding for the installation and interconnection of electrical and electronic equipment therein, and lightning protection.	
MIL-B-5423B 5 August 1960 Superseding MIL-B-005423 (ASG) 26 April 1956 MIL-B-5423 23 November 1949 MIL-B-19257 (ships) 11 June 1956	MILITARY SPECIFICATION BOOTS, DUST AND WATER SEAL (FOR TOGGLE AND PUSH BUTTON SWITCHES AND ROTARY-ACTUATED PARTS), GENERAL SPECIFICATION FOR This specification has been approved by the Department of Defense and is mandatory for use by the Departments of the Army, the Navy, and the Air Force.	This specification covers the general requirements for molded silicone-rubber rf-proof boots which can be used on toggle and push button switches and rotary-actuated parts such as rotary switches, variable resistors, capacitors, inductors, and transformers. The boots protect the switch-actuating mechanism from sand, dust, water, and other contaminants, and seal the panel on which the switches are mounted.	
MIL-C-5541B 30 June 1970 Superseding MIL-C-5541A 31 March 1964	MILITARY SPECIFICATION CHEMICAL CONVERSION COATINGS ON ALUMINUM AND ALUMINUM ALLOYS This specification is mandatory for use by all Departments and Agencies of the Department of Defense	This specification covers the requirements for chemical conversion coatings formed by the reaction of chemical conversion materials and the surfaces of aluminum and aluminum alloys. Chemical coatings resulting from treatment of aluminum and aluminum alloy surfaces with chemical conversion coating materials shall be of the following classes, as specified: Class 1A - For maximum protection against corrosion, painted or unpainted. Class 3 - For protection against corrosion where low electrical resistance is required.	Aluminum Alloy 2024, Plate and Sheet Aluminum Alloy 6061, Plate and Sheet Anodic Coatings, For Aluminum and Aluminum Alloys
MIL-C-7078C 9 August 1971 Superseding MIL-C-7078B 17 March 1964	MILITARY SPECIFICATION CABLE, ELECTRIC, AEROSPACE VEHICLE, GENERAL SPECIFICATION FOR This specification is mandatory for use by all Departments and Agencies of the Department of Defense	This specification covers electric cable for use in aerospace vehicles and other applications for which its performance characteristics are suitable.	The cable will be as described in the applicable military specification sheets and will be of the following types: Unshielded-unjacketed: Two or more spirally laid coded wires with no overall jacket or shield. Jacketed: Two or more spirally laid coded wires with no shield over the spirally laid wires but with an overall jacket. Shielded: A single wire or two or more spirally laid coded wires with an overall shield but with no jacket over the shield. Shielded and Jacketed: A single wire or two or more spirally laid coded wires with an overall shield and jacket over the shield.

Table 10-1 (cont'd)

Specification	Title	Scope	Comments
MIL-C-11693C 8 October 1971 Superseding MIL-C-11693B 9 March 1960	CAPACITORS, FEED THROUGH, RADIO-INTERFERENCE REDUCTION AC AND DC (HERMETICALLY SEALED IN METAL CASES), ESTABLISHED AND NON-ESTABLISHED RELIABILITY GENERAL SPECIFICATION FOR	This specification covers the general requirements for established reliability (ER) and non-ER capacitors designed for operation with alternating current (ac) and direct current (dc), feed-through capacitors, hermetically sealed in metal cases for use primarily in broadband, radio-interference suppression applications.	Capacitors meeting the established reliability requirements specified have a maximum failure rate of 1%/1000 hours. This failure rate is established with 90% confidence limit and maintained at a 10-percent producer's risk. An acceleration factor of 5:1 has been used to relate the life test data obtained at 140 percent of rated voltage at the applicable high test temperature. Styles which are of metallized construction should be used only in circuitry in which high values of insulation resistance are not essential, and in which occasional momentary breakdowns can be tolerated.
MIL-C-12889A 30 August 1957 Superseding MIL-C-12889 (Sig C) 10 July 1955	CAPACITORS, BY-PASS, RADIO-INTERFERENCE REDUCTION, PAPER DIELECTRIC, AC AND DC, (HERMETICALLY SEALED IN METALLIC CASES), GENERAL SPECIFICATION FOR	This specification covers the performance and general material requirements for bypass, radio-interference-reduction, alternating-current (ac) and direct-current (dc), paper-dielectric capacitors, hermetically sealed in metallic cases, for use primarily in broadband, radio-interference suppression applications. In addition, this specification indicates the ambient test conditions within which the capacitors must operate satisfactorily and reliably. These capacitors are suitable for operation over a temperature range of -55° to +85°C.	
MIL-C-13909B 3 January 1963 Superseding MIL-C-13909A (ORD) 28 April 1958	MILITARY SPECIFICATION CONDUIT, METAL, FLEXIBLE; ELECTRICAL, SHIELDED This specification has been approved by the Department of Defense and is mandatory for use by the Departments of the Army, the Navy, and the Air Force.	This specification covers shielded, electrical, flexible metal conduit. The conduit consists of a core of flexible metal tubing with a covering of wire braid.	Conduit shall be furnished in the following types, as specified: Type I - Waterproof Type II - Nonwaterproof. Conduit shall be furnished in the following grades, as specified: Grade A - Double wire braid. Grade B - Single wire braid.
MIL-C-25200B (USAF) 1 October 1971 Superseding MIL-C-25200A (USAF) 9 April 1962	MILITARY SPECIFICATION CABLE ASSEMBLIES, SPECIAL WEAPONS, ELECTRICAL, GENERAL REQUIREMENTS FOR	This specification establishes the detailed requirements for the design, manufacture and testing of Air Force special weapons cable assemblies, including interconnecting cables.	
MIL-C-81511C 5 June 1970 Superseding MIL-C-81511B (NAVY) 1 May 1968	MILITARY SPECIFICATION CONNECTORS, ELECTRICAL, CIRCULAR, HIGH DENSITY, QUICK DISCONNECT, ENVIRONMENT RESISTING: AND ACCESSORIES GENERAL SPECIFICATION FOR This specification is mandatory for use by all Departments and Agencies of the Department of Defense.	This specification covers two series of electrical connectors (plugs and receptacles) with removable crimp and nonremovable solder contacts and accessories. Electrical, mechanical, and environmental features of these connectors include: (1) Environment resisting at sea level and high altitude. (2) Quick disconnect. (3) RFI/EMI protection. (4) High density insert arrangements. (5) Low level circuit capabilities. (6) Scoop proof. (7) Fluid resistant class provided. (8) High temperature class provided. (9) Several voltage service ratings available.	The connectors are identified as series 1 or series 2. Both series are designed to prevent inadvertent electrical contact and to provide contact protection during mating. They can be mounted in the same size panel cutout but are not intermateable. They have the following features: Series 1 - Long shell. Scoop-proof when pins are installed in either the plug or the receptacle (formerly known as 100 percent no-scoop). Series 2 - Light-weight, short shell. Scoop-proof when pins are installed in the receptacle. Not scoop-proof when pins are installed in the plug (formerly known as 50 percent no-scoop).

Table 10-1 (cont'd)

Specification	Title	Scope	Comments
			These connectors are rated for operation within a temperature range specified for each class. The upper temperature limit is the maximum internal hot-spot temperature resulting from any combination of electrical load and ambient temperature. These connectors are rated for continuous operation, 1000 hours at the specified maximum internal hot-spot temperature.
MIL-E-6051D 7 September 1967 Superseding MIL-E-6051C 17 June 1960	MILITARY SPECIFICATION ELECTROMAGNETIC COMPATIBILITY REQUIREMENTS, SYSTEMS This specification is mandatory for use by all Departments and Agencies of the Department of Defense	This specification outlines the overall requirements for systems electromagnetic compatibility, including control of the system electromagnetic environment, lightning protection, static electricity, bonding and grounding. It is applicable to complete systems. including all associated subsystems and equipments.	
MIL-E-16400F (NAVY) 24 February 1966 Superseding MIL-E-16400E (NAVY) 15 June 1962	MILITARY SPECIFICATION ELECTRONIC EQUIPMENT, NAVAL SHIP AND SHORE: GENERAL SPECIFICATION All interested bureaus of the Navy Department and the Marine Corps have concurred in the use of this specification.	This specification covers the general requirements applicable to the design and construction of electronic equipment and associated and auxiliary electronic apparatus furnished as part of a complete system intended for Naval ship or shore applications. The intent of this specification is to set forth the ambient conditions within which equipment must operate satisfactorily and reliably; the general material, the process for selection and application of parts, and to detail the means by which equipment as a whole will be tested to determine whether it will so operate. Throughout the design and manufacture of the equipment, maximum effort shall be made to attain the basic design objectives in that the equipment will meet the needs of the Naval service. Requirements applicable to individual equipment specification.	• Unitized modular construction • Plug-in techniques • Light weight • Modular maintenance testing • Commercially acceptible metals • Planned for maintenance
MIL-E-45782B 28 December 1967 Superseding MIL-E-45782A (MI) 16 January 1964	MILITARY SPECIFICATION ELECTRICAL WIRING, PROCEDURES FOR This specification is mandatory for use by all Departments and Agencies of the Department of Defense.	This specification covers the fabrication and termination of shielded electrical harness and cable assemblies and the wiring of electrical and electronic circuits of subassemblies used in missile systems.	• Shield Termination • Securing of harnessess • Use of ferrules
MIL-F-15733E 24 March 1965 Superseding MIL-F-15733D 22 March 1960	MILITARY SPECIFICATION FILTERS, RADIO INTERFERENCE, GENERAL SPECIFICATION FOR This specification is mandatory for use by all Departments and Agencies of the Department of Defense	This specification covers the general requirements for current-carrying filters, alternating-current (ac) and direct-current (dc), for use primarily in the reduction of broadband radio interference.	The type designation shall be in the following form, and as specified: • Style • Current rating • Insertion-loss characteristic • Terminal identification • Operating temperature range • Grade

Table 10-1 (cont'd)

Specification	Title	Scope	Comments
MIL-F-18327C 25 May 1966 Superseding MIL-F-18327B 6 April 1962	MILITARY SPECIFICATION FILTERS; HIGH PASS, LOW PASS, BAND PASS: BAND SUPPRESSION, AND DUAL FUNCTIONING, GENERAL SPECIFICATION FOR This specification is mandatory for use by all Departments and Agencies of the Department of Defense	This specification covers the general requirements for passive frequency-selective networks, such as dual functioning, band suppression, band pass, low pass and high pass (or any combination thereof) electric-wave filters, including those employing electromechanical and piezoelectric elements, for use over the frequency range of 0 to 150 megahertz. Filters covered by this specification are intended for use where operation under various environmental conditions is required. This specification covers filters weighing not more than 50 pounds and requiring root-mean-square test voltage ratings not greater than 5,000 volts.	The type designation shall be in the form appearing below, and as specified. The type designation does not describe a discrete item. For complete identification, the part number and type designation should be referenced. • Component • Grade • Class • Life Expectancy • Family • Case and Mounting Dimensions • Composition
MIL-F-19207B/36A 31 August 1966 Superseding MIL-F-19207A (Ships) 23 May 1962	MILITARY SPECIFICATION FUSEHOLDERS, EXTRACTOR POST TYPE, BLOWN FUSE INDICATING AND NONINDICATING GENERAL SPECIFICATION FOR	This specification covers inclosed, panel mounted, extractor post type electrical fuseholders, blown fuse nonindicating, EMI shielded.	• Gaskets and o-rings, conductive: Knitted monel impregnated with silicone rubber, ZZ-R-765. • Shield cap, collar & associated hardware: Brass, w/lusterless black finish (collar & chain retainer rings exc.). • Rating: 30 amp., 250 volts max. • Nonindicating:
MIL-I-6181D 25 November 1959 Superseding MIL-I-006181C (USAF) 6 June 1957 MIL-I-6181B 29 May 1953	MILITARY SPECIFICATION INTERFERENCE CONTROL REQUIREMENTS, AIRCRAFT EQUIPMENT This specification has been approved by the Department of Defense and is mandatory for use by the Departments of the Army, the Navy, and the Air Force.	This specification covers design requirements, interference test procedures and limits for electrical and electronic aeronautical equipment to be installed in or closely associated with aircraft. The test procedures which are specified cover the following types of tests: (a) Interference tests: Conducted and radiated tests which measure the magnitude of the interference signals emanating from the equipment under test. (b) Susceptibility tests: Conducted, radiated, intermodulation and front-end rejection tests which determine whether an equipment will operate satisfactorily when exposed to external interference signals.	Electrical and electronic equipment shall operate satisfactorily, not only independently but also in conjunction with other equipment which may be installed nearby. This requires that the operation of such equipment shall not be adversely affected by interference voltages and fields reaching it from external sources, and also requires that such equipment shall not, in itself, be a source of interference which might adversely affect the operation of other equipments. The limits specified herein are established to insure that the air vehicles will meet the requirements of Specification MIL-I-6051 or other applicable system specification.
MIL-I-16165D (SHIPS) 22 August 1961 Superseding MIL-I-16165C (Ships) 23 November 1955	MILITARY SPECIFICATION INTERFERENCE SHIELDING, ENGINE ELECTRICAL SYSTEMS	This specification covers requirements for interference shielding items and shielded harnesses for engine electrical systems aboard Naval ships, at advance bases, and in the vicinity of electronic installations. It includes the allowable interference limits for such items and the permissible limits for auxiliary devices normally installed on electrical wiring systems associated with these engines.	The interference shielding electrical systems furnished under this specification shall be a product which has been tested and passed the qualification tests specified herein and has been listed or approved for listing on the applicable qualified products list. Reliability of operation shall be stressed throughout the design and manufacture of the shielding.

Table 10-1 (cont'd)

Specification	Title	Scope	Comments
MIL-P-24014 (WEP) 30 January 1964 Superseding WR-27 12 February 1962	MILITARY SPECIFICATION PRECLUSION OF HAZARDS FROM ELECTROMAGNETIC RADIATION TO ORDNANCE, GENERAL REQUIREMENTS FOR	This specification establishes general requirements for weapon systems to preclude hazards from environmental electromagnetic fields in the frequency range of 1- kilocycles per second to 40 gigacycles per second. These requirements apply to all Navy weapon systems, including safety and emergency devices and other ancillary equipment which contain electrically initiated, explosive or pyrotechnic components.	
MIL-W-6858C 20 October 1964 Superseding MIL-W-6858B 2 November 1960	MILITARY SPECIFICATION WELDING, RESISTANCE: ALUMINUM, MAGNESIUM, NONHARDENING STEELS OR ALLOYS, NICKEL ALLOYS, HEAT-RESISTING ALLOYS, AND TITANIUM ALLOYS; SPOT AND SEAM This specification is mandatory for use by all Departments and Agencies of the Department of Defense	This specification covers requirements for resistance spot and seam welding of the following non-hardening materials Group (a) - Aluminum, aluminum alloys and magnesium alloys. Group (b) - Steels, austenitic and ferritic and precipitation hardening steels, nickel and cobalt base alloys. Group (c) - Titanium and titanium alloys.	Classification shall be based on function of the spot and seam welded joint, as follows: Class A - Used in joints, the single failure of which during any operating condition would cause loss of the weapons system or one of its major components, loss of control, unintentional release of, or inability to release any armament store, failure of gun installation components, or which may cause significant injury to occupants of the manned weapons system. Class B - Used in joints, the failure of which would reduce the overall strength of the weapons system or preclude the intended functioning or use of equipment. Class C - A weld which is considered noncritical and for which no stress analysis is required.
MIL-STD-202D 14 April 1969 Superseding MIL-STD-202C 12 Sept. 1963	MILITARY STANDARD TEST METHODS FOR ELECTRONICS AND ELECTRICAL COMPONENT PARTS	The standard specifies test conditions obtainable in the laboratory which give results equivalent to actual service conditions existing in the field. All of the test methods of a similar character which appeared in the various joint or single-service electronic and electrical component-parts specifications; those newly-developed test methods which are feasible for use in several specifications; and the recognized extreme environments, particularly temperatures, barometric pressures, etc., at which component parts will be tested under some of the presently-standardized testing procedures have been consolidated.	This standard establishes uniform methods for testing electronic and electrical component parts, including basic environmental tests to determine resistance to natural elements and conditions surrounding military operations, and physical and electrical tests. This standard is intended to apply only to small parts weighing up to 300 pounds or having a root-mean-square test voltage up to 50,000 volts unless otherwise specifically invoked.

Table 10-1 (cont'd)

Specification	Title	Scope	Comments
MIL-STD-220A 15 December 1959	MILITARY STANDARD METHOD OF INSERTION-LOSS MEASUREMENT	This standard covers a method of measuring, in a 50-ohm system, the insertion loss of feed-through suppression capacitors, and of single- and multiple-circuit, radio-frequency (RF) filters at frequencies up to 1000 megahertz (MHz).	
MIL-STD-285 25 June 1956	ATTENUATION MEASUREMENTS FOR ENCLOSURES, ELECTROMAGNETIC SHIELDING, FOR ELECTRONIC TEST PURPOSES, METHOD OF	This standard covers a method of measuring the attenuation characteristics of electromagnetic shielding enclosures used for electronic test purposes over the frequency range 100 kilohertz to 10,000 megahertz.	Attenuation is the ratio, expressed in decibels (db), of the received powers on opposite sides of a shield when the shield is illuminated by electromagnetic radiation; and as used in this standard, is the figure of merit to designate the shielding effectiveness of electromagnetic enclosures.
MIL-STD-449C 1 March 1965	MILITARY STANDARD RADIO FREQUENCY SPECTRUM CHARACTERISTICS, MEASUREMENT OF	This technical standard establishes uniform measurement techniques that are applicable to the determination of the spectral characteristics of radio-frequency transmitters and receivers. The ultimate goal is to ensure the compatibility of present and future systems.	The data obtained from the measurements described in this standard will comprise one of the principal aids for (a) predicting the performance of equipment and systems in an operational electromagnetic environment, and (b) predicting the effect of a particular equipment or system on the electromagnetic environment of other equipments or systems. This radio-frequency-measurement standard will apply to all equipments and systems that are designed to emit or will respond to electromagnetic energy in the radio-frequency range of from 0.014 to 12,000 megahertz (MHz). This standard does not necessarily apply to all parameters of all systems.
MIL-STD-454C	STANDARD GENERAL REQUIREMENTS FOR ELECTRONIC EQUIPMENT	Requirements applicable to electronic equipment. This standard covers some of the common requirements to be used in military specifications for electronics equipment.	The requirements contained herein are intended to provide uniform requirements applicable to electronic equipment and shall be incorporated by reference in general equipment specifications.
MIL-STD-461A 1 August 1968	ELECTROMAGNETIC INTERFERENCE CHARACTERISTICS REQUIREMENTS FOR EQUIPMENT	This standard covers the requirements and test limits for the measurement and determination of the electromagnetic interference characteristics (emission and susceptibility) of electronic, electrical and electromechanical equipment. The requirements shall be applied for general or multi-service procurements and single service procurements, as specified in the individual equipment specification, or the contract or order. The requirements specified in this standard are established to: (a) Insure that interference control is considered and incorporated into the design of equipment. (b) Enable compatible operation of the equipment in a complex electromagnetic environment. This standard shall be used in conjunction with MIL-STD-463 and MIL-STD-462.	This standard shall be used in conjunction with MIL-STD-463 and MIL-STD-462.

Table 10-1 (cont'd)

Specification	Title	Scope	Comments
MIL-STD-462 31 July 1967	ELECTROMAGNETIC INTERFERENCE CHARACTERISTICS, MEASUREMENT OF	This standard establishes techniques to be used for the measurement and determination of the electromagnetic interference characteristics (emission and susceptibility (of electrical, electronic, and electromechanical equipment, as required by MIL-STD-461.	
MIL-STD-463 9 June 1966	DEFINITIONS AND SYSTEM OF UNITS, ELECTROMAGNETIC INTERFERENCE TECHNOLOGY	This standard contains general interference definitions, abbreviations, and acronyms used in MIL-STD-461 and MIL-STD-462. Definitions of abbreviations and terms are limited to statements of meaning as related to this and referenced standards, rather than encyclopedia or textbook discussions.	
MIL-STD-469 1 December 1966	RADAR ENGINEERING DESIGN REQUIREMENTS, ELECTROMAGNETIC COMPATIBILITY	The engineering design requirements are established to control the spectral characteristics of all new radar systems operating between 100 and 40,000 megahertz (MHz) in an effort to achieve electromagnetic compatibility and to conserve the frequency spectrum available to Military radar systems.	
MIL-STD-704A	ELECTRIC POWER, AIRCRAFT, CHARACTERSITICS AND UTILIZATION OF	This standard delineates the characteristics of electric power supplied to airborne equipment at the equipment terminals and the requirements for the utilization of such electric power by the airborne equipment.	The purpose of this standard is to foster compatibility between aircraft electric systems or ground support electric systems and airborne utilization equipment to the extent of confining the aircraft and ground support electric power characteristics within definitive limits and restricting the requirements imposed on the electric power by the airborne utilization equipment.
MIL-STD-0001310B (SHIPS) 15 April 1971	SHIPBOARD BONDING, GROUNDING AND OTHER TECHNIQUES FOR ELECTROMAGNETIC COMPATIBILITY AND SAFETY	This standard provides shipboard bonding, grounding and other techniques for electromagnetic compatibility and safety.	The requirements of this standard apply to all new shipboard installations and to that part of existing installations receiving modifications. It is not the intent of this standard to retrofit existing installations not programmed for modification or to retrofit work accomplished according to previous requirements. The procedures and methods specified in this standard shall be utilized wherever it is required to bond, ground, insulate or use nonmetallic materials for the following applications, as specified: (a) To provide electromagnetic compatibility. (b) To provide personnel safety from electrical shock hazards. (c) To safeguard electrical transmission of classified information. (d) To provide a d.c. reference ground.

Table 10-1 (cont'd)

Specification	Title	Scope	Comments
MIL-STD-1377 (NAVY)	MILITARY STANDARD EFFECTIVENESS OF CABLE, CONNECTOR, AND WEAPON ENCLOSURE SHIELDING AND FILTERS IN PRECLUDING HAZARDS OF ELECTROMAGNETIC RADIATION TO ORDNANCE; MEASUREMENT OF	This standard is intended to provide a weapon developer or designer with shielding and filter effectiveness test methods for determining whether the particular weapon design requirements of MIL-P-24014 have been properly implemented. It is not intended to be a substitute for full-scale electromagnetic hazards evaluation tests of the weapon system, but rather an aid in developing a weapon system with a high probability of successfully passing such environmental tests.	This standard covers the methods of shielding effectiveness of weapon enclosures, cables, and cable connectors over the frequency ranges of from 100 kilohertz (kHz) to 30 megahertz (MHz) and from 1000 MHz to 10 gigahertz (GHz). The shielding effectiveness test methods are conducted at two separate frequency ranges because of hardware limitations. This standard also covers the methods of measuring the filtering effectiveness of radio frequency (RF) suppression devices (filters) for weapon firing circuits over the frequency range of from 100 kHz to 10 GHz.
MIL-STD-1385 (NAVY) April 6, 1972	MILITARY STANDARD PRECLUSION OF ORDNANCE HAZARDS IN ELECTROMAGNETIC FIELDS; GENERAL REQUIREMENTS FOR	This standard establishes the general requirements to preclude hazards resulting from ordnance having electroexplosive devices when exposed to electromagnetic fields. The nominal frequency range covered by this standard is from 10 kilohertz (10^4 Hertz) by 40 gigahertz (4×10^{10} Hertz).	These requirements apply to all Naval Weapon systems, safety and emergency devices and other ancillary equipment containing electrically initiated explosives, propellant or pyrotechnic components. The trend in radar and communications equipment towards greater radiated power has resulted in growing concern with electromagnetic radiation hazards to ordnance. These hazards stem from the use of electroexplosive divices which can be initiated spuriously by means of electromagnetic energy. In addition to the hazards that could occur, reliability factors are also considered.
MSFC-SPEC-279 June 1, 1964	GEORGE C. MARSHALL SPACE FLIGHT CENTER NATIONAL AERONAUTICS AND SPACE ADMINISTRATION SPECIFICATION, ELECTROMAGNETIC COMPATIBILITY	This specification provides system electromagnetic compatibility test requirements, subsystem interference test requirements, and subsystem susceptibility test requirements. System test requirements are applicable to entire space vehicles and all associated support equipment connected to the vehicle. Subsystem test requirements are applicable to individual electronic and electrical equipment to be installed in or associated with either the space vehicle or checkout facility.	The test requirements specified herein shall be applicable for the following types of tests: (a) System electromagnetic compatibility tests: Radiated and conducted interference and susceptibility tests performed to determine the ability of all individual vehicle subsystems to operate simultaneously without performance degradation (b) Subsystem interference tests: Radiated, conducted, tests performed to measure the magnitude of interfering undesired signals emanating from individual vehicle or facility equipments. (c) Subsystem susceptibility tests: Radiated, conducted, intermodulation, transient, cross-modulation and receiver input rejection tests performed to determine whether individual vehicle or facility equipments will satisfactorily operate when subjected to undesired external signals.

11.0 TABLES AND NOMOGRAPHS

11.1 INTRODUCTION

This chapter provides a convenient reference for miscellaneous data useful in the design of electromagentically compatible GSE. The chapter contains tables for various conversions of ratio units to decibels, as well as tables, equations and nomographs for convenient calculation of EM wave shielding parameters, field strength and power density, cable separation distance, transmission path loss, and various useful characteristics.

11.2 ORDER OF CONTENTS

Page

DBM	u/volts 50 ohm	u/volts 72 ohm	u/volts 600 ohm	uu/watts
0	223,607.0	268,328.0	774,596.7	1,000,000,000.0
-3	158,314.0	189,976.0	548,379.4	501,200,000.0
-6	112,094.0	134,513.0	388,265.4	251,250,000.0
-9	79,358.0	95,230.0	274,845.4	125,900,000.0
-12	56,192.0	67,431.0	194,576.5	63,100,000.0
-15	39,780.0	47,736.0	137,738.9	31,620,000.0
-18	28,174.0	33,809.0	97,519.2	15,850,000.0
-21	19,932.0	23,919.0	69,034.8	7,943,000.0
-24	14,112.0	16,934.0	48,873.3	3,981,000.0
-27	9,990.0	11,988.0	34,597.7	1,995,000.0
-30	7,073.0	8,487.0	24,494.9	1,000,000.0
-33	5,009.0	6,011.0	17,341.3	501,200.0
-36	3,546.0	4,256.0	12,276.8	251,200.0
-39	2,511.0	3,013.0	8,691.4	125,900.0
-42	1,776.0	2,132.0	6,153.0	63,100.0
-45	1,258.0	1,509.0	4,355.7	31,620.0
-48	890.0	1,068.0	3,083.8	15,850.0
-51	630.0	756.0	2,183.1	7,943.0
-54	446.0	536.0	1,545.5	3,981.0
-57	316.0	379.0	1,094.0	1,995.0
-60	223.607	268.328	774.597	1,000.0
-63	158.314	189.976	548.379	501.2
-66	112.094	134.513	388.265	251.25
-69	79.358	95.230	274.845	125.9
-72	56.192	67.431	194.576	63.1
-75	39.780	47.736	137.739	31.62
-78	28.174	33.809	97.519	15.85
-81	19.932	23.919	69.035	7.943
-84	14.112	16.934	48.873	3.981
-87	9.990	11.988	34.598	1.995
-90	7.073	8.487	24.495	1.0
-93	5.009	6.011	17.341	0.5012
-96	3.546	4.256	12.277	0.2512
-99	2.511	3.013	8.691	0.1259
-102	1.776	2.132	6.153	0.0631
-105	1.257	1.509	4.356	0.03162
-107	0.999	1.199	3.460	0.01995

Figure 11-1

Voltage Into Different Impedances and Power to dBm

dBm	dBμV	Volts
+53	160	100
+48	155	56
+43	150	32
+38	145	18
+33	140	10
+28	135	5.6
+23	130	3.2
+18	125	1.8
+13	120	1.0
+ 8	115	.56
+ 3	110	.32
- 2	105	.18
- 7	100	.10
- 8	99	89 mV
- 9	98	79 mV
-10	97	71 mV
-11	96	63 mV
-12	95	56 mV
-13	94	50 mV
-14	93	45 mV
-15	92	40 mV
-16	91	35 mV
-17	90	32 mV
-18	89	28 mV
-19	88	25 mV
-20	87	22 mV
-21	86	20 mV
-22	85	18 mV
-23	84	16 mV
-24	83	14 mV
-25	82	13 mV
-26	81	11 mV
-27	80	10 mV
-28	79	8.9 mV
-29	78	7.9 mV
-30	77	7.1 mV
-31	76	6.3 mV
-32	75	5.6 mV
-33	74	5.0 mV
-34	73	4.5 mV
-35	72	4.0 mV
-36	71	3.5 mV
-37	70	3.2 mV
-38	69	2.8 mV
-39	68	2.5 mV
-40	67	2.2 mV
-41	66	2.0 mV
-42	65	1.8 mV
-43	64	1.6 mV
-44	63	1.4 mV
-45	62	1.3 mV
-46	61	1.1 mV
-47	60	1.0 mV
-48	59	.89 mV
-49	58	.79 mV
-50	57	.71 mV
-51	56	.63 mV
-52	55	.56 mV
-53	54	.50 mV
-54	53	.45 mV
-55	52	.40 mV
-56	51	.35 mV
-57	50	.32 mV
-58	49	.28 mV
-59	48	.25 mV
-60	47	.22 mV
-61	46	.20 mV
-62	45	.18 mV
-63	44	.16 mV
-64	43	.14 mV
-65	42	.13 mV
-66	41	.11 mV
-67	40	.10 mV

dBm	dBμV	Microvolts
- 68	39	89
- 69	38	79
- 70	37	71
- 71	36	63
- 72	35	56
- 73	34	50
- 74	33	45
- 75	32	40
- 76	31	35
- 77	30	32
- 78	29	28
- 79	28	25
- 80	27	22
- 81	26	20
- 82	25	18
- 83	24	16
- 84	23	14
- 85	22	13
- 86	21	11
- 87	20	10
- 88	19	8.9
- 89	18	7.9
- 90	17	7.1
- 91	16	6.3
- 92	15	5.6
- 93	14	5.0
- 94	13	4.5
- 95	12	4.0
- 96	11	3.5
- 97	10	3.2
- 98	9	2.8
- 99	8	2.5
-100	7	2.2
-101	6	2.0
-102	5	1.8
-103	4	1.6
-104	3	1.4
-105	2	1.3
-106	1	1.1
-107	0	1.0
-108	- 1	.89
-109	- 2	.79
-110	- 3	.71
-111	- 4	.63
-112	- 5	.56
-113	- 6	.50
-114	- 7	.45
-115	- 8	.40
-116	- 9	.35
-117	-10	.32
-118	-11	.28
-119	-12	.25
-120	-13	.22
-121	-14	.20
-122	-15	.18
-123	-16	.16
-124	-17	.14
-125	-18	.13
-126	-19	.11
-127	-20	.10
-132	-25	.056
-137	-30	.032
-142	-35	.018
-147	-40	.010
-152	-45	.006

Figure 11-2

dB Above 1 Microvolt and Volts Into 50 Ohms to dBm

dB	Voltage or Current Ratio	Power Ratio
0	1.000	1.000
1.0	1.122	1.259
2.0	1.259	1.585
3.0	1.413	1.995
4.0	1.585	2.512
5.0	1.778	3.162
6.0	1.995	3.931
7.0	2.239	5.012
8.0	2.512	6.310
9.0	2.818	7.943
10.0	3.162	10.000
10.5	3.350	11.22
11.0	3.548	12.59
11.5	3.758	14.13
12.0	3.981	15.85
12.5	4.217	17.78
13.0	4.467	19.95
13.5	4.732	22.39
14.0	5.012	25.12
14.5	5.309	28.18
15.0	5.623	31.62
16.0	6.310	39.81
17.0	7.079	50.12
18.0	7.943	63.10
19.0	8.913	79.43
20.0	10.000	10^2
25.0	17.8	3.16×10^2
30.0	31.62	10^3
35.0	56.2	3.16×10^3
40.0	100.0	10^4
45.0	177.8	3.16×10^4
50.0	316.2	10^5
60.0	1000	10^6
70.0	3162.0	10^7
80.0	10,000	10^8
90.0	31,620	10^9
100.0	100,000	10^{10}

Figure 11-3
Conversion Ratios, dB to Voltage, Current and Power

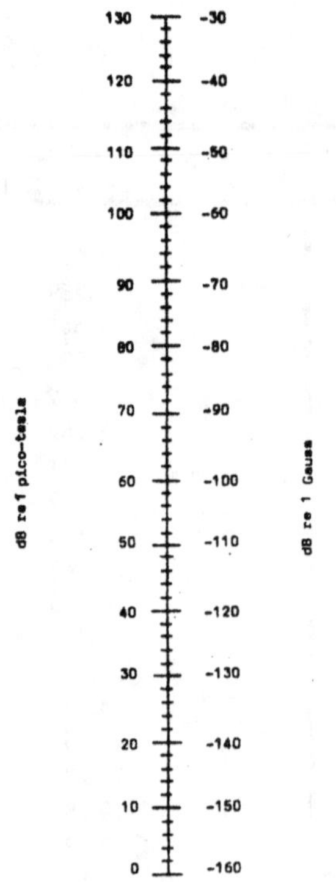

Figure 11-4

Converting Flux Density From Pico-Tesla to Gauss

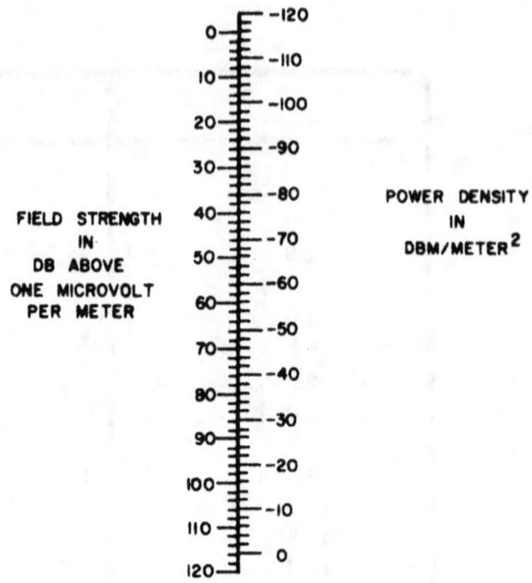

Figure 11-6

Converting dB Above 1 Microvolt per Meter to dBm per Square Meter

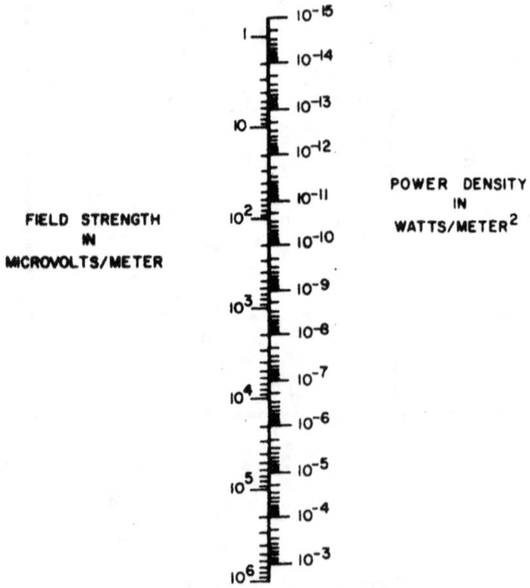

Figure 11-5

Converting Microvolts-per-Meter to Watts-per-Square-Meter

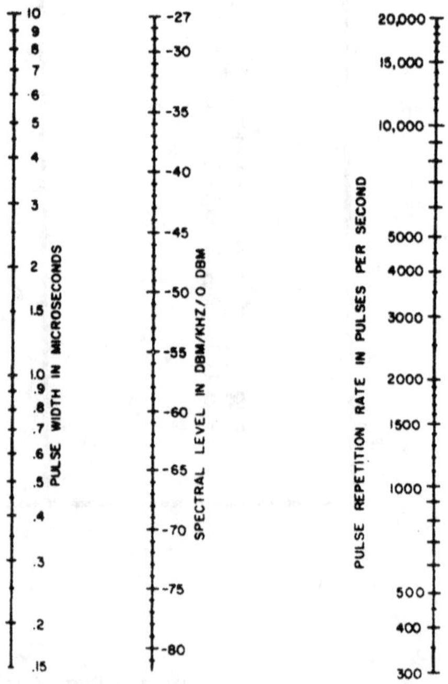

Figure 11-7

Determining Spectral Level from Pulse Width and Pulse Repetition Rate

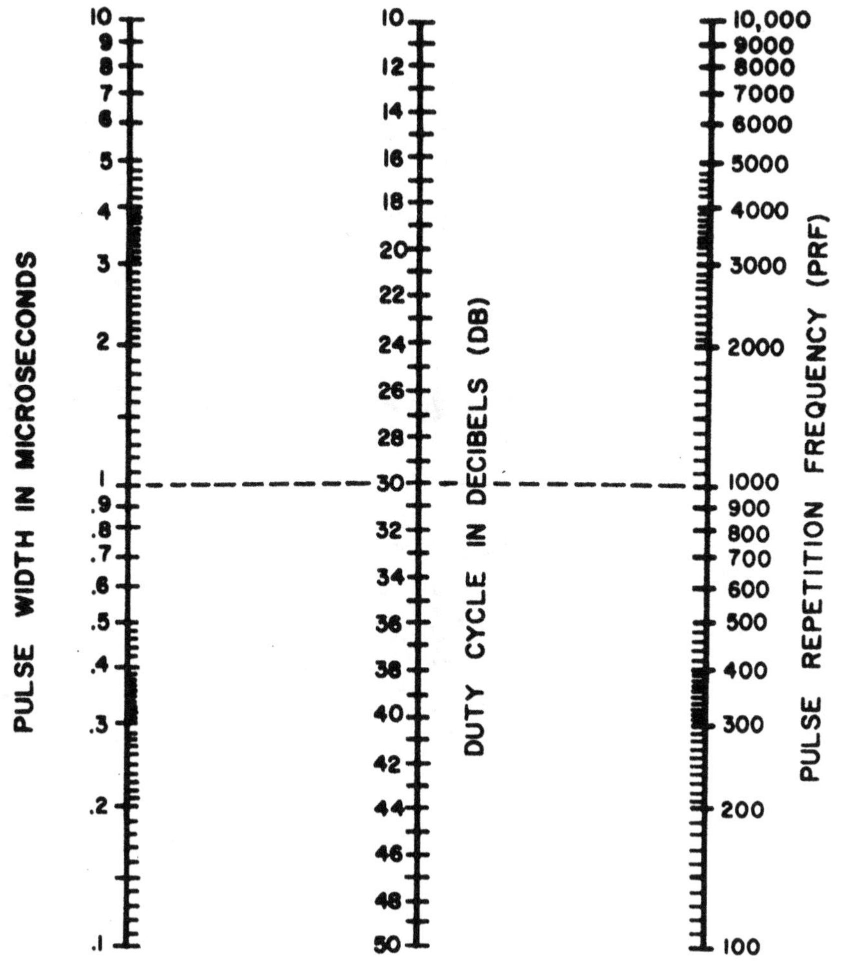

Figure 11-8

Relationship Between Pulse Width, Pulse Repetition Rate and Duty Cycle

Nomograph for the Relationship Between Pulse Width, Pulse Repetition Rate, and Duty Cycle

The duty cycle, expressed in decibels, is dB

$$dB = 10 \log_{10} \frac{T_{on}}{T_{on} + T_{off}}$$

where T_{on} is the on-time, and T_{off} the off-time.

As an example of the use of this nomograph, the dashed line shows that pulses of one microsecond width and a 30 dB (0.001) duty cycle correspond to a pulse repetition frequency of 1000.

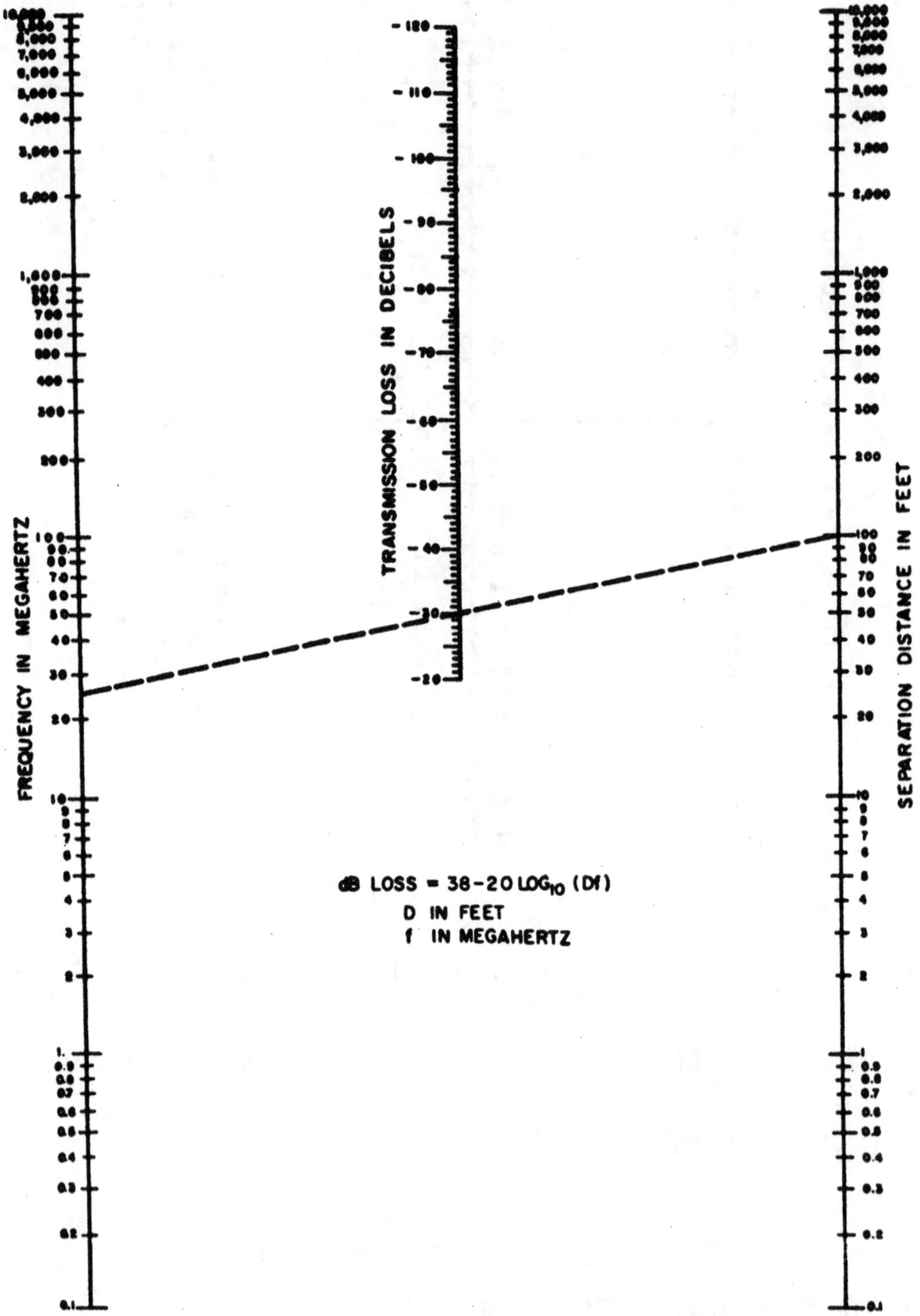

Figure 11-9
Free Space Nomograph (Feet And Megahertz)

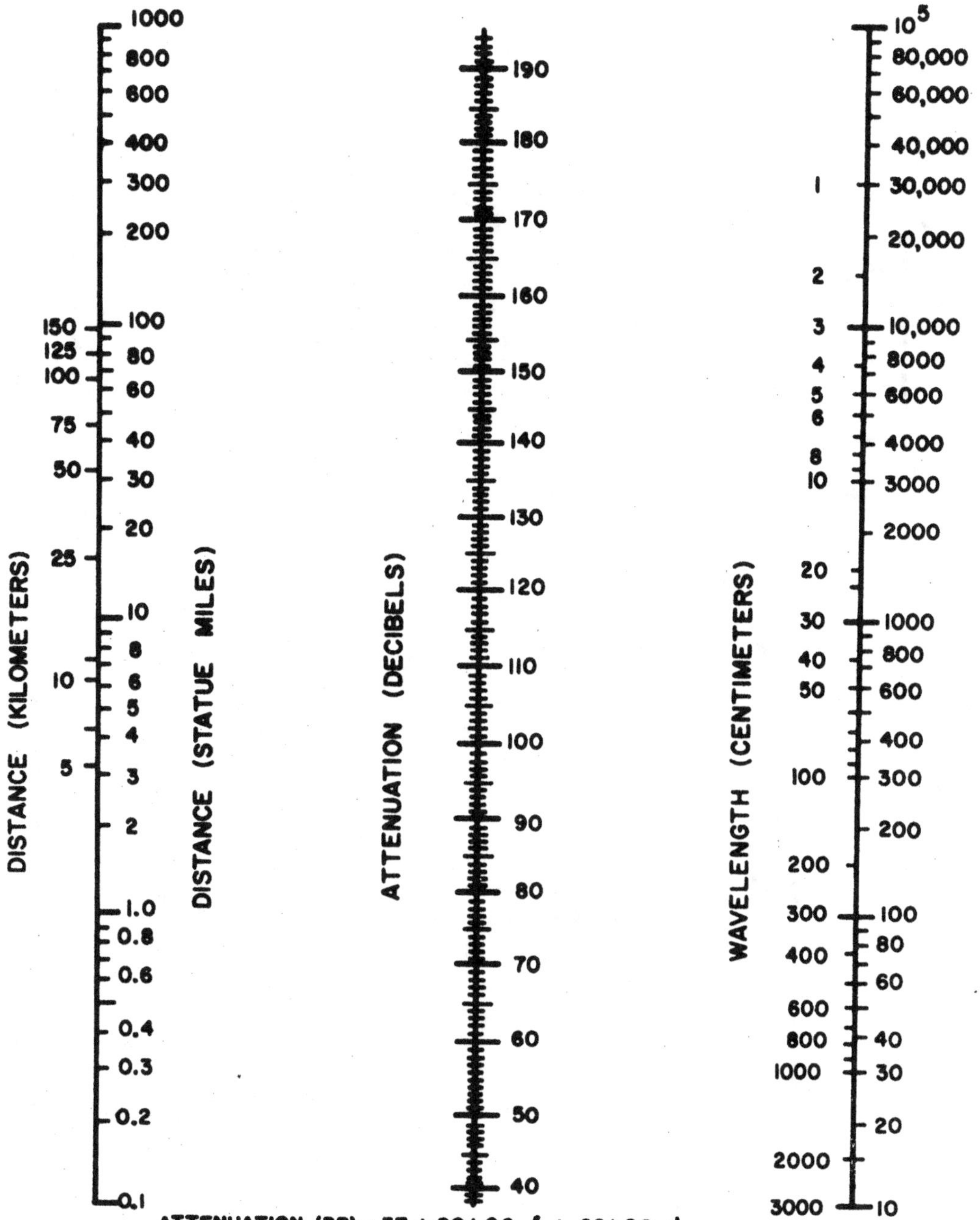

ATTENUATION (DB) = 37 + 20 LOG f + 20 LOG d
f = FREQUENCY IN MEGAHERTZ
d = SEPARATION IN MILES

Figure 11-10

Free Space Nomograph (Miles and Megahertz)

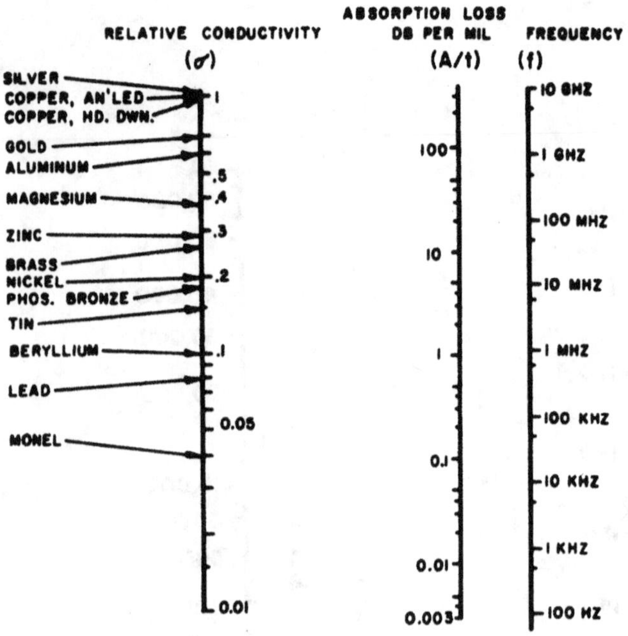

Figure 11-11

Absorption Loss for Nonmagnetic Materials

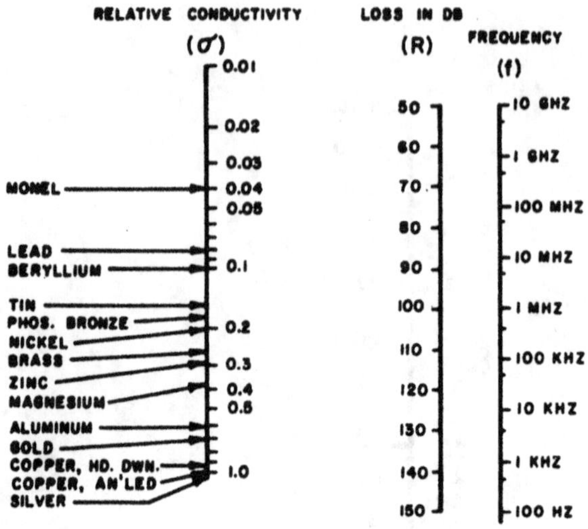

Figure 11-13

Plane Wave Reflection Loss for Nonmagnetic Materials

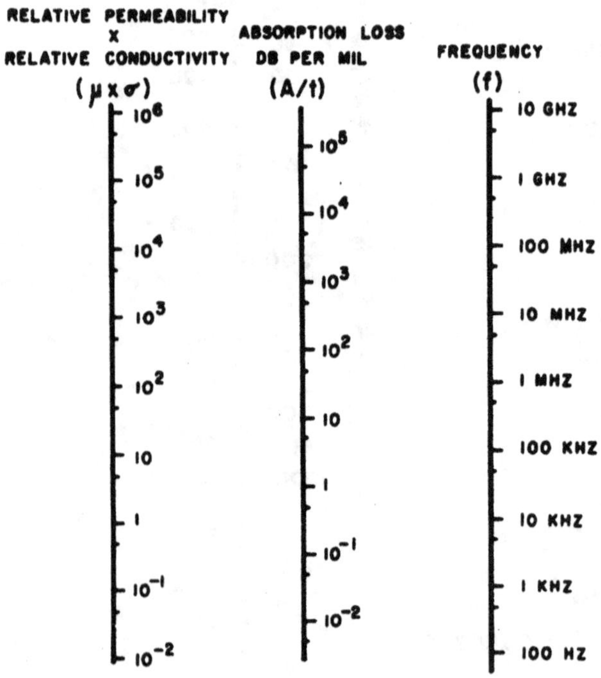

Figure 11-12

Absorption Loss for Magnetic Materials

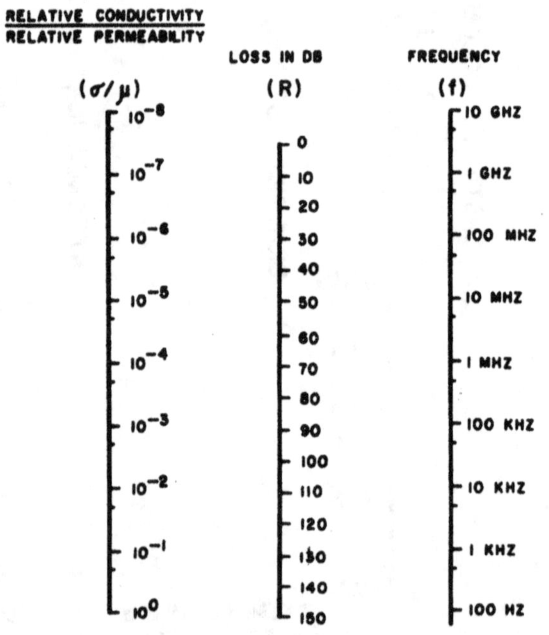

Figure 11-14

Plane Wave Reflection Loss for Magnetic Materials

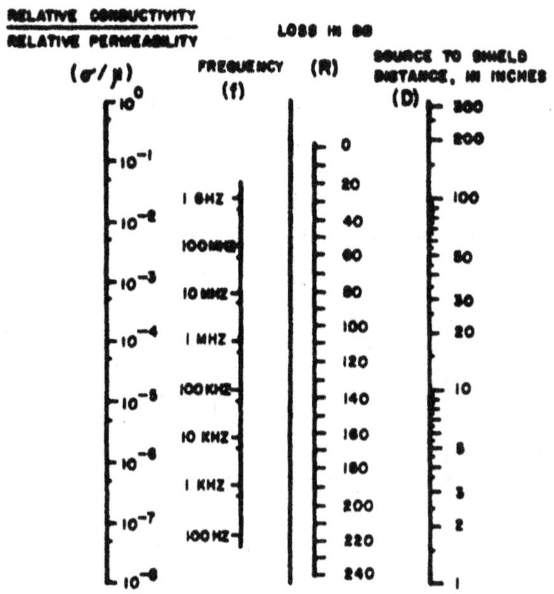

Figure 11-15

Electric Field Reflection Loss for Magnetic Materials

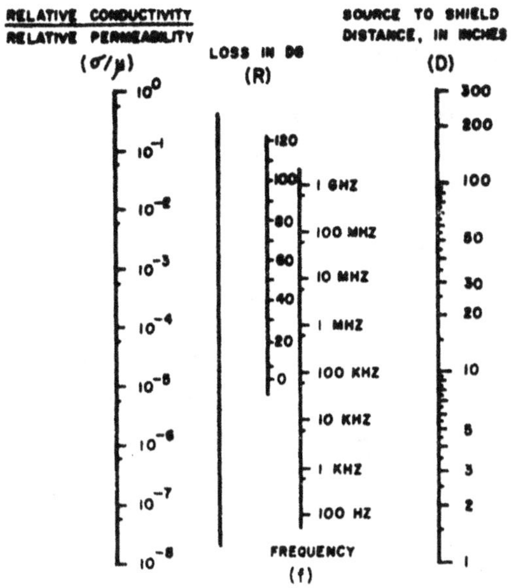

Figure 11-17

Magnetic Field Reflection Loss for Magnetic Materials

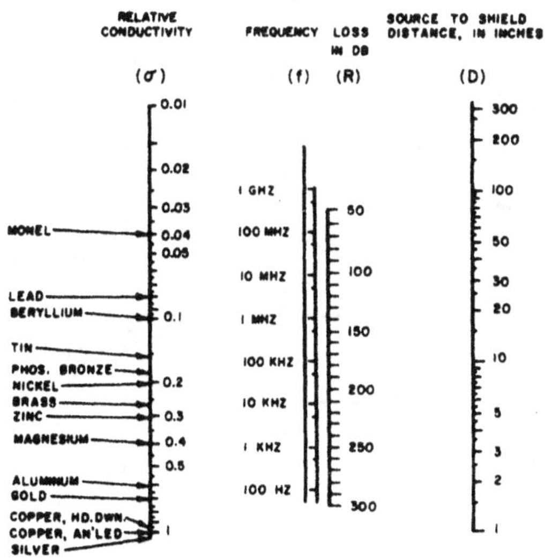

Figure 11-16

Magnetic Field Reflection Loss for Nonmagnetic Materials

Nomograph For Estimating Allowable Distance Between Parallel Conductors (Figure 11-18)

At frequencies up to where a cable run is approximately one-sixteenth of a wavelength, the nomograph given below can be used to estimate the amount of distance allowable between a radiating line and a susceptible line.

Example:

A = level in radiating cable = -25 dBmW

B = susceptible level of receiving cable = -30 dBmW

C = length of cable run = 20 ft.

D = frequency = 1000 Hz

E = elevation or distance from ground plane or shield

X = compensated difference between emitting cable and susceptible cable = (A) - (B) + (C) + (D), in dB

R = required cable separation, in inches

Procedure:

1. From C, Length Factor = +4.5 dB
2. From D, Frequency Factor = +8.0 dB
3. Solve for X = (-25) - (-30) + 4.5 + 8.0 = + 17.5 dB
4. Find R from nomograph, entering at X = 17.5 and E = 2 inches. Find R = 4 inch required separation.

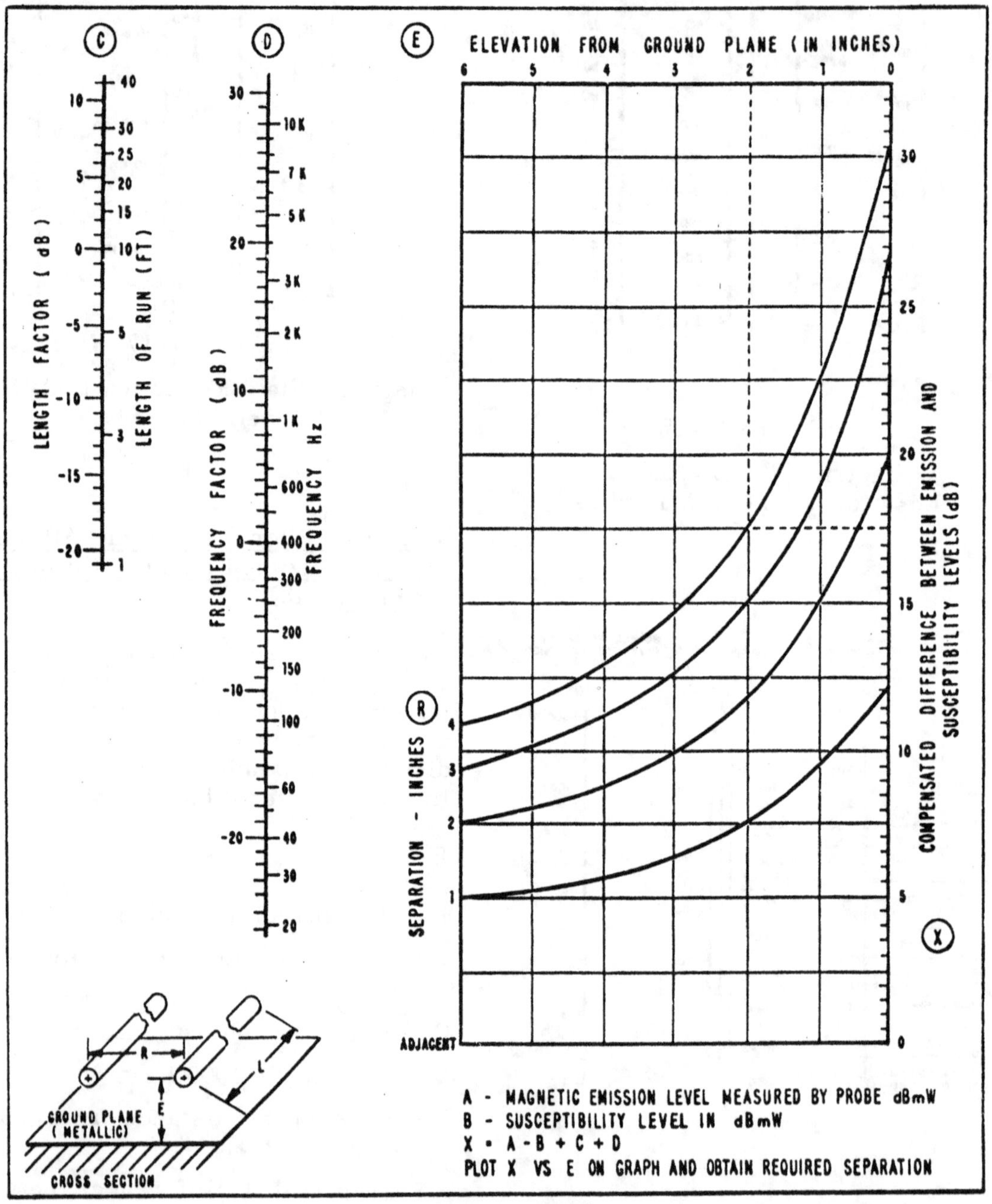

Figure 11-18

Nomograph for Estimating Allowable Distance Between Parallel Conductors

Estimation of Capacitance in Unshielded Cable Harnesses and Runs

The capacitance between unshielded wiring is given below. The chart may be used to estimate total capacity for a given length of cable harness.

The capacitance between wires in a long cable run can exceed circuit requirements; standard coaxial and shielded cable can be used to control capacitance. The table below, "Capacitance of Various AN Cables", is given to assist the designer in selecting the proper cable configuration for his needs.

Capacitance vs. Wire Size for Various Grouping of Wires

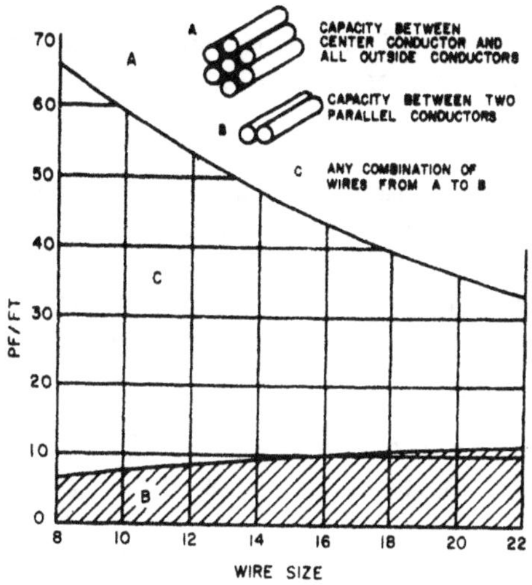

Figure 11-19

Estimation of Capacitance in Unshielded Cable Harnesses and Runs

CAPACITANCE IN PICOFARAD PER FOOT												
TYPE	ONE WIRE SHIELDED				TWO WIRE SHIELDED				THREE WIRE SHIELDED		ONE WIRE DOUBLE SHIELDED	
CONFIGURATION												
WIRE SIZE	16	18	20	22	16	18	20	22	20	22	22	22
CONDUCTOR TO SHIELD pf/ft	89	91	74	98	68	65	64	62.5	60	52	98 (CONDUCTOR TO INNER SHIELD)	340 (INNER SHIELD TO OUTER SHIELD)
					(EITHER CONDUCTOR TO SHIELD)				(ANY CONDUCTOR TO SHIELD)			
CONDUCTOR TO CONDUCTOR pf/ft					42	39.5	38	36.5	36	30		

Figure 11-20

Capacitance of Various Shielded AN Wires

Insertion Loss and Parameter Value Calculations for Single Element Filters

See Equations (7.2) and (7.3) of this publication for the single element filter design equations.

To find the insertion loss at any frequency for a given shunt capacitance, using the following chart, proceed as follows:

a. Place a straight edge so that it lies parallel to the sloping guide lines and intersects the given value of capacitance on the C scale. The straight edge now lies along the insertion loss characteristic for that ideal element.

b. At any frequency along the abscissa, note where the frequency intersects the straight edge. Read insertion loss in dB along the ordinate corresponding to that intersection.

To find insertion loss at any frequency for a given series inductance, using the chart, proceed as above, but use the L scale instead of the C scale.

To find the amount of shunt capacitance or series inductance required to produce a desired amount of insertion loss at a given frequency, using the chart, proceed as follows:

a. Place a straight edge so that it lies parallel to the guide lines and passes through the desired amount of insertion loss at the given frequency.

b. Read the required capacitance on the C scale or the required inductance on the L scale.

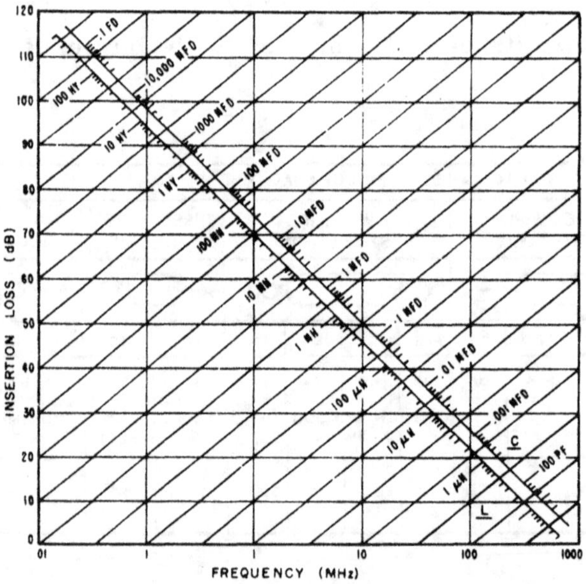

Figure 11-21

Chart for Computing Single-Element Filters

Insertion Loss and Parameter Value Calculations for Ideally-Damped Two-Element Filters

See Equation (7.4) of this publication for the two-element filter design equation.

To find the insertion loss of an ideally damped "L" section filter (d = 1) at any frequency when f_o is known,

a. Place a straight edge parallel to the guide lines and pass it through the known cutoff frequency. The straight edge now lies along the insertion loss characteristic for the ideally damped L section having that cutoff frequency.

b. At any frequency along the abscissa, read the insertion loss in dB along the ordinate.

To find the cutoff frequency (f_o) required to produce a desired amount of insertion loss at a given frequency,

a. Place a straight edge parallel to the guide lines and pass it through the desired insertion loss at the given frequency.

b. Read the cutoff frequency on the f_o scale.

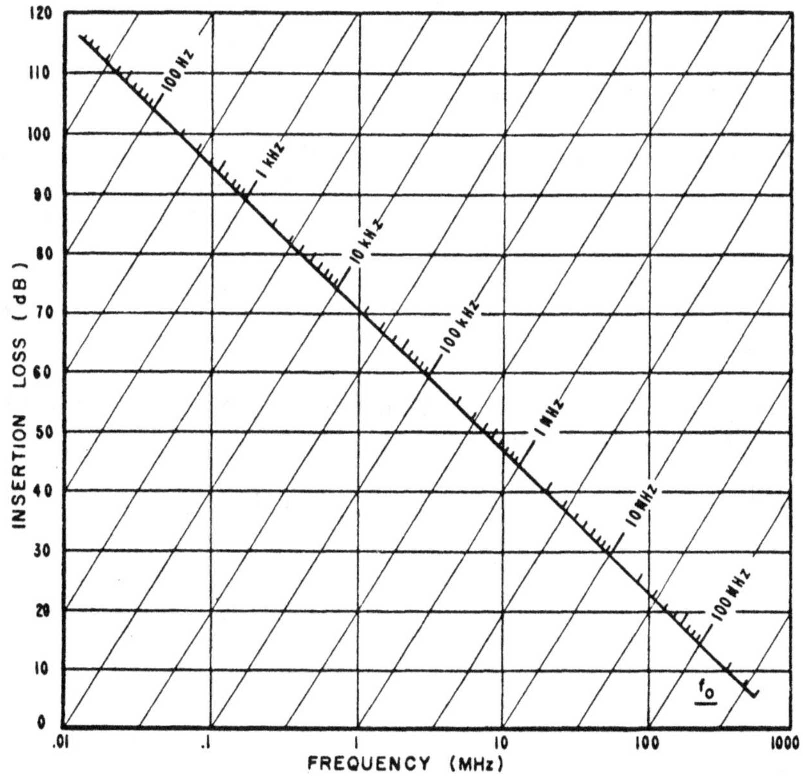

Figure 11-22

Chart for Computing Ideally-Damped Two-Element Filters

Insertion Loss and Parameter Value Calculations for Non-Ideally Damped Two-Element Filters

See Equation (7.4) of this publication for the two-element filter design equation.

To find the insertion loss of a non-ideally damped L section filter at any frequency when the cutoff frequency (f_o) and the damping ratio (d) are known, proceed as follows, using Chart A on the next page.

a. This chart is normalized for $f_o = 1$; thus, divide the frequency of interest by the known f_o to determine the normalized frquency of interest.

b. Locate the characteristic for which d is equal to the known damping ratio. Read the approximate value of d.

c. At the normalized frequency of interest, read the insertion loss in dB.

To establish the relationship between inductance, capacitance, cutoff frequency, and damping factor of two-element filters, use Chart B on the second following page. To find the cutoff frequency and the damping ratio for a given inductance and capacitance:

a. Place a straight edge so that it passes through the known L and C values.

b. Read the cutoff frequency (f_o) on the center scale.

c. The damping ratio slope line which lies parallel to the straight edge indicates the damping ratio. Note that the point at which the straight edge intersects the damping ratio scale is irrelevant; the "d" sloped line parallel to the straight edge determines the damping ratio.

To find the circuit elements required to produce a desired cutoff frequency at a given damping ratio, use Chart B and:

a. Place a straight edge so that it passes through the desired cutoff frequency and lies parallel to the given damping ratio line.

b. Read the inductance and capacitance on the L and C scales.

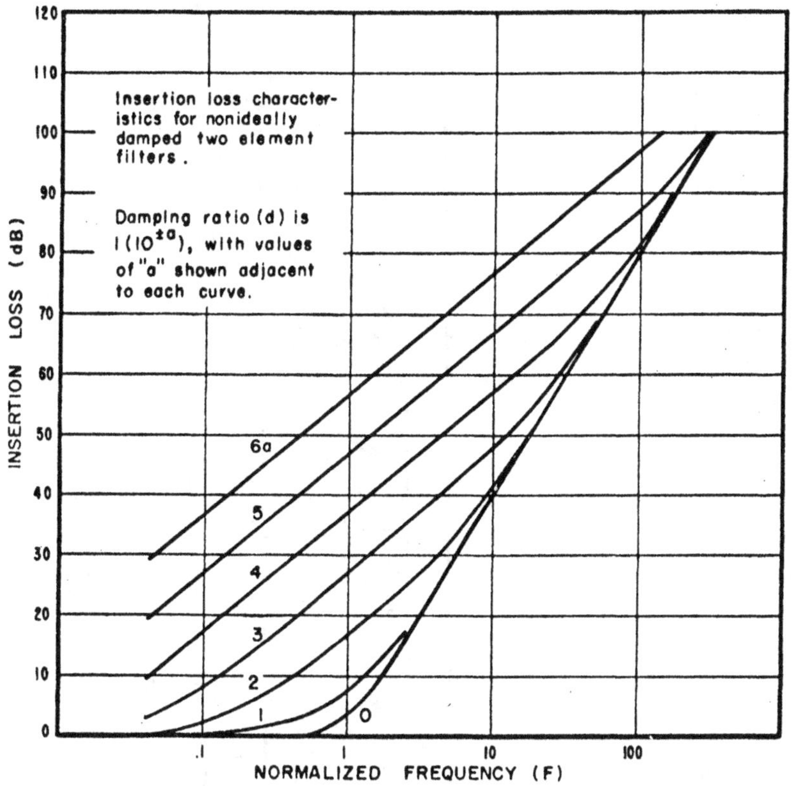

Figure 11-23

Chart A for Computing Non-Ideally Damped Two-Element Filters

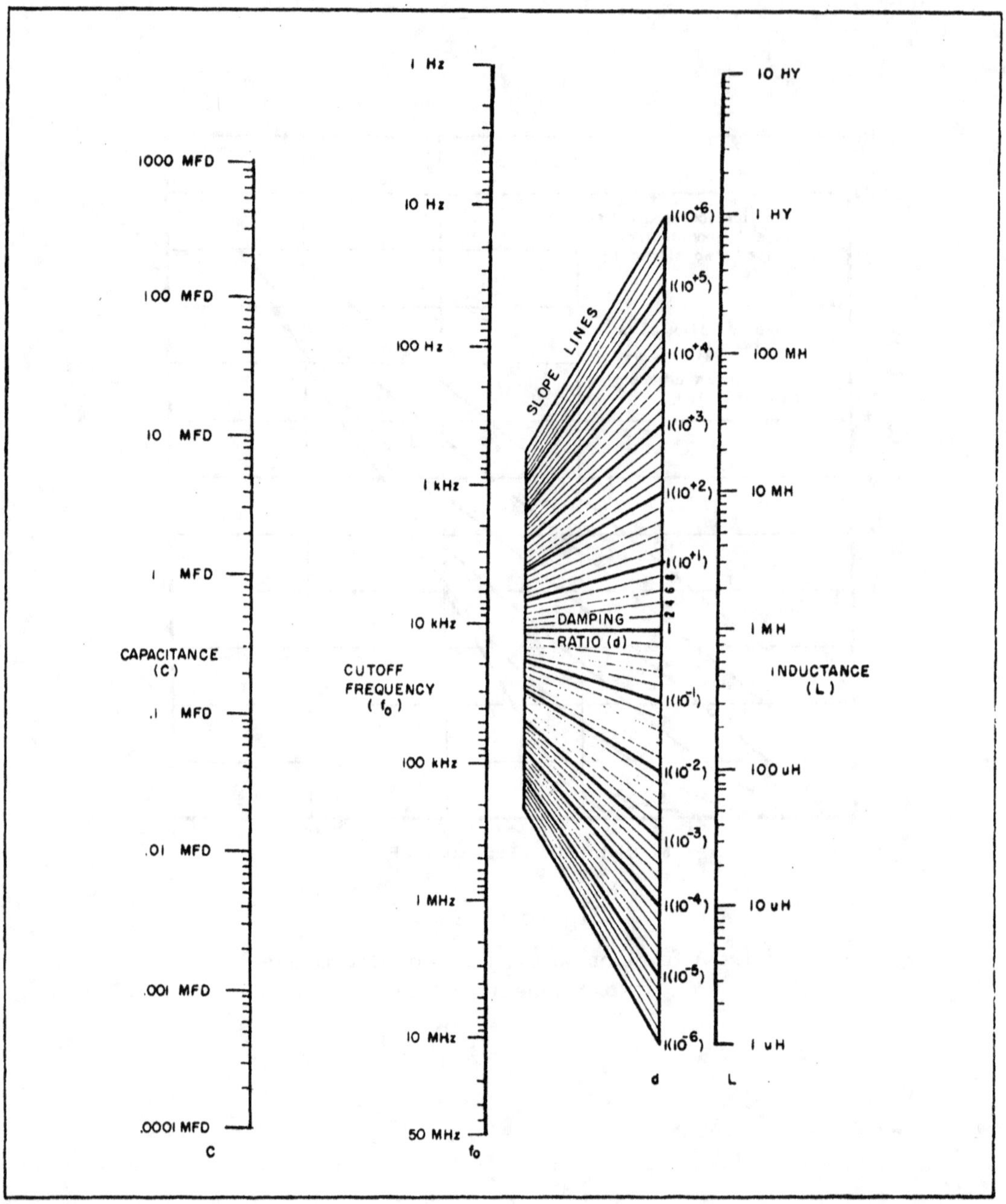

Figure 11-24

Chart B for Computing Two-Element Filters For Any Damping Ratio

Insertion Loss and Parameter Value Calculations for Ideally-Damped Three-Element Filters

See Equations (7.6) and (7.7) of this publication for the three-element filter design equations

To find the insertion loss of an ideally-damped "T" or π section filter ($d = 1$) at any frequency when f_o is known,

a. Place a straight edge parallel to the guide lines and pass it through the known cutoff frequency. The straight edge now lies along the insertion loss characteristic for the ideally-damped L section having that cutoff frequency.

b. At any frequency along the abscissa, read the insertion loss in dB along the ordinate.

To find the cutoff frequency (f_o) required to produce a desired amount of insertion loss at a given frequency,

a. Place a straight edge parallel to the guide lines and pass it through the desired insertion loss at the given frequency.

b. Read the cutoff frequency on the f_o scale.

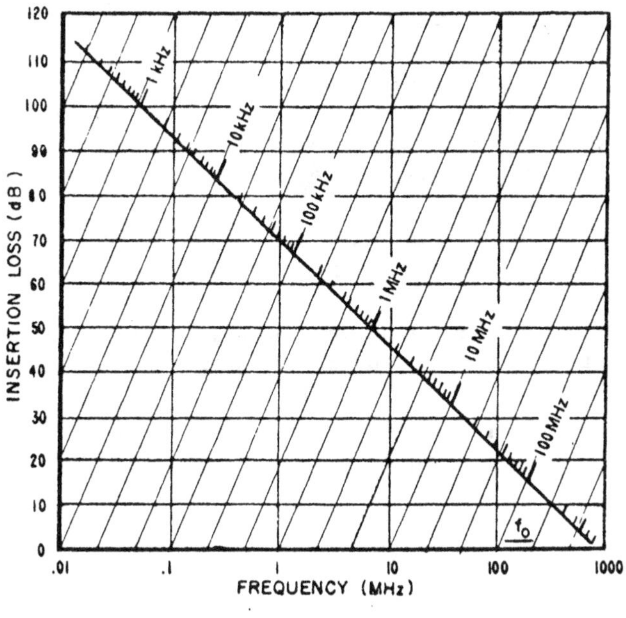

Figure 11-25
Chart for Computing Ideally Damped Three-Element Filters

Insertion Loss and Parameter Value Calculations for Non-Ideally Damped Three-Element Filters

See Equations (7.6) and (7.7) of this publication for the three-element filter design equations.

To find the insertion loss of a non-ideally damped "T" or π section filter at any frequency when the cutoff frequency (f_o) and the damping ratio (d) are known, use Chart A on the next page for the overdamped case (d > 1) or Chart B on the next page for the underdamped case (d < 1). Proceed as follows:

a. These charts are normalized for $f_o = 1$; thus, divide the frequency of interest by the known f_o to determine the normalized frequency of interest.

b. Locate the characteristic for which d is equal to the known damping ratio. Read the approximate value of d.

c. At the normalized frequency of interest, read the insertion loss in dB.

To establish the relationship between inductance, capacitance, cutoff frequency, and damping factor of three-element filters, use Charts C or D that follow. Chart C applies to π section filters, while Chart D applies to "T" section filters. To find the cutoff frequency and the damping ratio for a given inductance and capacitance:

a. Place a straight edge so that it passes through the known L and C values.

b. Read the cutoff frequency (f_o) on the center scale.

c. The damping ratio slope line which lies parallel to the straight edge indicates the damping ratio. Note that the point at which the straight edge intersects the damping ratio scale is irrelevant; the "d" sloped line parallel to the straight edge determines the damping ratio.

To find the circuit elements required to produce a desired cutoff frequency at a given damping ratio, use Chart C or D and:

a. Place a straight edge so that it passes through the desired cutoff frequency and lies parallel to the given damping ratio line.

b. Read the inductance and capacitance on the L and C scales.

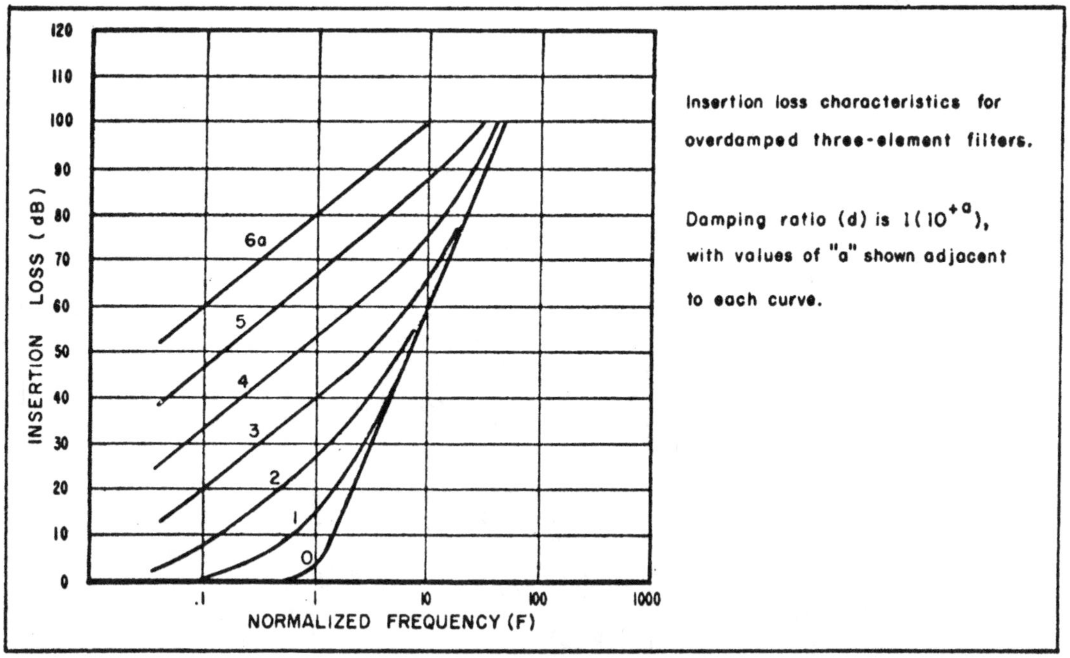

Figure 11-26

Chart A For Computing Overdamped Three-Element Filters

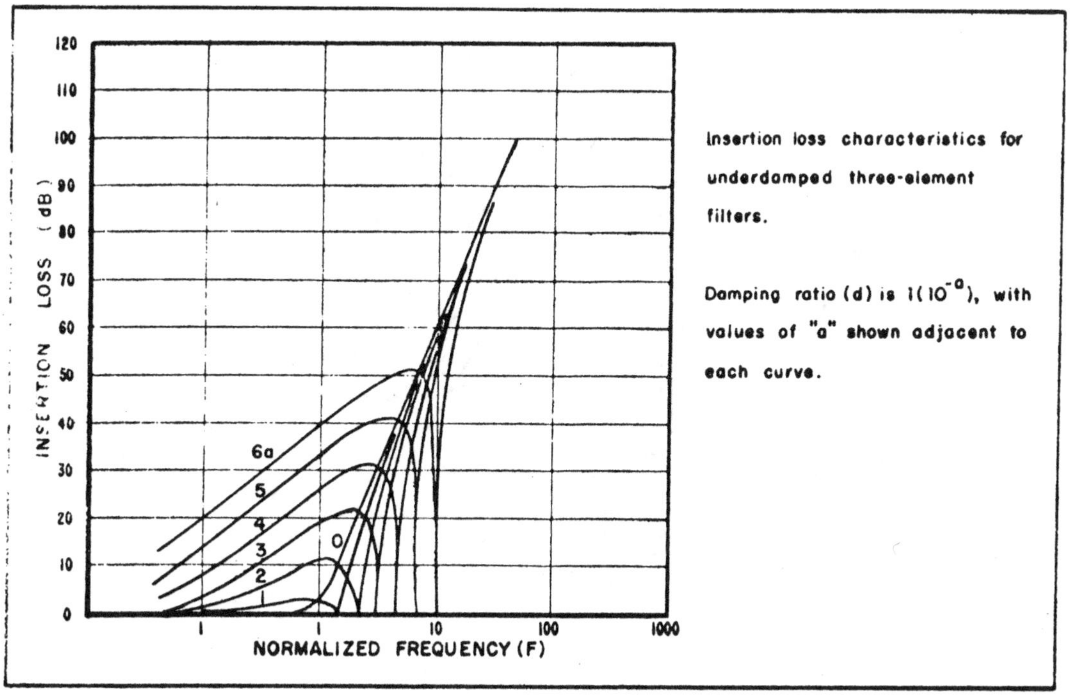

Figure 11-27

Chart B For Computing Underdamped Three-Element Filters

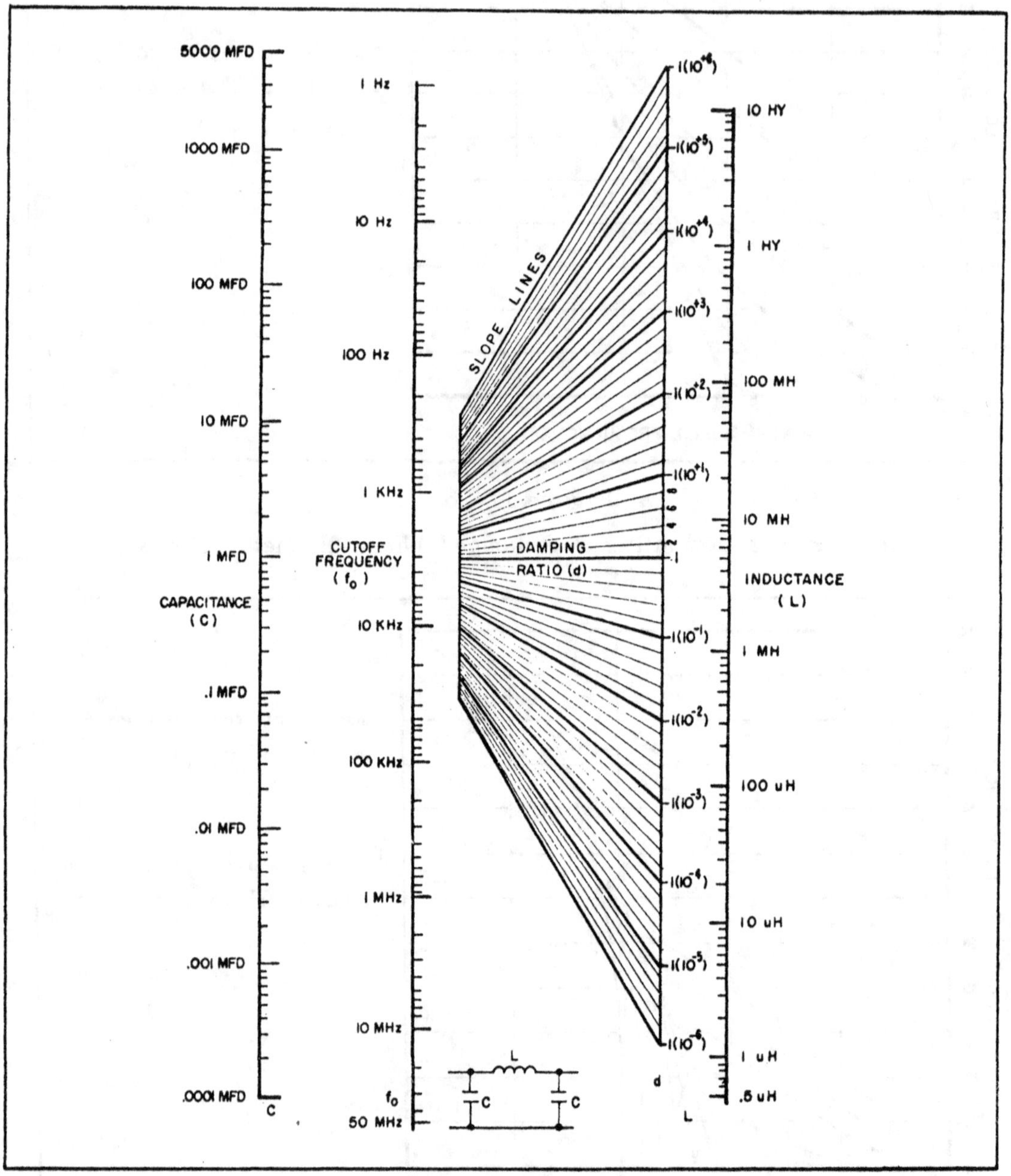

Figure 11-28
Chart C For Computing π Filters For Any Damping Ratio

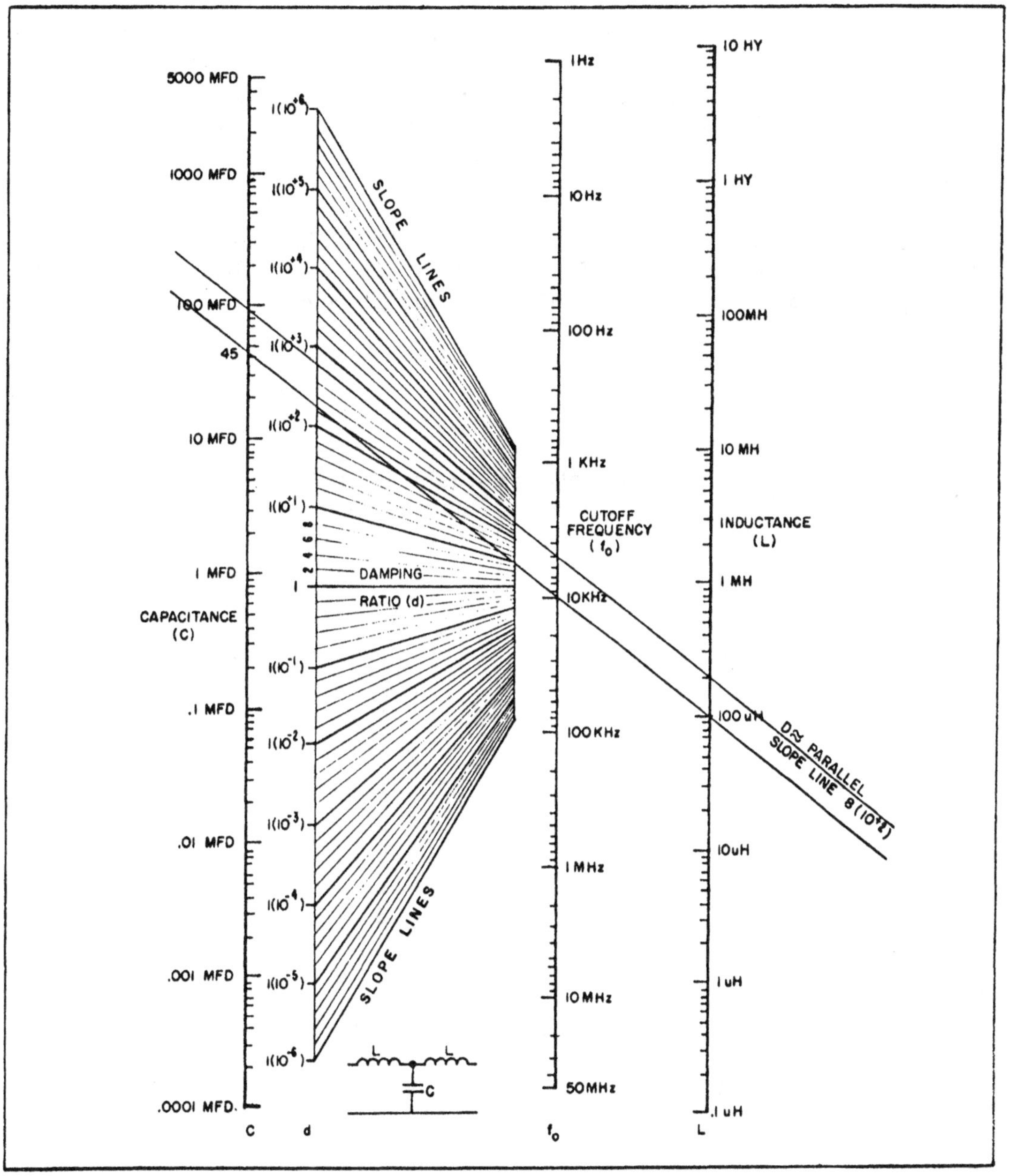

Figure 11-29

Chart D For Computing T Filters For Any Damping Ratio

Parameter Calculations for Twin-T Notch Filters for K = 1 or 2

See Equation (7.10) of the publication for the twin-T notch filter design equation.

The following nomographs are provided for rapidly computing the parameters of twin-T notch filters when K = 1 or 2. Nomograph A is used to relate the source resistance, load resistance and R, while Nomograph B identifies the relationship between f_o, R and C.

The example shown assumes a source resistance of 300 ohms, a load resistance of 100,000 ohms, a notch frequency of 60 Hertz, and K = 1. Under these circumstances, R from Nomograph A is 7500 ohms, and C from Nomograph B is 0.4 μf.

From J. F. Sorado, "Twin-T Network Design" 1958
Electrical Design Manual, Rogers Printing Co., Englewood, Cal., 1958, pp. 27-30.

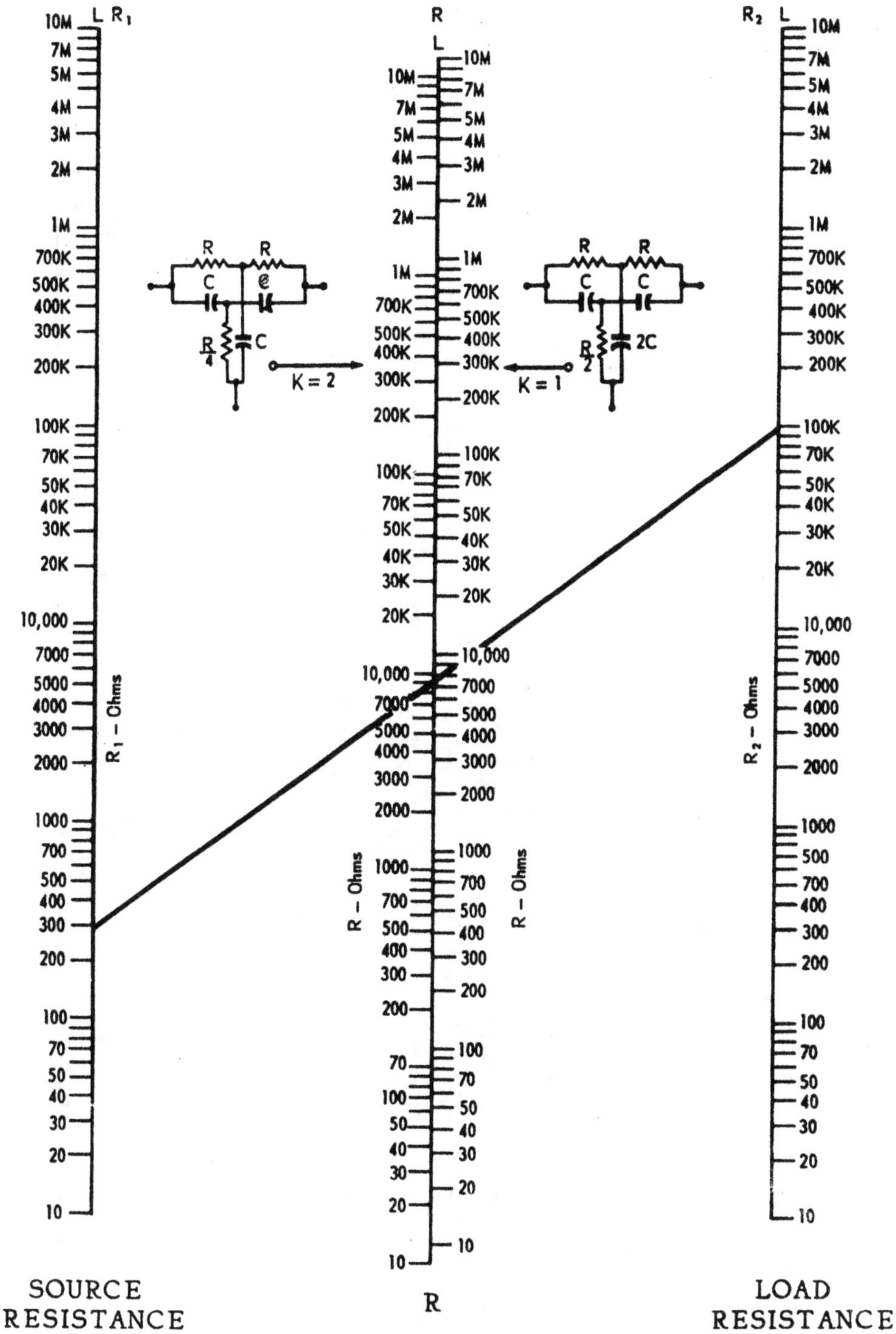

Figure 11-30

Nomograph A for Determining Resistance Values for Twin-T Notch, K = 1 and K = 2

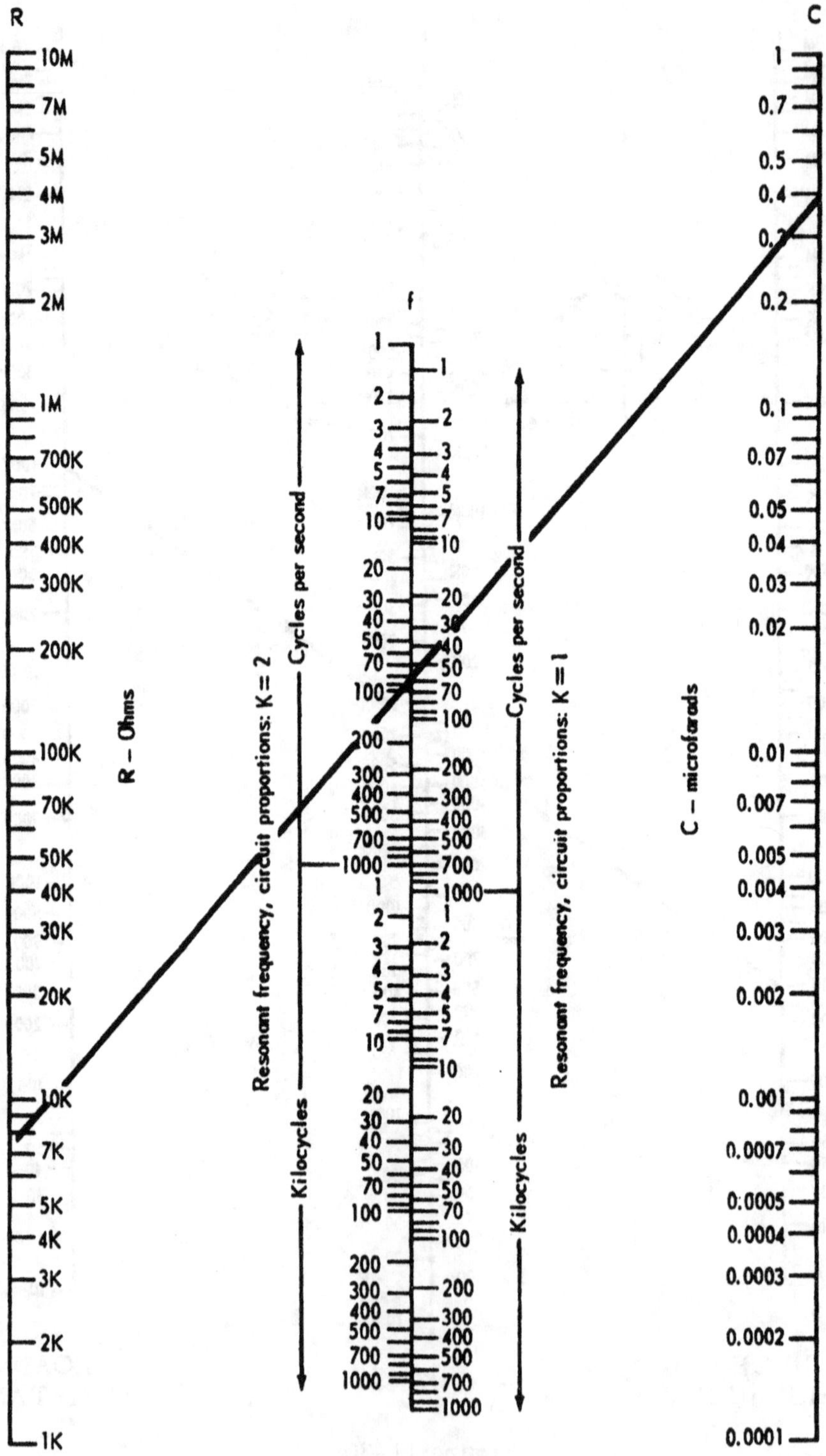

Figure 11-31

Nomograph B for Determining Capacitor Values for Twin-T Notch, K = 1 and K = 2

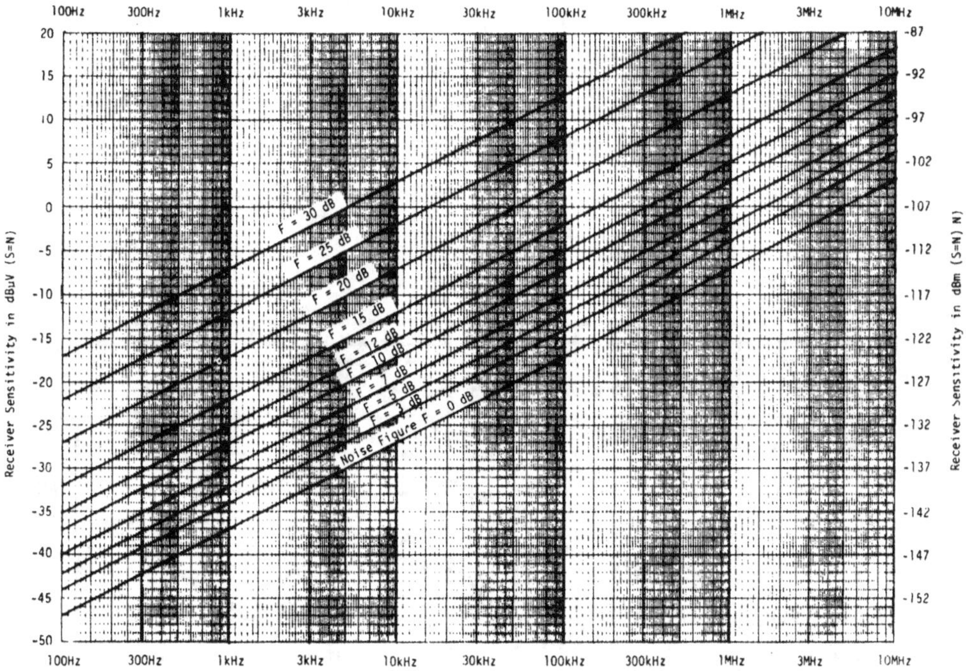

Figure 11-32 Courtesy of DON WHITE CONSULTANTS, INC.

Receiver Narrowband Sensitivity as a Function of Noise Figure and Bandwidth

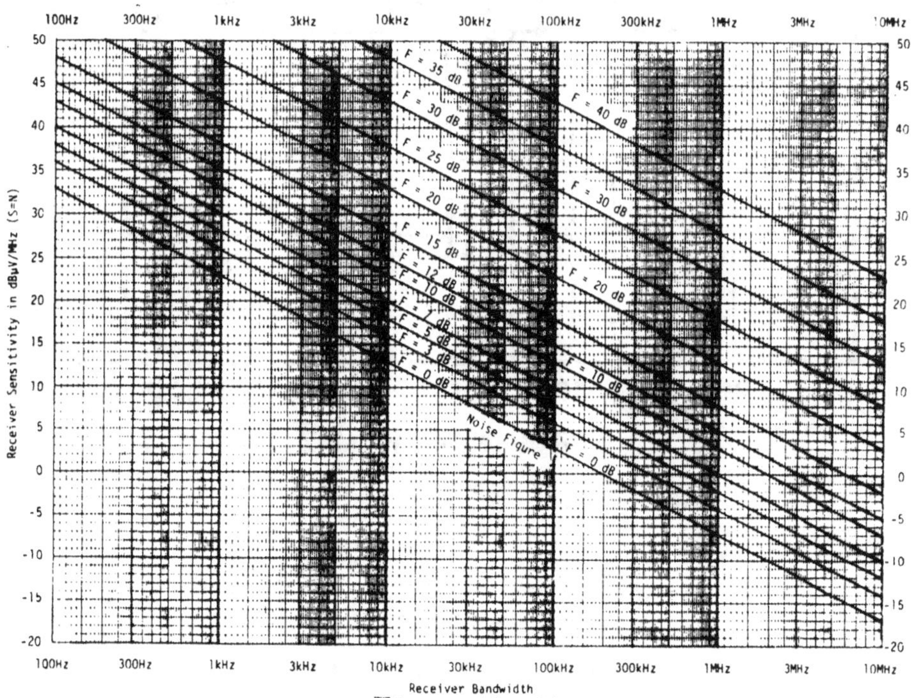

Figure 11-33

Receiver Broadband Impulse Noise Sensitivity as a Function of Noise Figure and Bandwidth

Courtesy of DON WHITE CONSULTANTS, INC.

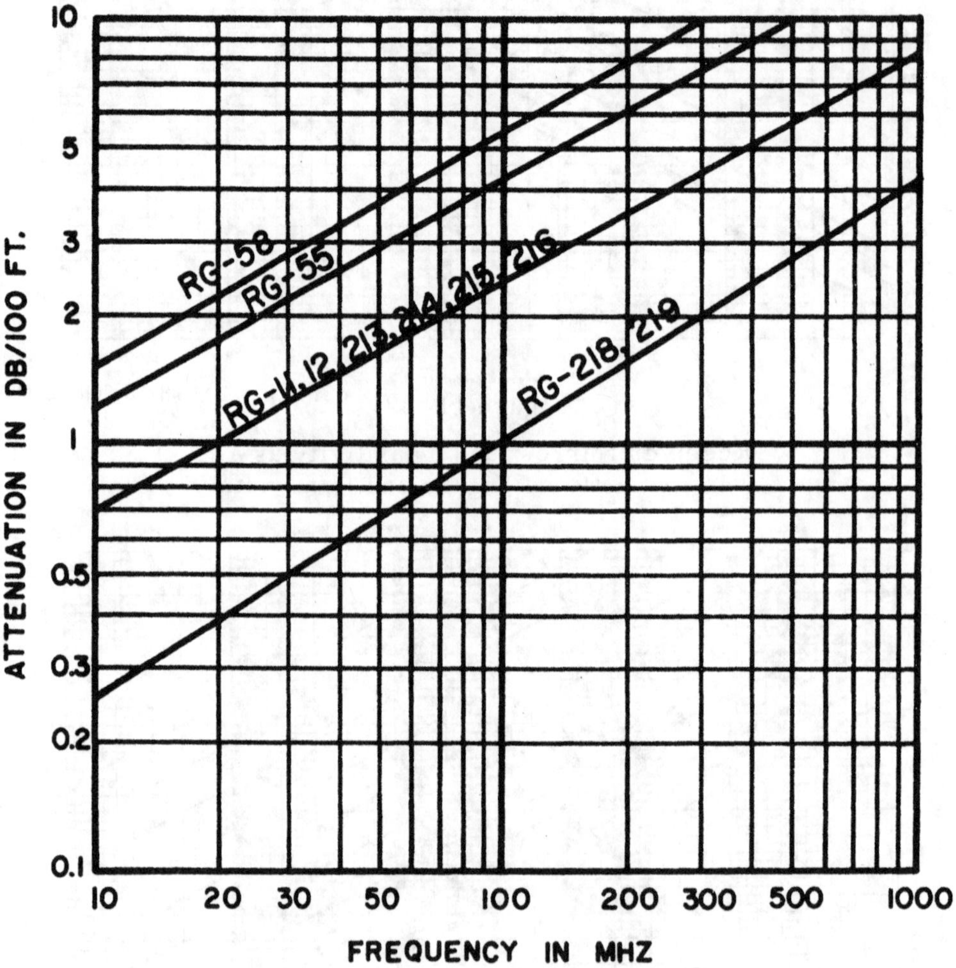

Figure 11-34
Attenuation of Coaxial Cable

MISCELLANEOUS CONVERSIONS

λ (meters) $= \frac{300}{f_{(mc)}}$

λ (feet) $= \frac{984}{f_{(mc)}}$

1 Meter = 3.28 feet = 6.214×10^{-4} statute miles

1 Foot = .3048 meters

1 Mile (statute) = 5280 feet = 1609 meters

Statute Miles = Nautical Miles x 1.1508

Free Space Impedance, Z = 377 ohms

Diameter of the Earth = 7917 nautical miles

= 9112 statute miles

Maximum line of sight distance over smooth earth in statute miles, $D_{(St.\ miles)} = \sqrt{2h_1} + \sqrt{2h_2}$

h_1 is the height of one antenna, in feet

h_2 is the height of the second antenna, in feet

BASIC NOISE POWER IN A SYSTEM

$P_N = KT\beta$

$K = 1.380 \times 10^{-23}$ joule/degree Kelvin

degrees Kelvin = °C + 273

room temperature is taken as 68° F = 20° C = 293° K

$P_N = 1.380 \times 10^{-23} \times 273 \times 1$ for $\beta = 1$ Hertz and P_N in watts

$P_N = 4.05 \times 10^{-21}$ watts/Hertz or 4.05×10^{-18} milliwatts/Hertz

$P_N = -173$ dBm/Hertz

To convert to wider bandwidths, add 10 log β in Hertz

Example: for β = 1 MHz 10 log 10^6 = 60

$P_N = -173 + 60 = -113$ dBm/MHz

To compute real noise level in a receiver, add the noise figure in dB to the computed noise power.

$P_N = P_N + NF$ (dB)

For a typical 1 MHz receiver, NF = + 9 dB

Sensitivity = -113 + 9 = -104 dBm

POWER DENSITY RELATIONSHIPS

$dBm/m^2 = dBm/ft^2 + 10.3$

$dBm/m^2 = dBm/cm^2 + 40$

(Personnel Hazard of .01 watts/cm^2 = + 50 dBm/m^2)

POWER-VOLTAGE CONVERSIONS

For a 50 ohm load:

$$P_{dBm} = E_{dB\mu} - 107$$

For a 72 ohm load:

$$P_{dBm} = E_{dB\mu} - 108.6$$

For a 300 ohm load:

$$P_{dBm} = E_{dB\mu} - 114.8$$

For a 377 ohm load (free space):

$$\xi_{(dB\mu/m)} = P_{d\ (dBm/m^2)} + 116$$

PATH LOSS AND PROPAGATION LOSS

Path Loss in dB (Free space propagation between isotropic antennas)

(1) for R in feet $L_i = 20 \log f_{MHz} + 20 \log R_{(feet)} - 37$

(2) for R in yards $L_i = 20 \log f_{MHz} + 20 \log R_{(yards)} - 28$

(3) for R in meters $L_i = 20 \log f_{MHz} + 20 \log R_{(meters)} - 26.8$

(4) for R in K yards $L_i = 20 \log f_{MHz} + 20 \log R_{(K\ yards)} + 32$

(5) for R in miles $L_i = 20 \log f_{MHz} + 20 \log R_{(miles)} + 37$

(6) for R in Nautical miles $L_i = 20 \log f_{MHz} + 20 \log R_{(N\ miles)} + 38.6$

Propagation Loss in dB (Free space propagation)

Gain-Coupling Loss

(1) for R in feet $L_{gc} = 20 \log f_{MHz} + 20 \log R_{(feet)} - G_t - G_r - 37$

(2) for R in yards $L_{gc} = 20 \log f_{MHz} + 20 \log R_{(yards)} - G_t - G_r - 28$

(3) for R in meters $L_{gc} = 20 \log f_{MHz} + 20 \log R_{(meters)} - G_t - G_r - 26.8$

(4) for R in K yards $L_{gc} = 20 \log f_{MHz} + 20 \log R_{(K\ yards)} - G_t - G_r + 32$

(5) for R in miles $L_{gc} = 20 \log f_{MHz} + 20 \log R_{(miles)} - G_t - G_r + 37$

REDUCTION OF POWER DENSITY TO TRANSMITTER POWER OUTPUT

(1) $P_{t(dBm)} = P_{d(dBm/m^2)} - G + 20 \log R + 11$ (R in meters)

(2) $P_{t(dBm)} = P_{d(dBm/m^2)} - G + 20 \log R + 0.7$ (R in feet)

(3) $P_{t(dBm)} = P_{d(dBm/m^2)} - G + 20 \log R + 75$ (R in miles)

(1, 2, 3, hold for free space propagation with P_d measured in the far field.)

REDUCTION OF FIELD STRENGTH TO EFFECTIVE RADIATED POWER (P_{erp}, dBm)

(1) $P_{erp\ (dBm)} = \xi_{(dB\mu v/m)} + 20 \log R = 105$ (R in meters)

(2) $P_{erp\ (dBm)} = \xi_{(dB\mu v/m)} + 20 \log R - 115$ (R in feet)

(3) $P_{erp\ (dBm)} = \xi_{(dB\mu v/m)} + 20 \log R - 41$ (R in miles)

(1, 2, 3, hold for free space propagation only.)

ANTENNA CONVERSIONS

For a 50 ohm load

$$A_e = 1.885\ L_e^2$$

$$A_{e(dB)} = G_{(dB)} - 20 \log f_{(MHz)} + 38.6$$

$$G_{(dipole)} = G_{(isotropic)} + 2.15 \text{ dB}$$

$$G_{(dipole)} = G_{(isotropic)} \times 1.64 \text{ dB}$$

POWER, VOLTAGE, FIELD STRENGTH AND POWER DENSITY RELATIONSHIPS

1. Effective area A_e, and effective length L_e, for the same antenna in a constant field:

$$A_e = L_e^2\ \frac{377}{4R_L}$$

where R_L is the antenna matched load

2. Field Strength and Received Voltage:

$$\xi_{(dB\mu/m)} = E\ (dB\mu) - 10 \log G_r + 20 \log f_{MHz} - 29.8$$

3. Power and Voltage in a 50 ohm load

$$P_{(dBm)} = E_{(dB\mu\ )} - 107$$

4. Power Density and Field Strength (Free Space Z = 377)

$$P_d(dBm/m^2) = \xi\ (dB\mu/m) - 116$$

5. Power Density and Received Voltage (for P_d at the Rx antenna)

$$P_d(dBm/m^2) = E_R(dB\mu) - G_R(dB) + 20 \log f_{MHz} - 146$$

6. Power Density and Received Power (for P_d at the Rx antenna)

$$P_d(dBm/m^2) = P_R(dBm) - G_R(dB) + 20 \log f_{MHz} - 38.5$$

FREE SPACE PROPAGATION VELOCITY

186,280 Statute Miles/Second

161,300 Nautical Miles/Second

984×10^6 Feet/Second

328×10^6 Yards/Second

299.8×10^6 Meters/Second

Coupling of EM Fields Into Wire Loops

The two nomographs (Figures 11-35 and 11-36) allow an EMC engineer to determine the electromagnetic field induced open circuit voltage in a rectangular loop (hxL) such as is found in common mode coupling.

Analysis of the magnitude of these signals can be performed by determining the equivalent loop aperture of the wire. An isolated single wire can be analyzed as a loop by considering the end-to-end capacitance as the return path. The wire would therefore act as an unintentional antenna. A pair of wires in a cable could similarly be considered as a loop when the capacitive and inductive coupling between them is taken into account. These effects can be differential mode and common mode. The loop for differential mode signals between two wires would be defined by the common run length and spacing between the wires. The signal would be 180° out-of-phase between the wires, therefore, "differential-mode". The loop for common mode signals would be defined by the wires and the closest ground plane. The induced signal would be in-phase on both wires, hence, "common-mode". The open circuit voltage induced in a loop by either an electric or magnetic field is a function of the area of the loop and its orientation in the field.

For an electric field;

$$Vi/E = 2L \cos 0 \sin \left(\frac{\pi h}{\lambda} \cos \alpha\right) \text{ (meters) for h and L} \ll \alpha$$

where Vi = open circuit induced voltage (volts)
E = electric field (V/M)
h = height of loop
L = length of loop
λ = wavelength
α = angle between plane of loop and direction of propagation of field
0 = angle between the wire and the E-field polarization

The maximum value would occur when the loop is orthogonal to the direction of propagation and the wire and field polarization are in alignment. In such a condition, $\cos 0 = \cos \alpha = 1$.

Therefore,

$$\left(\frac{Vi}{E}\right)_{max.} = 2L \sin \left(\frac{\pi h}{\lambda}\right) \text{ for h and L} \ll \lambda$$

$$\sim \frac{2\pi Lh}{\lambda}$$

$$\sim 2.1\ (10)\ LhF_{Hz}$$

Similarly, magnetic coupling,

$$Vi/B = 2LC \cdot \sin 0 \cdot \sin \left(\frac{\pi h}{\lambda} \cos \alpha\right), \text{ meters for h and L} \ll \lambda$$

where where V_i = open circuit induced voltage (Volts)
B = magnetic flux density (Gauss)
λ = wavelength
C = velocity of propagation = 3 (10) 8M/S
h = height of loop
L = length of loop
α = angle between plane of the loop and the direction of propagation of the field
0 = angle between the wire and a plane orthogonal to the field

The maximum, therefore, is

$$\left(V_i/B\right)_{max.} = 2LC \sin\left(\frac{\pi h}{\lambda}\right)_{radians}, \text{ for h and L} << \lambda$$

$$\sim \frac{2\pi LhC}{\lambda}, \text{ for h and L} << \lambda$$

$$\sim 6.28\,(10)^{-4} hLF_{Hz}$$

The maximum coupling values in dB,

$$20 \log \left(V_i/E\right)_{max.}{}^{dB(Meter)} \text{ and } 20 \log \left(V_i/B\right)_{max.}{}^{dBVT}$$

as a function of loop area (h · L) can be determined using the nomographs.

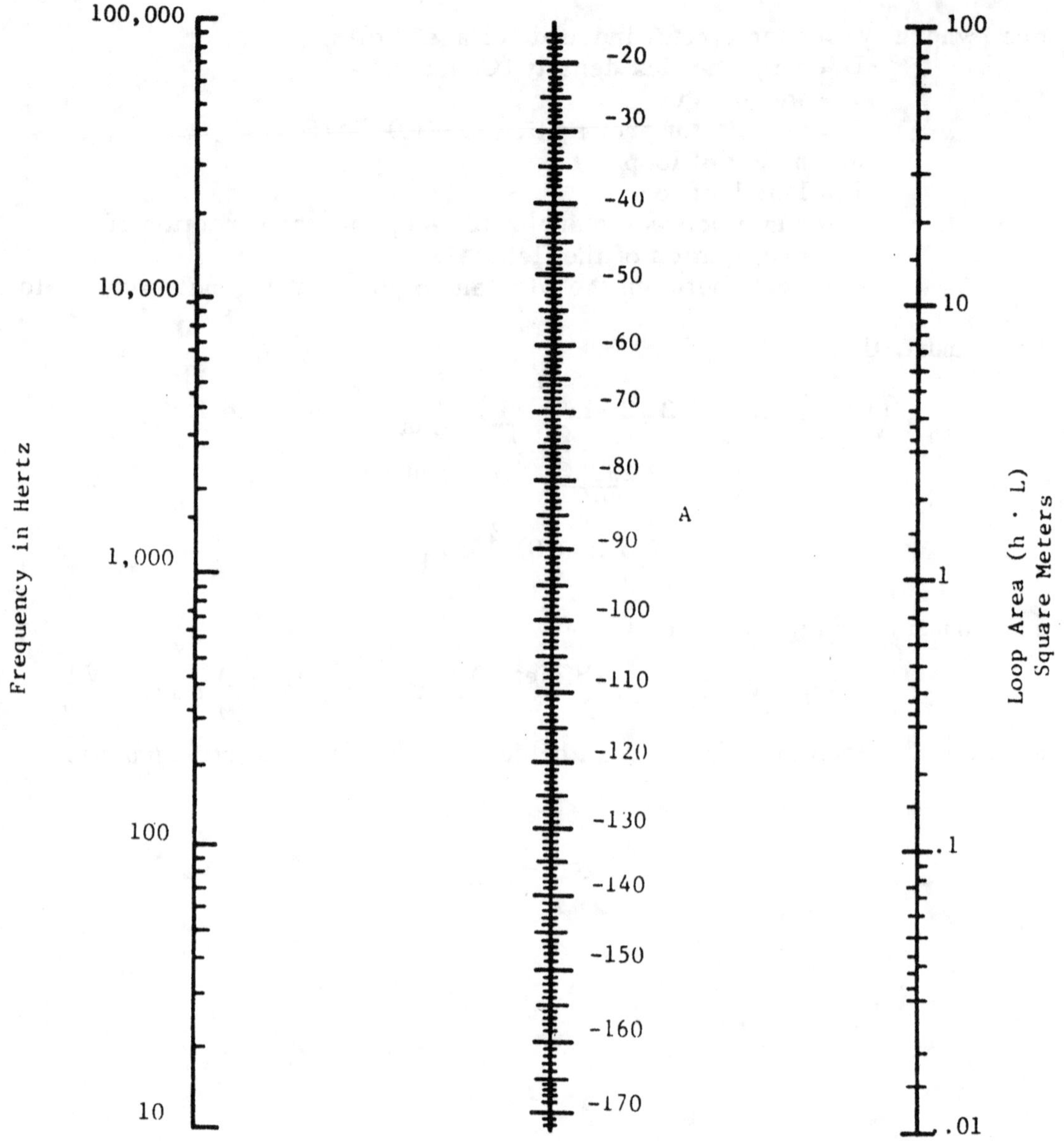

A(dB Meter) = 153.56 + 20 log (hL) + 20 log F_{Hz}, h and L

V (dBV) = A + E (dBV/M)

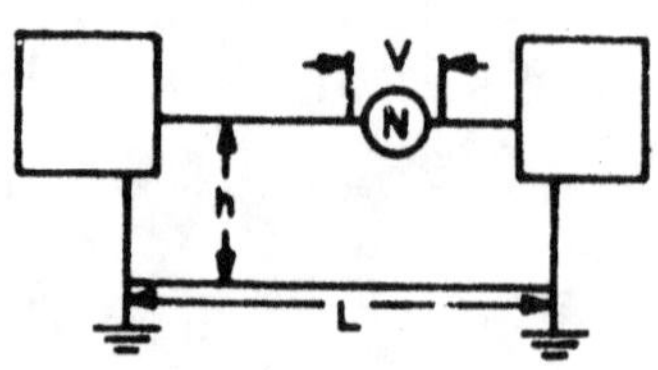

Figure 11-35
Maximum Open Circuit Loop Voltage

Due to Electric Field

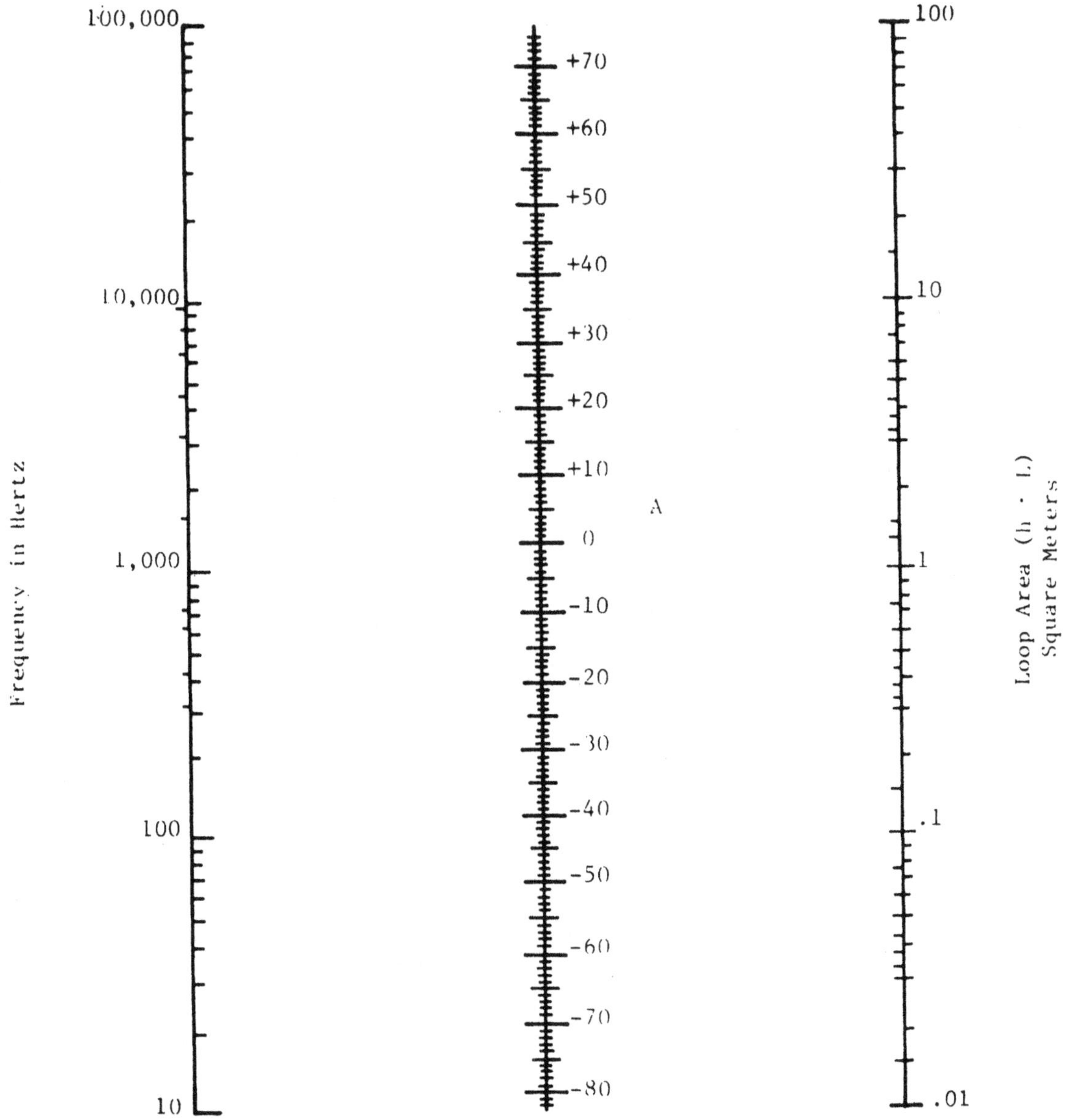

$$A(dBV/G) = -64.04 + 20 \log (hL) + 20 \log F_{Hz}, \text{ h and L}$$

$$V\ (dBV) = A + B\ (dBGauss)$$

Figure 11-36

Maximum Open Circuit Loop Voltage

Due to Magnetic Field

INDEX

www.ingramcontent.com/pod-product-compliance
Lightning Source LLC
LaVergne TN
LVHW080035210926
848481LV00011B/3199